Data Modelling and Analytics for the Internet of Medical Things

The emergence of the Internet of Medical Things (IoMT) is transforming the management of diseases, improving diseases diagnosis and treatment methods, and reducing healthcare costs and errors. This book covers all the essential aspects of IoMT in one place, providing readers with a comprehensive grasp of IoMT and related technologies.

Data Modelling and Analytics for the Internet of Medical Things integrates the architectural, conceptual, and technological aspects of IoMT, discussing in detail the IoMT, connected smart medical devices, and their applications to improve health outcomes. It explores various methodologies and solutions for medical data analytics in healthcare systems using machine learning and deep learning approaches, as well as exploring how technologies such as blockchain and cloud computing can further enhance data analytics in the e-health domain. Prevalent IoMT case studies and applications are also discussed.

This book is suitable for scientists, design engineers, system integrators, and researchers in the field of IoMT. It will also be of interest to postgraduate students in computer science focusing on healthcare applications and a supplementary reading for IoMT courses.

Data Modelling and Analytics for the Internet of Medical Things

Edited by
Rajiv Pandey, Pratibha Maurya,
and Raymond Chiong

CRC Press
Taylor & Francis Group
Boca Raton London New York

CRC Press is an imprint of the
Taylor & Francis Group, an **informa** business

Designed cover image: © Shutterstock

MATLAB® is a trademark of The MathWorks, Inc. and is used with permission. The MathWorks does not warrant the accuracy of the text or exercises in this book. This book's use or discussion of MATLAB® software or related products does not constitute endorsement or sponsorship by The MathWorks of a particular pedagogical approach or particular use of the MATLAB® software.

First edition published 2024
by CRC Press
2385 NW Executive Center Drive, Suite 320, Boca Raton FL 33431

and by CRC Press
4 Park Square, Milton Park, Abingdon, Oxon, OX14 4RN

CRC Press is an imprint of Taylor & Francis Group, LLC

ISBN: 978-1-032-41423-2 (hbk)
ISBN: 978-1-032-41831-5 (pbk)
ISBN: 978-1-003-35995-1 (ebk)

DOI: 10.1201/9781003359951

Typeset in Times
by codeMantra

Contents

About the Editors xiv
Contributors xv
Preface... xviii

SECTION 1   IoMT Datasets and Storage

Chapter 1   Remote Health Monitoring in the Era of the Internet of Medical
Things... 3

Dalal Alshehri, Nasimul Noman, Raymond Chiong, and Shah J Miah

1.1   Introduction 3
1.2   Types of Remote Health Monitoring... 4
1.3   IoMT Architectures.. 5
1.4   IoMT Devices for Remote Health Monitoring 7
1.5   Technologies for Remote Health Monitoring 8
1.6   Benefits of IoMT-Based RHM 11
1.7   Challenges for IoMT-Based RHM 12
1.8   Security and Privacy Requirements for RHM 13
1.9   Remote Monitoring Case Studies.. 14
        1.9.1   Elderly Patent Monitoring...................................... 14
        1.9.2   Monitoring of Infectious Diseases 15
        1.9.3   Intensive Care Unit Patient Monitoring 16
1.10   Conclusion 16
References 17

Chapter 2   Diabetic Healthcare Data Analytics and Application 19

*Dev Kapadia, Pratyush Mohanty, Anupama Namburu,
and Aravapalli Rama Satish*

2.1   Introduction 19
2.2   Architectures for Diabetic Healthcare..................................... 20
        2.2.1   IoT Architecture for Healthcare 20
        2.2.2   Fog Computing and Edge Computing
                  Architecture for Healthcare...................................... 21
        2.2.3   Machine Learning Architecture for Healthcare......... 22
        2.2.4   Big Data Architectures for Healthcare..................... 23
        2.2.5   Architecture Based on Deep Learning 23
2.3   Sensor, Devices and Applications for Diabetic Healthcare..... 25
        2.3.1   Continuous Glucose Meters 25
        2.3.2   Wearable Devices... 25

2.3.3   IoT and Sensor Devices 26
2.3.4   Decision Support Systems...................................28
2.3.5   Telemedicine 28
2.3.6   Advanced Telediabetology Methods Based on
        New Technologies 28
2.3.7   Mobile Applications 28
2.3.8   Smart Pens..28
2.3.9   Decision Support Systems for Insulin Therapy 29
2.4   Diabetic Data Processing and Mining...........................29
        2.4.1   Use of Machine Learning Algorithms 29
        2.4.2   Feature Reduction 30
2.5   Diabetic Complications and Its Monitoring 31
        2.5.1   Challenges in Diabetic Health Monitoring 32
2.6   Conclusion 33
References 34

Chapter 3   Blockchain for Handling Medical Data 38

T. Vigneswari, Farjana Farvin Sahapudeen, and G. Kalaiselvi

3.1   Introduction 38
3.2   Blockchain—An Overview 40
3.3   Blockchain Framework for IoMT............................42
3.4   Blockchain and Data Management..............................44
        3.4.1   Data Storage in Blockchain.....................................44
        3.4.2   Data Access 46
        3.4.3   Data Preservation System.......................................49
        3.4.4   Privacy Protection 51
3.5   Conclusion 54
References 54

Chapter 4   Cloud Computing for Complex IoMT Data 58

Akrati Sharma and Priti Maheshwary

4.1   Cloud Introduction..58
        4.1.1   IoMT System Architecture.......................................61
        4.1.2   Technologies in IoMT 61
        4.1.3   Wireless Sensor Network (WSN)............................62
        4.1.4   Middleware...64
        4.1.5   The Comparative Experimental Study between
                IoMT-Based Sensors 64
        4.1.6   IoMT Implementations..65
4.2   Application of IoMT along with Cloud................................66
4.3   Conclusion 69
References 69

Chapter 5   The Potential of IoMT Devices in Early
               Detection of Suicidal Ideation..72

               Archana Sahai and Farooqui UK

       5.1   Introduction 72
       5.2   Internet of Things (IoT)..73
       5.3   Internet of Medical Things (IoMT)................................73
       5.4   IoMT Devices and Mental Health..................................74
       5.5   Parameters Measured by IoMT Devices 75
       5.6   IoMT Devices in Predicting Suicide Rates 77
       5.7   IoMT Devices Specifically for Mental Health in India..........82
       5.8   Application of IoMT Devices in India 83
       5.9   Conclusion 85
       References 85

SECTION 2   *Machine Learning for Medical Things*

Chapter 6   Artificial Intelligence and Internet of Medical Things in the
               Diagnosis and Prediction of Disease..91

               *Narendra Kumar Sharma, Shahnaz Fatima, and
               Alok Singh Chauhan*

       6.1   Introduction 91
       6.2   IoMT...91
             6.2.1   IoMT and Machine Learning....................................92
             6.2.2   Importance of IoMT in Healthcare 92
       6.3   Artificial Intelligence 93
             6.3.1   Artificial Intelligence and Machine Learning 93
       6.4   AI/ML Framework for Disease Detection and Modeling.......94
             6.4.1   Artificial Intelligence and Disease Diagnostic..........95
             6.4.2   Machine Learning and Disease Diagnosis.................96
       6.5   Implementation of AI/ML Algorithms in the Diagnosis
             and Prediction of Disease 97
             6.5.1   Materials and Methods..97
             6.5.2   Decision Tree...97
             6.5.3   K-Nearest Neighbor (KNN) 98
             6.5.4   Naïve Bayes 98
             6.5.5   Random Forest 98
             6.5.6   Python 99
             6.5.7   Graphical User Interface 99
             6.5.8   Dataset..99

          6.5.9   Working Methodology 100
          6.5.10 Experimental Evaluation 100
          6.5.11 Result Analysis 101
    6.6   Application of AI/ML Algorithms in the Diagnosis and
          Prediction of Disease 102
          6.6.1   Disease Symptoms and Diagnostic Difficulties 102
          6.6.2   Technological Perspective of IoMT in Disease
                  Diagnosis and Prediction 104
          6.6.3   Need to Overcome IoMT Constraints 105
    6.7   Opportunities and Challenges 106
    6.8   Conclusion and Future Prospects 106
    References 107

Chapter 7   Predicting Cardiovascular Diseases Using Machine Learning:
A Systematic Review of the Literature 109

*Abhay Kumar Pathak, Bhupendra Kumar Dewangan, and
Manjari Gupta*

    7.1   Introduction 109
    7.2   Methodology 110
          7.2.1   Study Selection 110
          7.2.2   Year-Wise Distribution of Selected Articles 111
    7.3   Comparison Based on Models and Datasets Used 112
          7.3.1   Datasets Used 112
    7.4   Comparison Based on Performance Metrics 115
    7.5   Conclusion 124
    References 124

Chapter 8   Identification of Unipolar Depression Using
Boosting Algorithms 126

Parul Verma, Roopam Srivastava, and Shikha Srivastava

    8.1   Introduction 126
    8.2   Related Work 130
    8.3   Methodology 132
          8.3.1   Dataset 132
          8.3.2   Clean/Explore Data 133
          8.3.3   Feature Categorization and Data Visualization 135
    8.4   Train/Test Split Method 138
    8.5   Calculating Accuracy Using Classifiers 138
    8.6   Conclusion 139
    References 139

Chapter 9   Development of EEG-Based Identification of Learning
Disability Using Machine Learning Algorithms 141

Nitin Ahire, R. N. Awale, and Abhay Wagh

9.1   Introduction 141
      9.1.1   Electroencephalography 141
9.2   Method.. 142
      9.2.1   Literature Review 142
      9.2.2   Limitations of Existing Systems 145
      9.2.3   Problem Statement 145
      9.2.4   Proposed System 146
9.3   Results 147
9.4   Discussion and Conclusions 147
      9.4.1   Discussion... 147
      9.4.2   Innovation... 148
      9.4.3   Conclusion.. 150
      9.4.4   Future Scope.. 150
      9.4.5   Declarations.. 151
References 151

Chapter 10 Deep Learning Approaches for IoMT.. 153

*Farjana Farvin Sahapudeen, T. Vigneswari, and
S. Krishna Mohan*

10.1   Introduction 153
10.2   DL Frameworks for IoMT 155
       10.2.1   Convolutional Neural Network (CNN) 156
       10.2.2   Recurrent Neural Network (RNN)........................... 156
       10.2.3   Deep Autoencoder (DAN) 156
       10.2.4   U-Net.. 157
       10.2.5   Generative Adversarial Network (GAN)................. 157
10.3   Deep Learning Healthcare Monitoring............................. 157
10.4   Deep Learning Healthcare Diagnosis System (DL-HDS) 159
10.5   Deep Learning in Security 161
       10.5.1   Security Essentials for IoMT 162
       10.5.2   Security Threats in IoMT Domain........................... 163
       10.5.3   Deep Learning in Intrusion Detection 166
10.6   Performance Metrics 168
10.7   Conclusion 168
References 168

Chapter 11 Machine Learning and Deep Learning Techniques to Classify
Depressed Patients from Healthy, by Using Brain Signals from
Electroencephalogram (EEG) 171

Gosala Bethany, Gosala Emmanuel Raj, and Manjari Gupta

11.1   Introduction 171
11.2   Related Work 173
11.3   Material and Methods 174
       11.3.1   Dataset 174
       11.3.2   Preprocessing 174
       11.3.3   Feature Extraction 177
       11.3.4   Supervised Learning Algorithm 177
       11.3.5   Performance Metrics 178
       11.3.6   Experiment Settings 180
11.4   Results and Discussions 180
       11.4.1   Results of Logistic Regression 181
       11.4.2   Results of Support Vector Machine 181
       11.4.3   Results of Convolutional Neural Network 181
11.5   Conclusion and Future Work 182
Acknowledgments 184
List of Abbreviation 184
References 184

Chapter 12 Dimensionality Reduction for IoMT Devices Using PCA 186

*Rajiv Pandey, Radhika Awasthi, Archana Sahai,
and Pratibha Maurya*

12.1   Introduction 186
12.2   Internet of Things 188
       12.2.1   IoT Architecture 189
       12.2.2   Applications of IoT 191
12.3   Internet of Medical Things (IoMT) 193
       12.3.1   Architecture of IoMT 194
       12.3.2   Technologies Integrated with IoMT 195
       12.3.3   Categories of IoMT Devices 198
12.4   Experimentation 199
       12.4.1   Dataset and Features 199
       12.4.2   Principal Component Analysis 200
       12.4.3   Applying PCA to Dataset 204
       12.4.4   Result 207
12.5   Conclusion 208
References 209

Chapter 13 Face Mask Detection System 211

> *Rajiv Pandey, Radhika Awasthi, and Archana Sahai*

13.1 Introduction 211
13.2 Methodology.. 212
    13.2.1 Database Used 212
    13.2.2 Convolutional Neural Network.............................. 212
    13.2.3 Topology and Results 218
13.3 Conclusion 220
Acknowledgements 222
References 222

SECTION 3   IoMT: Data Analytics and Use Cases

Chapter 14 An IoT-Based Real-Time ECG Monitoring Platform for
Multiple Patients.. 227

> *Vishnu S, Lova Raju K, Shanmugam M, and Sharma S*

14.1 Introduction 227
14.2 Literature Survey... 229
14.3 System Model 230
    14.3.1 Home Unit 230
    14.3.2 ECG-Data Server 231
14.4 Evaluation of PWN... 233
14.5 Conclusion 234
14.6 Current and Future Developments 234
References 235

Chapter 15 Study of Anomaly Detection in Clinical Laboratory
Data Using Internet of Medical Things............................... 237

> *Richa Singh and Nidhi Srivastava*

15.1 Introduction 237
15.2 Literature Survey... 238
15.3 Types of Anomalies in Clinical Laboratory.......................... 242
    15.3.1 Detection of Specimen-Based Anomalies.............. 242
15.4 Detection of Anomalies in Patients..................................... 244
15.5 Detection of Population-Based Anomalies 244
15.6 Types of Algorithms for Anomaly Detection....................... 246
15.7 Smart E-Healthcare and IoMT.. 246
    15.7.1 Architecture... 247
    15.7.2 Intelligent e-Healthcare System 247
    15.7.3 IoMT Systems in Electronic Healthcare 248

15.8   Components Used in IoMT 248
        15.8.1 IoMT Monitoring and Data Acquisition (MDA)......248
        15.8.2 IoMT Central Detection (CD) 250
15.9   Conclusion and Future Scope 251
References 251

Chapter 16 Computational Intelligence Framework for Improving Quality
               of Life in Cancer Patients..256

               Perla Sunanda and Dwaram Kavitha

16.1   Introduction 256
        16.1.1 Big Data...256
        16.1.2 Difference between Artificial Intelligence and
                Computational Intelligence 256
        16.1.3 Fuzzy Logic...257
        16.1.4 Expert Systems..257
        16.1.5 Artificial Neural Networks....................................257
        16.1.6 Genetic Algorithms 257
        16.1.7 Machine Learning..258
16.2   Quality of Life...258
        16.2.1 Characteristics of QoL Instruments 258
        16.2.2 Quality of Life Measurement Instruments..............259
        16.2.3 Generic versus Disease-Specific Instruments.........259
        16.2.4 Calculation of QoL Scores 261
        16.2.5 Types of Interventions 262
16.3   Methods and Design..262
        16.3.1 Search and Selection Criteria................................262
        16.3.2 Results 264
16.4   Computational Intelligence Framework 268
        16.4.1 Data Collection..271
        16.4.2 Data Preparation..271
        16.4.3 Study Design 272
16.5   Development of Prediction Model....................................272
        16.5.1 Building the Model...273
        16.5.2 Model Evaluation 274
        16.5.3 Data Visualisation 274
16.6   Conclusion 274
References 275

Chapter 17 Major Depressive Disorder Detection Using Data Science and
               Wearable Connected Devices..285

               Umar Khalid Farooqui and Archana Sahai

17.1   Introduction 285
17.2   Detecting Suicidal Behaviour...288

17.2.1 Use of Computerised Speech and Facial
Emotions Analysis.. 289
17.2.2 Use of Smart Phone and Computer Therapy............ 290
17.2.3 Use of Wearable IoT (WIoT).................................... 290
17.3 General Methods and Classification Used in Predicting
the Suicidal Tendency.. 292
17.3.1 Content Analysis 292
17.3.2 Feature Engineering 293
17.3.3 Deep Learning.. 294
17.4 Summary 297
17.5 Conclusion 297
References 298

Index.. 303

About the Editors

Dr. Rajiv Pandey, Senior Member of IEEE, is a faculty member at Amity Institute of Information Technology, Amity University, Uttar Pradesh, Lucknow Campus, India. He possesses a diverse background experience of around 38 years, including 15 years of industry and 23 years of academic experiences. His research interests lie in contemporary technologies such as blockchain and cryptocurrencies, information security, semantic Web provenance, cloud and big data, and data analytics. He has published more than 60 research papers. He has also been a session chair and a technical committee member for various IEEE and other conferences.

Dr. Pandey has served on technical committees of various government and private universities. He is intellectually involved in supervising Doctorate research scholars and postgraduate students. He is also an active contributor to professional bodies like the IEEE, IET, and Lucknow Management Association. He is a member of Machine Intelligence Research Labs.

Dr. Pratibha Maurya is currently an assistant professor at Amity Institute of Information Technology, Amity University, Uttar Pradesh, Lucknow Campus, India. She has more than 16 years of experience in academia. Her research interests include machine learning, data analytics, natural language processing, and information retrieval. She has published numerous journal and conference papers in these areas. She has been a session chair and a technical committee member for various international conferences and an editorial member for various journals.

Prof. Raymond Chiong is currently an associate professor with the University of Newcastle, Australia. His research focuses on using automated intelligent computing methods, including machine learning and optimization algorithms, to drive societal change and improve quality of life. For example, he uses a wide range of machine learning and optimization methods for prediction and data analytics, with applications in areas such as depression detection and suicide prevention. He has published more than 220 papers and attracted over 3 million dollars in research and industry funding. He is the Editor-in-Chief of the *Journal of Systems and Information Technology* and an Editor of *Engineering Applications of Artificial Intelligence.*

Contributors

Nitin Ahire
VJTI
Mumbai, Maharashtra, India

Dalal Alshehri
The University of Newcastle
Callaghan, NSW, Australia

R.N. Awale
VJTI
Mumbai, Maharashtra, India

Radhika Awasthi
Amity University
Lucknow, Uttar Pradesh, India

Gosala Bethany
Banaras Hindu University
Varanasi, Uttar Pradesh, India

Alok Singh Chauhan
Galgotias University
Greater Noida, Uttar Pradesh, India

Raymond Chiong
The University of Newcastle
Callaghan, NSW, Australia

Bhupendra Kumar Dewangan
DST-CIMS BHU
Varanasi, Utter Pradesh, India

Shahnaz Fatima
Amity University
Lucknow, Uttar Pradesh, India

Gaurav Gupta
Shoolini University
Solan, Himachal Pradesh, India

Manjari Gupta
DST-CIMS BHU
Varanasi, Utter Pradesh, India

Varun Jaiswal
Shoolini University Solan
Solan, Himachal Pradesh, India

Shashi Kala
Jaypee Institute of Information
    Technology
Noida, Uttar Pradesh, India

G. Kalaiselvi
Anjalai Ammal Mahalingam
    Engineering College
Thiruvarur, Tamil Nadu, India

Dev Kapadia
School of Engineering
JNU
New Delhi, India
Amaravati, Andhra Pradesh, India

Dwaram Kavitha
G. Pulla Reddy Engineering College
Kurnool, Andhra Pradesh, India

Umar Khalid Farooqui
Amity University
Lucknow, Uttar Pradesh, India

Shanmugam, M.
Department of Computer Science
Pondicherry University
Pondicherry, Tamil Nadu, India

Priti Maheshwary
Rabindranath Tagore University
Bhopal, Madhya Pradesh, India

Pratibha Maurya
Amity University
Lucknow, Uttar Pradesh, India

Shah J. Miah
The University of Newcastle
Callaghan, NSW, Australia

S. Krishna Mohan
E.G.S. Pillay Engineering College
Nagapattinam, Tamil Nadu, India

Pratyush Mohanty
School of Engineering
JNU
New Delhi, India

Anupama Namburu
School of Engineering
JNU
New Delhi, India

Nasimul Noman
The University of Newcastle
Callaghan, NSW, Australia

Rajiv Pandey
Amity University
Lucknow, Uttar Pradesh, India

Abhay Kumar Pathak
DST-CIMS BHU
Varanasi, Utter Pradesh, India

Gosala Emmanuel Raj
Osmania University
Secunderabad, Telangana, India

K. Lova Raju
Electronics and Communication
  Engineering
VFSTR
Guntur, Andhra Pradesh, India

Sharma, S.
Besoins Technologies
Chennai, Tamil Nadu, India

Vishnu, S.
Department of Computer Science and
  Engineering, Amrita School of
  Computing
Amrita Vishwa Vidyapeetham
Coimbatore, Tamil Nadu, India

Archana Sahai
Amity University
Lucknow, Uttar Pradesh, India

Farjana Farvin Sahapudeen
Anjalai Ammal Mahalingam
  Engineering College
Thiruvarur, Tamil Nadu, India

Aravapalli Rama Satish
School of Engineering
JNU
New Delhi, India

Ajay Sharma
Lovely Professional University
Bangaluru, Karnataka, India

Akrati Sharma
Rabindranath Tagore University
Bhopal, Madhya Pradesh, India

Narendra Kumar Sharma
Amity University
Kanpur, Uttar Pradesh, India

Shamneesh Sharma
Up-Grad Education Private Ltd
Bangaluru, Karnataka, India

Richa Singh
KIET Group of Institutions
Ghaziabad, Uttar Pradesh, India

Nidhi Srivastava
Amity University
Lucknow, Uttar Pradesh, India

Roopam Srivastava
Mahatma Gandhi Post Graduate
   College
Gorakhpur, Uttar Pradesh, India

Shikha Srivastava
Mahatma Gandhi Post Graduate
   College
Gorakhpur, Uttar Pradesh, India

Perla Sunanda
G. Pulla Reddy Engineering College
Kurnool, Andhra Pradesh, India

Parul Verma
Amity University
Lucknow, Uttar Pradesh, India

T. Vigneswari
Anjalai Ammal Mahalingam
   Engineering College
Thiruvarur, Tamil Nadu, India

Abhay Wagh
DTE Maharashtra
Mumbai, Maharashtra, India

Preface

The Internet of Medical Things (IoMT) refers to a collection of connected medical equipment, software applications, and health services. These medical devices generate, gather, examine, and transmit health data by establishing connections with healthcare provider networks and transmitting data to internal servers on the cloud. With the help of IoMT, medical organizations can transform the healthcare sector in novel ways, bringing significant changes in terms of facilitating disease management, improving disease diagnosis and treatment methods, and reducing healthcare costs and errors.

According to a leading market research provider, *AllTheResearch*, the global IoMT market was valued at $44.5 billion in 2018; it is expected to grow to $254.2 billion in 2026. The IoMT environment comprises hardware, software and service components, platforms for device management, application management, and cloud management, connectivity devices (either wired or wireless), different types of applications (e.g., on-body devices, healthcare providers, home-use medical devices, community), and end users (e.g., hospitals, clinics, research institutes and academics, homecare). With the integration of artificial intelligence into the environment, which makes the connected devices more capable of real-time, remote analysis of patient data, the IoMT market is expected to grow further in the near future. According to mHealthIntelligence reports, 88% of healthcare providers are investing in remote patient monitoring solutions.

The IoMT helps patients to avoid unnecessary hospital visits by connecting patients to their doctors and transferring medical data over a secure network, thereby scaling up virtual consultation. According to a survey conducted by Accenture, if a patient is given a choice, about 62% of them are likely to opt for virtual consultation and 57% of consumers are positive about home monitoring through medical devices. The IoMT also played a very vital role during the Covid-19 pandemic.

Nevertheless, technology always comes with the threat of data security. The healthcare providers are concentrating on the identification of these threats and minimizing the risks associated with them. IoMT devices are more vulnerable to data breaches and cyber-attacks compared to other wireless equipment. The *American Journal of Managed Care* states that the healthcare industry is the most targeted industry for cyber criminals. Thus, appropriate security solutions are needed to protect the IoMT ecosystem from cyber security threats.

This volume integrates the architectural, conceptual, and technological aspects of IoMT. This book discusses in detail the IoMT, connected smart medical devices, and their applications to improve health outcomes. This book explores various methodologies and solutions for medical data analytics in healthcare systems using machine learning and deep learning approaches, and looks into how technologies such as blockchain and cloud computing can further enhance data analytics in the e-health domain. This book also discusses prevalent IoMT use cases and applications. This edited volume can provide a high level of understanding to researchers, academicians, and learners who wish to delve into the IoMT domain.

Section 1 – IoMT Datasets and Storage: This section discusses the importance of IoMT in the present context and the role of advanced technologies for handling complex e-health data.

"Remote Health Monitoring in the Era of the Internet of Medical Things", authored by Dalal Alshehri, Nasimul Noman, Raymond Chiong, and Shah J. Miah, aims to provide readers with the necessary background information on the IoMT in Remote Health Monitoring (RHM). This chapter introduces the general architecture of IoMT for RHM and discusses the major RHM types, namely, remote diagnosis, remote healthcare consultation, and remote patient monitoring. A brief review of the infrastructure and software technologies, such as 5G, cloud, fog, artificial intelligence, machine learning, and blockchain, is provided, along with the discussion of various hardware-based technologies that are playing a key role in RHM, such as medical sensors and smart devices. Subsequently, this chapter discusses the benefits that the IoMT can bring to RHM and challenges to overcome in order to take full advantage of IoMT in RHM. Finally, three representative case studies of RHM developed with the help of IoMT are presented to give readers a better understanding of the sophisticated technologies used in such environments to deliver enhanced and efficient services.

"Diabetic Health Care Data Analytics and Application", by Dev Kapadia, Pratyush Mohanty, Anupama Namburu, and Aravapalli Rama Satish, comprehensively discusses the data analytics and applications used for analysing diabetes. This chapter introduces various architectures of diabetic health data modelling and analysis, devices, sensors and applications for different stages of diabetic data acquisition, processing, prediction, and treatment planning. This chapter also discusses the secondary complications and associated analytics available for the diagnosis of diabetes.

"Blockchain for Handling Medical Data", by T. Vigneswari, Farjana Farvin Sahapudeen, and G. Kalaiselvi, deals with the application of blockchain in handling medical data. Blockchain offers several features such as tamperproof, confidentiality, immutability, traceability, data integrity, security, and privacy, which make it more convenient to use along with the IoMT. The major data management components are data storage, access control, data preservation, privacy preservation, and data auditing. This chapter offers a detailed review of how the enlisted data management components are achieved efficiently by integrating blockchain with the IoMT to provide better healthcare to users of the IoMT framework.

"Cloud Computing for Complex IoMT Data", by Akrati Sharma and Priti Maheshwary, discusses how the cloud can be used to manage data that have been collected via various medical devices and sent to the concerned end without any manual intervention. This chapter also illustrates another vibrant use of the cloud, which is based on data collection and how it helps to diagnose diseases and assists in drug formation.

"Potential of IoMT Devices in Early Detection of Suicidal Ideation", by Archana Sahai and Umar Khalid Farooqui, aims to provide a comprehensive overview of the current state of research on the use of IoMT devices in mental health and discusses how the IoMT devices have been used for various purposes, including tracking and monitoring symptoms, providing real-time feedback, and facilitating communication with healthcare professionals to improve mental health outcomes, particularly in the areas of mood disorders, anxiety, and stress.

Section 2 – Machine Learning for Medical Things: This section deals with scalable machine learning and deep learning approaches that can provide solutions for health data analytics.

"Application and Challenges of Machine Learning in Healthcare", by Ajay Sharma, Shashi Kala, Gaurav Gupta, Varun Jaiswal, and Shamneesh Sharma, provides an overview of the applications of machine learning in the healthcare industry, including cancer detection, automatic computer-assisted diagnosis, and the development of a cloud-based platform for diseases such as Alzheimer and Parkinson. This chapter also presents case studies on the application of artificial intelligence in healthcare.

"Artificial Intelligence and Internet of Medical Things in the Diagnosis and Prediction of Disease", by Narendra Kumar Sharma, Shahnaz Fatima, and Alok Singh Chauhan, focuses on artificial intelligence methods that are used with IoMT devices in disease diagnosis and discusses the implementation of artificial intelligence and machine learning algorithms in the diagnosis and prediction of diseases. This chapter also explores the applications of artificial intelligence and highlights the diagnostic difficulties in disease prediction. The technological perspective of IoMT and challenges in disease diagnosis and prediction are also included.

"Predicting Cardiovascular Diseases Using Machine Learning: A Systematic Review of the Literature", by Abhay Kumar Pathak, Bhupendra Kumar Dewangan, and Manjari Gupta, aims to provide a comprehensive literature review by considering research work in the field of cardiovascular diseases that was published between 2019 and 2022. This chapter also outlines in detail the different datasets that were considered by different researchers to predict cardiovascular diseases using machine learning techniques. This chapter ends with a critical comparison of the results of all machine learning techniques used in this field.

"Identification of Unipolar Depression Using Boosting Algorithms", by Parul Verma, Roopam Srivastava, and Shikha Srivastava, discusses depression as the depressive disorders that affect 280 million people globally. The current system of mental healthcare is inadequate to deal with it. Smart healthcare is making an impact in different healthcare areas. This chapter discusses the huge potential of IoMT in the identification of depression through data collected by wearable devices that do provide continuous data on various behavioural patterns. This chapter aims to detect depression based on the actigraph watch measures by using two machine learning algorithms, namely, AdaBoost and XGBoost, and show how XGBoost outperforms the other.

"Development of EEG-Based Identification of Learning Disability Using Machine Learning Algorithms", by Nitin Ahire, R.N. Awale, and Abhay Wagh, aims to identify the most efficient machine learning algorithms and data pre-processing techniques for diagnosing learning impairments. Learning disability is impacting a large number of school-aged children who are at risk of experiencing lifelong low self-esteem and deprived academic performances. This chapter utilizes EEG signals from people with learning disabilities to interpret and deliver the most accurate results in the shortest amount of time, by exploring machine learning models including

Principal Component Analysis (PCA), Independent Component Analysis (ICA), and Linear Discriminant Analysis (LDA).

"Deep Learning Approaches for IoMT", by Farjana Farvin Sahapudeen, T. Vigneswari, and S. Krishna Mohan, discusses the role of deep learning algorithms in assessing patient data maintained in IoMT servers and forecasting the existence of deadly illnesses. Security has been identified as a critical aspect that should be seriously considered in such IoMT applications to ensure secured data access, as the medical servers and healthcare records need to be secured from attackers by using an intrusion detection system. This chapter discusses the various deep learning technologies applied in healthcare monitoring, healthcare diagnosis, and the intrusion detection system of IoMT systems.

"Machine Learning and Deep Learning Techniques to Classify Depressed Patients from Healthy, by Using Brain Signals from Electroencephalogram (EEG)", by Bethany Gosala, Emmanuel Raj Gosala, and Manjari Gupta, points out that depression is one of the most common and serious mental disorders and it is likely to be on top among mental disorders by 2030. Detecting depression at an early stage needs an easy, patient-friendly, and inexpensive method. This chapter provides an artificial intelligence solution for this problem based on machine learning and deep learning methods, using EEG signals for the detection of depression. The authors extracted 11 statistical features from the signals using three different criteria: 1) when the eyes of the subject are closed; 2) when the eyes of the subject are open; and 3) when the subject is doing some tasks. Their results show that deep learning achieves the highest accuracy when the subject is doing some tasks and that it is better than some of the available state-of-the-art methods.

"Dimensionality Reduction for IoMT Devices Using PCA", by Rajiv Pandey, Radhika Awasthi, Archana Sahai, and Pratibha Maurya, assesses the significance of utilizing PCA in reducing the dimensions of the dataset used in an IoMT-enabled system, coupling their research with a previously proposed framework called "Prenatal Healthcare System of Remote Mother and Fetal Surveillance via IoMT". The prenatal device increases the likelihood of a safe and healthy birth while lowering pregnancy risks. This chapter discusses the implementation of PCA to highlight variations and identify significant patterns in the dataset to accurately forecast outcomes.

"Face Mask Detection System: CNN-Based Implementation", by Rajiv Pandey, Radhika Awasthi, and Archana Sahai, discusses the World Health Organization declaration of face mask use as a mandatory measure to prevent the spread of infection during the Covid-19 pandemic. This motivated the invention of an automated system to monitor persons wearing face masks. This chapter presents a mask detector that uses a machine learning facial categorization algorithm to determine if a person is wearing a mask or not. The mask detector may then be connected to a CCTV system to guarantee that only those wearing masks are allowed to enter the premise.

Section 3 – IoMT: Data Analytics and Use Cases
"An IoT-Based Real-Time ECG Monitoring Platform for Multiple Patients", by Vishnu, S., Lova Raju, K., Shanmugam, M., and Sharma, S., presents an IoMT-based RHM system for cardiovascular patients to monitor their ECG signals from their homes themselves. Each patient is allotted a node, namely, the Patient Wireless Node (PWN), which consists of three lead ECG electrodes, a signal conditioning

chip (AD8232), and a wireless microcontroller, to transmit the patient's ECG to the ECG-data server. The ECG-data server authenticates each patient's credentials and stores the respective ECG signals, which will be viewed by the doctor or clinician to monitor and analyse the ECG signals for abnormalities. This system overcomes the limitations of existing methods, such as Holter monitors and event monitors, and does not require smartphones to provide Internet connectivity to the PWN.

"Study on Anomaly Detection in Clinical Laboratory Data Using the Internet of Medical Things", by Richa Singh and Nidhi Srivastava, discusses the existing literature on anomaly identification, identifies causes for unusual outputs inside a laboratory, and aims to propose potential fixes for typical flaws in present laboratory procedures. This is much required as it is important to reduce the clinical effect of abnormalities due to the rising volume of laboratory tests. The authors also describe the concepts of IoMT, its architecture, and how to translate the processes into an architectural design.

"Computational Intelligence Framework for Improving Quality of Life in Cancer Patients", by Perla Sunanda and Dwaram Kavitha, analyzes how various physical, emotional, societal, and symptom domains affect the quality of life in cancer patients. This chapter presents a computational intelligence conceptual framework that provides valuable information to monitor health status and quality of life after cancer treatment.

"Major Depressive Disorder Detection Using Data Science and Wearable Connected Devices", by Umar Khalid Farooqui and Archana Sahai, illustrates how IoMT and wearable connected devices are used for collecting, analysing, and processing data of individuals. The authors further examine prominent methods and techniques used to detect major depression disorders by using data science techniques like content analysis, deep learning, and feature engineering.

To wrap up, we would like to extend our gratitude to all the reviewers who have evaluated the submissions and helped improve the ones that have been selected for inclusion in this volume. Thank you for your hard work and support.

Dr. Rajiv Pandey,
Amity University Uttar Pradesh, Lucknow, India.

Dr. Pratibha Maurya,
Amity University Uttar Pradesh, Lucknow, India.

Prof. Raymond Chiong,
The University of Newcastle, Australia.

Section 1

IoMT Datasets and Storage

1 Remote Health Monitoring in the Era of the Internet of Medical Things

Dalal Alshehri, Nasimul Noman,
Raymond Chiong, and Shah J Miah
NSW

1.1 INTRODUCTION

With the tremendous progress in medical science, we can successfully treat and cure many diseases that were impossible to cure even a decade ago. The growing world population and increasing life expectancy are creating pressure on healthcare systems, because it is a legal obligation of states to provide all people with the highest attainable standard of health (Belova & Pavlov, 2020). To tackle the mounting demands, healthcare systems are adopting intelligent and technology-dependent solutions to deliver quick, quality services. The Internet of Medical Things (IoMT), in particular, is leading the way in building 'smart healthcare' by revolutionising the structure, operations and services in healthcare systems.

The Internet of Things (IoT), a significant technological term in this era, has offered numerous benefits for various sectors. According to a report by Grand View Research (2019), the IoT healthcare market is anticipated to grow globally to USD 534.3 billion by 2025, increasing at an annual compound growth rate of 19.9% from 2019. Before addressing the IoT in healthcare specifically, the term IoT must be defined. Currently, a commonly accepted definition of IoT is 'a network of physical objects or "things" that can interact with each other to share information and take action' (Mandal, 2020).

IoT applications range widely from small private networks, such as in smart homes, to large corporate networks, such as in cloud-based business applications. The IoT has been used in different application areas, such as manufacturing, transportation, supply chain and logistics, environmental monitoring, home automation (smart homes), agriculture, aquaculture and healthcare. As one of the most prominent application areas of IoT, the IoMT refers to a hardware and software infrastructure in which medical devices can communicate and share information across a network without requiring any interactions between the users or between users and computers (Vishnu et al., 2020). In an IoMT environment, numerous smart medical devices and sensors, connected over the Internet, collect various types of data from patients and the environment. The gathered data are transferred over the network for storing, analysis and prediction; sophisticated computation methods, such as artificial intelligence (AI) and machine learning (ML), are used to make interpretation, analysis and prediction from

DOI: 10.1201/9781003359951-2

3

the collected data. From that medical diagnosis, treatment prescriptions and additional service requests might be triggered (Srivastava et al., 2022). An IoMT can be deployed in a hospital, for example, for critical patient monitoring in intensive care units (ICUs); at home, for example, for detecting falls or critical medical emergencies such as heart attacks; in a patient's body, for example, for alerting fluctuations in glucose levels of diabetic patients; or in the community, for example, for tracking patients during transit in an ambulance. Although the IoMT is a special application domain of IoT, because of the sensitivity of the information managed, it differs from the general IoT in terms of its design, accuracy, reliability and security requirements, as will become clear in the subsequent sections of this chapter.

1.2   TYPES OF REMOTE HEALTH MONITORING

With the development of IoMT, healthcare services and professionals have turned to activating and using various types of remote health monitoring (RHM), namely, remote diagnosis, remote healthcare consultation and remote patient monitoring.

Remote diagnosis aids in prevention of illness through early and accurate diagnosis even when the doctor and patient are in different geographic locations (Kavitha et al., 2021). It could be costly and time-consuming for people to attend hospitals on a regular basis for diagnosis and check-ups, and there are occasions when they cannot even go to the nearest hospital. Moreover, for many chronic conditions, such as cancer, patients need regular blood testing to monitor the progression of their conditions and the effectiveness of treatment. The development of portable diagnostic platforms allows clinicians to treat patients without the need for planned or specialised hospital visits, which also lowers expenses and improves efficiency of the treatment and convenience of both the clinician and patients. Remote diagnosis can be beneficial in many situations such as for monitoring patients after their discharge from hospitals, for early detection and treatment of heart diseases, and for providing daily reminders to patients on medication. Using this technology, healthcare providers can monitor patients' data, record them, analyse them using AI- and ML-powered algorithms and send back important metrics (Arora, 2020).

The IoMT assists in *remote health consultations* by using the data collected by various sensors and monitoring devices at the patient's end. In the IoMT era, remote healthcare consultation goes far beyond tele-appointments, and specialised consultations can take place during surgery, while transporting through ambulance and so on. Remote healthcare consultation can lower the cost of medical care and offer an early diagnosis of health issues. Additionally, quick and specialised medical advice from a doctor who is located far away can aid people in case of an emergency (Tiwari et al., 2021).

Finally, *remote patient monitoring*, i.e., monitoring and following up patients' vital signs while they are at home, is one of the greatest benefits of technology in the healthcare industry. People who are elderly or who suffer from chronic illnesses require ongoing care from their families. In addition, some people with disabilities frequently live completely alone or are alone for a significant portion of the day when other family members are at work. An IoT-assisted remote patient monitoring system can offer a reliable and satisfactory solution for these patients and can have a significant impact by saving priceless human lives and enhancing their standard of living. The COVID-19 pandemic has catalysed the research and interest in RHM, which otherwise would have developed at a slower pace.

In providing different kinds of RHM services, hospitals become the centre of the IoMT environment and ensure seamless integration across these technologies. In smart hospitals, humans, equipment, places and systems are all connected in real time. A huge amount of data produced by this interconnectedness are shared from edge to cloud securely. Using AI and other technologies, smart hospitals ensure streamlined workflows for key operations, safe and individualised patient care, superior service and excellent patient experience (Dash et al., 2019).

1.3 IoMT ARCHITECTURES

Irfan and Ahmad (2018) and Srivastava et al. (2022) reviewed various IoMT architectures proposed by different researchers. Although the number of layers in those IoMT architectures ranges between two and five, based on their analysis, these authors agreed that the complexities of different IoMT architectures can be abstracted into three basic layers or tiers. The general three-layered architecture of IoMT, which can be used for various types of RHM, is shown in Figure 1.1. First, the *sensors* or *things*

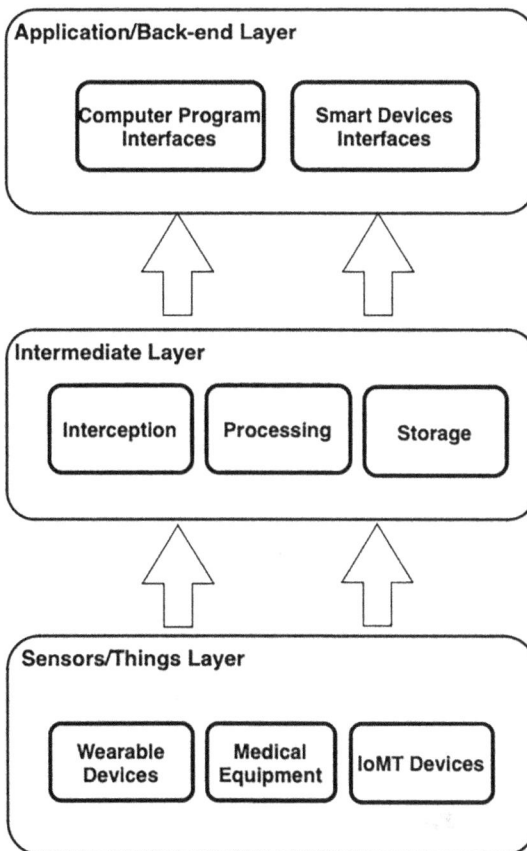

FIGURE 1.1 IoMT architecture.

layer contains different physical entities, such as medical equipment, sensors and smart devices, that can interact using different communication protocols. Second, the *intermediate* or *perceptual* layer is responsible for supporting the interaction among heterogeneous devices, implementing the security protocols and processes, and storing the data coming from the first layer locally in on-body devices such as smartphones or off-body devices such as routers. Third, the *application* or *back-end* layer stores the vast amount of data received from the intermediate layer, processes the data using several applications and provides an interface to multiple services to the end users, i.e., doctors and patients.

For example, the three layers in an IoMT-based patient monitoring system can be conceptualised as follows. The sensors/things layer uses a range of sensors to read and collect data (electrocardiogram [ECG], heart rate, blood pressure, glucose, location, etc.) from the patient. Then, using WiFi or Bluetooth for data transmission, the collected data are processed in the intermediate layer through a tablet or personal digital assistant before passing them to the centralised server. Then, to communicate the data and information gathered by the perception layer in real time, with high precision and reliability, the Internet, mobile communication networks and other specialised networks are used. The application layer, which comes last, is where all the patient data are stored, and it uses various intelligent programmes to analyse the patient data for diagnosis of possible medical conditions and prescribe proper medicines. Figure 1.2 describes a representative architecture for such a system.

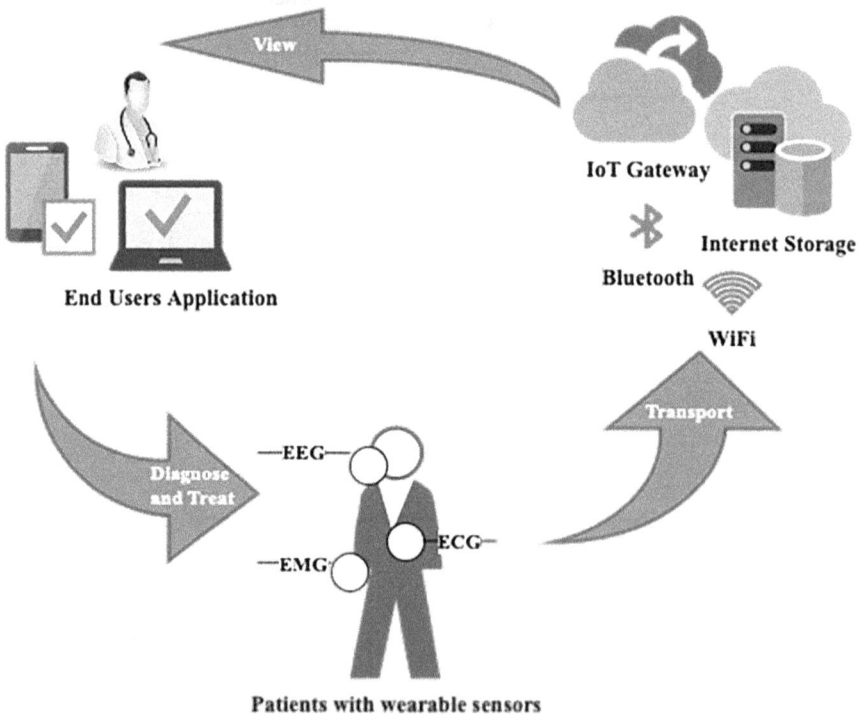

FIGURE 1.2   IoMT-based remote patient monitoring.

1.4   IoMT DEVICES FOR REMOTE HEALTH MONITORING

Different types of medical devices, monitoring systems, wearable hardware and medical sensors are being used in IoMT health monitoring. A smart inhaler is a good example of contemporary IoMT devices. It is used to alert users when it is time to take a dose, instruct users on how to operate the device correctly or extract data from the inhaler to a storage database so that doctors can view their patients' inhaler usage. Smart inhalers employ Bluetooth to track inhaler usage through a mobile app, so the patients, along with their doctors, can manage their asthma better. Some gadgets measure the quality of air in real time, sending users alerts to warn them of potentially dangerous situations and keeping track of the environmental triggers for asthma attacks (Yan, 2018). Many companies have created similar devices for different purposes, such as Respira, which is developing a single device with these three processes: breath-activated dose counters, customising particles to improve medication distribution, and smart technical capabilities (Yan, 2018). Additionally, this device eliminates the inaccuracy and unpredictability of therapies by employing designed molecules to separate the patient's inhalation technique from drug intake. According to NekoPoi (2020), the Propeller smart inhaler demonstrated a 50% increase in medication adherence in more than 15 clinical studies, which resulted in a 79% decrease in asthma episodes.

Several other IoMT devices, such as ingestible sensors, are available or under development to be used in the IoMT environment in the near future. Ingestible sensors are electronics in the shape of pills that, after being swallowed, ping your smartphone with data. Most often, they take measurements of pH, temperature and pressure or check to see whether patients have taken their medications (Molteni, 2018). Because they enable gastrointestinal (GI) information to be acquired with little to no invasiveness, ingestible devices are expected to become more popular as a tool for illness diagnosis, prevention and management. In addition to their small size and ease of use, they provide the benefit of easing the burden on patients and extending the range of in vivo testing. There are various types of ingestible sensors: wireless capsule endoscopy, pH sensors and haemoglobin sensors (Inami et al., 2022). By enabling examination of the digestive system without discomfort or anaesthesia, the wireless capsule endoscopy technique eliminates the dangers associated with traditional endoscopy and may persuade people to have GI tract examinations (Ciuti et al., 2011). The second one is the pH sensor, which can be used in wireless capsule systems. Microfabricated pH, pressure and temperature sensors are part of a wireless capsule that can pick up GI physiological information. The stomach's highly acidic pH values are measured using a microfabricated capacitive pH sensor (Arefin et al., 2018). The third example of ingestible sensors is the haemoglobin (Hb) sensor. Haemoglobin is the primary component of red blood cells, and it is responsible for transporting oxygen from the lungs to the tissues and reabsorbing carbon dioxide from tissues. Anaemia results from a decrease in haemoglobin concentration, and it can be brought on by factors such as blood loss, lack of iron, and illnesses that destroy red blood cells. A quick Hb test by haemoglobin sensors is crucial to detect anaemia, allowing for the most effective use of transfusions while also saving money on medical expenses (Bai & Kumar, 2021). From their functionalities, it is evident that these smart devices are suitable for various types of RHM in an IoMT environment.

1.5 TECHNOLOGIES FOR REMOTE HEALTH MONITORING

The IoMT uses a variety of technologies, including body sensor networks (BSNs), wireless sensor networks (WSNs) and radio frequency identification (RFID), to gather and then transmit sensor data to servers. BSNs, also known as body area networks, are a standard method for short-range wireless communication among devices and sensors placed in users' bodies or clothes. BSNs are primarily used in healthcare systems to monitor patient health remotely using a variety of wireless sensor nodes that are light and consume little electricity (Srivastava et al., 2022). WSNs, initially developed for military applications, became a key driving force for the development of IoT. In WSNs, the network's devices are equipped with sensors to sense the data from the environment and transmit them by actuators in an infrastructureless ad hoc network environment (Kotha & Gupta, 2018). WSNs can be used in RHM to track the physiological state of the subject of observation in real time. Additionally, there are sensors that can assess a patient's body temperature, rate of movement and perspiration to determine the pressure level (Srivastava et al., 2022). RFID, originally developed to use in transportation, became the foundational technology in the early stages of IoT, enabling users to send and receive data using radio frequency signals. RFID tags with a reader and a transmitter should be attached to the linked device. People can find and keep an eye on the devices with the use of an RFID reader (Jia et al., 2012). In RHM, RFID can be used for administration of medical assets and equipment, and gathering of ECG and blood pressure data and other vital sign data from patients (Srivastava et al., 2022).

To support and strengthen its growth, other cutting-edge technologies, such as cloud, blockchain, software-defined networking (SDN), physically unclonable function (PUF), AI and ML, 5G and fog, are being incorporated in IoMT environments. Traditional technologies struggle to guarantee the confidentiality, security and integrity of enormous amounts of data collected and processed in RHM. With their excellent sharing capability, scalability and reliability, cloud and blockchain computing technologies have recently been applied in the IoMT, promising a significant advancement in medical reform. The development of cloud technology has had a significant impact on the way medical data and information are processed, managed and shared, creating a huge prospect of realising smart hospitals. The development of cloud computing-based medical information systems significantly reduces the scarcity of medical supplies, increases the effectiveness and quality of medical services, and lowers the price of treatments. Establishing a platform for sharing medical information based on electronic health records is at the heart of cloud-based IoMT. Hospitals and patients transfer electronic health records comprising patients' vital signs, personal data and other associated information to the cloud for processing and storage. Researchers are primarily concerned with whether cloud-based electronic health data is completely secure and confidential. Using cloud computing, RHM systems transfer medical data from edge devices to the cloud. In the cloud, the received or stored data undergo various types of processing, and the results are stored or returned to the terminal devices. Nonetheless, if the massive quantity of data produced by the expanding medical technology for RHM is uploaded to the cloud, it will put a great deal of pressure on the cloud, causing excessive energy consumption and a significant amount of delay. Because of its centralised computer architecture, such a huge volume of data cannot be handled by cloud computing alone, nor can it give real-time responses. Hence, developing an effective medical cloud IoMT environment requires leveraging distributed computing facilities to enable the processing of computing tasks near the network's edge. In edge computing—an open platform—the essential functions, such as computing and storage, are realised on the edge of the network, locally through distributed computing, avoiding upload of data to the cloud as much as possible (Sun et al., 2020).

Blockchain is a decentralised ledger that records transactions across the entire network of computing nodes. Several security issues in the IoMT can be resolved using blockchain. One of the key challenges with IoMT data is ensuring that they are secure and cannot be tampered with or altered. This is where blockchain immutability comes in. In blockchain, the information passed between network nodes is stored as a sequence of blocks and can be used for cross-references. Blocks for which consensus cannot be achieved are discarded, which opens the door for using blockchain as an entrusted technique for sharing information in IoMT environments (Razdan & Sharma, 2022). Each block in the blockchain contains a cryptographic hash of the previous block, which makes it virtually impossible to alter the data in any block without changing the hash of all subsequent blocks (Saldamli et al., 2022). Hence, the blocks are immutable, and it is easy to track down the source of fraudulent or illegal activities in the network. By using a blockchain to store IoMT data, healthcare organisations can ensure that the data are secure and tamper-proof, and cannot be altered in an unauthorised way without leaving a trace (Pelekoudas-Oikonomou et al., 2022). This can be especially important for maintaining sensitive medical data, for example, remote patient monitoring data, where accuracy and privacy are of utmost importance. The distributed database electronically stores encrypted information in a decentralised fashion, ensuring the data security and authenticity. As a result, it fosters confidence without the help of a third party. Hence, using blockchain within an IoMT-based framework for RHM would provide enhanced security for digital health records and sensitive information of patients (Srivastava et al., 2022).

Data plane and control plane are two components that make up the network in general as well as in IoMT environments. The control plane carries out the necessary duties so that the data plane can decide how to proceed with the traffic to its target destination (Razdan & Sharma, 2022). SDN is a novel approach that separates the control and data planes and transforms the traditional network architecture into a more agile, controllable and dynamic system. SDN allows the network to be flexible and evolve into new architectures using high-level programming languages. Using SDN, more flexible and dynamic network structures can be created in WSNs and BSNs (introduced above). Additionally, SDN is used in multiple IoMT architectures for many reasons, namely, protecting the network by identifying threats and malware; enhancing energy consumption, network delay and effective throughput; and satisfying different quality of service (QoS) requirements. One example of SDN usage, as suggested by Razdan and Sharma (2022), is for IoMT devices to be linked to e-healthcare applications via an SDN control plane that gathers and sends data from the IoMT devices to the applications.

A PUF is a device that takes advantage of the randomness that is inevitably present during production of semiconductor devices to give a physical object a distinctive fingerprint. These devices have a wide range of possible uses, including complex protocols such as transfer of knowledge without awareness, key renovation, key exchange and virtual existence evidence, as well as anti-counterfeiting, identity, authentication and key creation (Gao et al., 2020). When it comes to device authentication in an IoMT environment, PUFs produce a distinct fingerprint for the IoMT-vulnerable hardware that cannot be cloned even with physical access to those devices. In the RHM environment, sensors and other devices in the 'things' layer are vulnerable to attacks aimed at tampering with the hardware; hence, these fingerprints can be used to generate cryptography codes that will safeguard those devices and the data they contain (Razdan & Sharma, 2022).

Among other technologies, AI and ML are offering various applications for the IoMT environment. AI and ML are assisting health practitioners in their duties by simplifying human interventions in disease diagnosis, medical image analysis, treatment plan or dose determination and many other decision-making processes. AI can expand and simulate the application, theory and technique of human intelligence and thought processes (Sun et al., 2020). From vast amounts of data, AI systems can glean relevant information and produce actionable insights that can be used in a variety of applications. ML is a subfield of AI; it uses algorithms for learning a model of data and making predictions on unseen data. In the RHM environment, ML plays a crucial role in identifying patients' condition, keeping track of their health and advising necessary precautions. In addition, ML has been successfully incorporated into paediatric care in recent years to provide the best, most personalised treatments for children. Since the emergence of COVID-19 pandemic, ML has come into the public spotlight (Javaid et al., 2022).

One of the benefits of applying ML in remote patient monitoring is that analysis of big data that aids in the diagnosis and prevention of diseases becomes possible. Analytics can be used on patients' big data to identify people who need to alter their lifestyles in order to prevent the deterioration of their health (Sampathkumar et al., 2022). For instance, individuals with heart failure in its early stages, which is frequently brought on by certain risk factors such as hypertension, should be able to gain from preventive therapy. To achieve an excellent result in the analysis, ML uses various techniques, including convolutional neural networks (CNNs). CNN is a neural network model used for identifying and classifying images based on deep learning knowledge. In their paper, Sampathkumar et al. (2022) proposed an IoMT system that uses a metaheuristic optimisation technique for large data analysis and employs CNN for data categorisation to predict diabetes.

Moreover, in recent years, implementation of 5G for communication in different fields has been growing quickly because of its higher speeds compared with other wireless networks. 5G has created the opportunity to provide many services in hospitals, such as speeding up the download of patient files, enabling the rapid transmission of huge image files and data analytics, and real-time patient monitoring (Barneih et al., 2021). For wearable and wireless communication, full-fledged implementation of IoMT in RHM will greatly benefit from 5G technology.

Fog is a decentralised infrastructure that was developed as an alternative to centralised cloud architecture. Fog architecture is a compromise between cloud and edge computing. To provide computation or storage services to IoMT devices, the fog nodes need to connect to end devices using a wireless connection or Ethernet. Fog nodes receive, gather and analyse patient data from sensors and medical equipment in real time (Mourse et al., 2021). Fog networks have to work with BSNs to reduce the cost of administration, increase scalability and improve the effectiveness of resource use.

1.6   BENEFITS OF IoMT-BASED RHM

The infrastructure and technology that constitute the IoMT environment are bringing fundamental changes to the way health systems operate, services are offered and stakeholders interact with the health system. In general, the IoMT has created opportunities to guarantee better services for everyone's health, an efficient and productive operating environment for healthcare professionals and a safe and secure environment for managing sensitive personal data. IoMT-based RHM systems offer patients urgent medical care wherever they are and whenever they need, providing whatever service is best for their needs, thus offering numerous benefits to both the patients and healthcare system. Using the following scenario, we will highlight some of the key advantages of an IoMT-based RHM system.

Sensors in the RHM system accumulate patient data and send the data via wireless networks. Through smartphones and other portable devices, healthcare providers can notify patients about impending medical appointments and medication. Next, medical records can be sent quickly and easily to responsible doctors and healthcare experts. This saves patients' time and eliminates the necessity of filling multiple forms in the hospital, and also offers the highest quality healthcare at a minimal cost. The time requirements and expenditures can be further reduced in RHM when doctors can consult, diagnose a medical condition and prescribe treatments for patients without the need for specialised or planned visits. Additionally, IoMT-based RHM has created a wider range of services for babies, the elderly and disabled patients. For example, a computerised pill dispenser can be used to track medicine consumption and relay usage data to doctors and medical providers for the benefit of the patient.

Moreover, to support patients who have a chronic disease and need ongoing monitoring and care for their well-being, IoMT sensing devices are able to contact and send alerts to ambulances and medical personnel simultaneously in case of an emergency. Using RFID, IoMT-based RHM systems can track and ensure that the patient takes the correct drug with suitable doses and thus help to prevent uncertain medication errors that may result in patient mortality. Furthermore, with the use of IoMT, the risk of treating infectious diseases (e.g. COVID-19) may be lowered, by making decisions remotely based on the information collected from wearable devices. Besides the benefits for the patients and healthcare providers, the IoMT helps in managing hospital equipment efficiently and effectively. Medical equipment and wearable sensors used by doctors and nurses may convey feedback and alter their configurations automatically in accordance with the needs of patients and professionals (Irfan &

Ahmad, 2018). In summary, it can be stated that IoMT-based RHM can significantly contribute to lowering healthcare costs, offering superior health services more efficiently, enhancing patient quality of life, improving sustainability and expanding the healthcare system by lowering the need for hospitalisation (Arora, 2020).

1.7   CHALLENGES FOR IoMT-BASED RHM

As a nascent technology, the IoMT has several challenges to overcome, including security, scalability, device limitations, reliability of IoT devices, inadequate data collection and low communication rate. First, using IoMT devices to collect highly sensitive data from patients and transfer them through the network will pose significant security challenges. Increasing the number of IoMT devices in RHM will open new doors for different types of potential attacks, such as confidentiality attacks against devices or data, tag tracking, replay attacks and denial-of-service attacks. Hence, network-based medical data transfer requires adequate security and protection from external interference or monitoring (Irfan & Ahmad, 2018). Additionally, user data are saved in the cloud for more accurate analysis and quick responses to the patients under the monitoring of IoMT devices, but this carries the risk of user data being accessed or used in an unauthorised manner. In recent history, cyber hackers have chosen to target healthcare facilities for personal information such as email addresses, social security numbers, blood type and other sensitive health records of patients. Because of their high sensitivity and private nature, these types of data are especially targeted by cybercriminals (Hireche et al., 2022). Therefore, the IoMT-based RHM environment should be able to ensure the confidentiality and privacy of all these data from any kind of unauthorised access or use.

Another challenge for IoMT-based RHM is ensuring QoS. Real-time BSN applications, such as ECGs, are extremely vulnerable to data loss and have tight deadlines. Therefore, it is crucial that the QoS handles such circumstances. Additionally, the network transmission overhead and data transfer overhead contribute to a high transmission latency. Moreover, the majority of IoMT components have limited energy and processing resource capabilities. The processing capabilities, in terms of central processing unit, random access memory, read-only memory, and so on, of medical devices employed in the IoMT environment are limited. Another issue in RHM is energy efficiency, which affects the size, lifespan and usability of the IoMT devices. As a result, the routing protocol should reduce the IoMT device's energy usage and monitor the device's battery life. The usability of wearable devices is affected by frequent battery changes or recharging (Srivastava et al., 2022). Maintaining the data integrity and reliability of IoMT devices is necessary because most IoT equipment used in healthcare is wireless in nature. Data integrity assures that patient health information is accepted accurately and without modification during transmission. Tampered or incorrectly collected data may result in inaccurate diagnoses, which may in turn result in the use of wrong medications and treatments that may be fatal (Hireche et al., 2022). Furthermore, the continuous provision of approved services only to authorised users, which is known as IoMT availability, must also be taken into consideration.

Additionally, scalability is one of the most pressing issues in an IoMT environment. The capacity of a healthcare IoT system to continue running efficiently even when the system load changes dynamically in terms of sensors, hardware or any other services, is known as its scalability (Moosavi et al., 2016).

1.8 SECURITY AND PRIVACY REQUIREMENTS FOR RHM

As mentioned in the previous section, security and privacy concerns are the key challenges in adopting the IoMT. Because of their significance and implications for RHM, we discuss them separately in this section. The most critical issues to be resolved are confidentiality, integrity, availability, non-repudiation and authentication. First, in IoMT environments, confidentiality focuses on safeguarding the medical data that patients disclose to their doctors or other medical personnel privately. Therefore, limiting network access and encrypting data are essential for ensuring the secrecy of the IoMT environment (Moosavi et al., 2016).

Second, privacy is concerned with shielding patient information from prying eyes and attempts to use it illegally. Many nations currently have laws in place governing the gathering and storage of patient health data for privacy regulations, e.g., the Health Insurance Portability and Accountability Act in the United States. These privacy laws are upheld by the IoMT systems. Next, for IoMT-based RHM systems to function properly, data integrity is essential. This is mostly related to ensuring that the data are not altered during wireless transmission and are delivered to the intended location without modification. It safeguards patients' data from being manipulated or processed by unauthorised users. Message authentication codes and a cyclic redundancy checksum, which are used to identify random errors during packet transmission, are often used to provide data integrity (Moosavi et al., 2016). Data integrity is crucial in healthcare since diagnoses, medications and health states are directed by data. Proper and adequate precautions must be in place to avoid any hostile attempts to alter the collected data.

Next, availability is a crucial part of an RHM system, particularly when a patient's health needs to be monitored continuously. The system therefore must guarantee to provide unquestionable data storage and a safe way to transmit if denial-of-service or distributed denial-of-service attacks occur. In addition, to ensure availability, the system's functionality and capability to tolerate and fast response to any faults should be increased.

Moreover, the capacity to hold any authorised users accountable for their actions is known as non-repudiation. This requirement prevents authorised users from rescinding previous system activities or responsibilities. This metric assesses a system's ability to verify whether an activity took place or not. The simplest way to meet this requirement is to employ digital signature methods. The use of blockchain can be another way to trace some types of user activities in an IoMT system.

Authentication means that users' identity should be confirmed when they log in to a system. The safest type of security is mutual authentication, which requires the client and server to first authenticate one another before exchanging secure keys or data. The lack of memory in some IoMT devices or a weaker processing unit that cannot handle the cryptographic operations needed by traditional authentication protocols has led to an increase in the use of lightweight authentication techniques. Authorisation refers to the necessity to safeguard sensitive medical data from illegal or unapproved access. As a result, in an RHM environment, only trusted parties should be granted access to carrying out specific tasks, such as configuring medical IoMT devices, upgrading their firmware or installing security patches. Finally, anonymity ensures that patients' and doctors' identities remain confidential when unauthorised users interact with the system. In certain scenarios, while a patient and a doctor are interacting, their identities should not be shared. Attacks that are passive can monitor only what a person does, but not who they are (Hireche et al., 2022).

1.9   REMOTE MONITORING CASE STUDIES

The financial and administrative pressure that healthcare organisations and professionals are undergoing can be ameliorated by remote patient monitoring and remote diagnosis systems. The growth of the world's ageing population and the desire of elderly people to maintain their independence have exacerbated this need. One crucial aspect of IoMT-based RHM is real-time data processing and management. In this section, we present three use case studies of RHM that were developed for the IoMT environment to give readers a better grasp of the advanced technologies used in such contexts to deliver enhanced and efficient service.

1.9.1   ELDERLY PATENT MONITORING

Our first case study is the RAMi platform, developed by Debauche et al. (2022) using an IoMT infrastructure to provide a real-time framework to monitor elderly patients. The RAMi platform is intended for the monitoring of patients and elderly people during their hospital stays and after their return to home. To avoid false-positive alerts, it contains a real-time early warning system that analyses real-time data using ML algorithms as well as an annotation system that specifically tags each patient's alert on historical data. Additionally, the system enables the training of AI algorithms using federated learning while maintaining data security and privacy. The system also uses AI to find patterns of intrusion.

To address the unique issues of IoMT, this cutting-edge architecture incorporates a things layer, where information is acquired using sensors or cell phones, a fog layer constructed on a smart gateway, blockchain, AI, mobile edge computing (MEC) and a cloud component. In accordance with the workload, resource needs and degree of secrecy, data are processed at the fog level, the MEC or the cloud. While a global blockchain uses smart contracts to safeguard transactions and data sharing, a local blockchain enables task management between the fog, the MEC and the cloud. To prevent and remain unaffected by potential connection problems, such as link congestion and momentary network connection outages, that could be disastrous for patients, the design connects a cloud and a fog level of treatment.

False positives, which frequently call for human intervention, are expensive for medical facilities. By reducing the number of false alarms, the architecture lowers the expenditures associated with them, thereby enabling the care structures to be relieved. To simplify deployment, scaling and management, they selected an architecture that is deployed using containers. The data are sent to the MQTT server and translated into messages that are then queued up. The data train is consumed and transformed in real time, and data anomalies are predicted. The relational database stores the patterns found so that they can be annotated. The eponymous service allows the nursing team to categorise patterns for each patient into real and fake positives. The time series database stores all the data produced by real-time processing, and deep storage is used to retain all data. However, relational database data must first be transformed into objects. An administrative model, a monitoring system, a log system and other software components that make up the architecture's core form the foundation of the cloud portion of the architecture.

1.9.2 Monitoring of Infectious Diseases

In the second case study, we will outline how the IoMT can offer a unique advantage over conventional medical practices in battling serious global health crises such as COVID-19. With its maturity, the IoMT will find more advanced applications and technologies in the coming years that will enable more accurate disease diagnosis, innovative facilities, effective cost reduction and quicker healthcare services (Hireche et al., 2022). The COVID-19 pandemic has demonstrated that with the IoMT, it is possible to deliver the necessary healthcare services to patients with infectious diseases in a faster, safer and more convenient way. Remote monitoring of COVID-19 patients is quite helpful and effective in non-severe cases. According to Sharma et al. (2021), there are many innovative IoMT devices and facilities for telemedicine in hospitals and medical centres that can enable seamless interactions between doctors and patients and thereby reduce in-person visits and protect both patients and doctors from infection.

VinCense is one example of such facilities. It is a B2B pre-screening tool from MedIoTek Health Systems. VinCense is an IoMT platform that can track key COVID-19 risk factors such as one's skin temperature, oxygen saturation, blood pressure and breathing rate. Before doing a COVID-19 test, the gadget is used to check for common signs of the disease. It should be noted that the device's results do not always translate effectively into a confirmation that a person is COVID positive. All the patient data are kept in the cloud and are accessible through the VinCense smartphone application. The device divides patients into four danger categories—low risk, moderate risk, moderate-severe risk and extreme risk—depending on the readings. Patients with low risk may be instructed to isolate themselves, while those with moderate risk or higher are advised to take the COVID-19 test, according to Sharmila Devadoss, the founder and managing director at MedIoTek Health Systems. By using such platforms, doctors and nurses can keep a safe social distance from the suspected patients, who can use the smartphone app to exchange data with healthcare experts. In addition, through the use of the data, which are updated on the app at predetermined regular intervals, close monitoring of the patients is possible while greatly reducing the risk for frontline health workers (Ganguly, 2020).

1.9.3 INTENSIVE CARE UNIT PATIENT MONITORING

The last example of the IoMT application in RHM is in the intensive care unit. Sarmiento-Rojas et al. (2021) proposed an architecture to prevent the development of pressure ulcers by using a new device that monitors the patient position changes from right lateral decubitus to left lateral. The architecture was created for the IoMT platform, which has a multilayered structural design. The hardware used for the system consists of a microcontroller, an inertial measurement unit sensor and communication cards that track and broadcast the status of the variables observed by the sensor. These devices work in the first layer, which is typically referred to as the perception layer. The second layer, known as the network layer, is responsible for data transmission and network connectivity. Using a WiFi communication card and the MQTT (Message Queuing Telemetry Transport) protocol, data are transmitted over the Internet and sent to the cloud for storage. The HTTPS protocol is used to send data from the development platform. The third layer is in charge of handling, displaying and storing the data.

The authors used a relational database as the primary storage method. A website was created where the positional information of the patients can be visualised. Various audible and visible notifications via on-screen messages are used by the same web application to send warnings. As a result, the notifications let the medical staff know when it is necessary to change the patient's position. The application layer, which is the final layer, is concentrated mostly on the health sector because it may be applied to medium- and high-complexity ICU services. The study showed that the usage of this device boosted healthcare professionals' compliance with position alterations and thereby significantly reduced pressure ulcers.

1.10 CONCLUSION

The IoMT's robustness and adaptability are sufficiently attested to by the explosion of successful applications in various sectors of healthcare. However, this field's ultimate objective is much broader and more universal than its extraordinary performance, in particular in health circumstances. We need to integrate and use the knowledge we have amassed from IoT and related technology scenarios to achieve the success of IoMT. Under the umbrella of IoMT, the area of RHM, which is quickly expanding, has made outstanding progress in a variety of real-time healthcare applications.

In this chapter, we presented a very general and preliminary introduction to RHM in the IoMT environment. The main objective was to walk readers through the IoMT background, technologies, benefits and challenges in remote healthcare systems. The architectures of IoMT were presented along with its layers and related devices used in them. The principal variations of different merging technologies used with the IoMT, such as cloud, blockchain, SDN, PUF, AI and ML, 5G and fog, were also introduced. In addition, the major benefits of RHM were briefly discussed. As with any other technology, the IoMT has some challenges, namely, security, limited energy, data integrity, device limitations, availability, inadequate data collection, and scalability, and these were presented with suggested solutions. Finally, we presented three use cases of IoMT in RHM used in hospitals and patients' homes.

REFERENCES

Arefin, M. S., Redouté, J. M., & Yuce, M. R. (2018). Integration of low-power ASIC and MEMS sensors for monitoring gastrointestinal tract using a wireless capsule system. *IEEE Journal of Biomedical and Health Informatics*, *22*(1), 87–97. https://doi.org/10.1109/JBHI.2017.2690965

Arora, S. (2020). IoMT (Internet of Medical Things): reducing cost while improving patient care. *IEEE Pulse*, *11*(5), 24–27. https://doi.org/10.1109/MPULS.2020.3022143

Bai, J. R., & Kumar, V. J. (2021). Use of magneto plethysmogram sensor for real-time estimation of hemoglobin concentration. *IEEE Sensors Journal*, *21*(4), 4405–4411. https://doi.org/10.1109/JSEN.2020.3029179

Barneih, F., Majali, E. A., Alshaltone, O., Nasir, N., & Al-Shammaa, A. (2021). 5G Ring Antenna for Internet of Medical Things (IoMT) Applications. In *Proceedings of the 14th International Conference on Developments in eSystems Engineering (DeSE)*, Sharjah, United Arab Emirates.

Belova, G., & Pavlov, S. (2020). Some comments on the highest attainable standard of health. *International Conference on Knowledge-Based Organization*, *26*(2), 134–140.

Ciuti, G., Menciassi, A., & Dario, P. (2011). Capsule endoscopy: from current achievements to open challenges. *IEEE Reviews in Biomedical Engineering*, *4*, 59–72. https://doi.org/10.1109/RBME.2011.2171182

Dash, S., Shakyawar, S. K., Sharma, M., & Kaushik, S. (2019). Big data in healthcare: management, analysis and future prospects. *Journal of Big Data*, 6(1), 54. https://doi.org/10.1186/s40537-019-0217-0

Debauche, O., Nkamla Penka, J. B., Mahmoudi, S., Lessage, X., Hani, M., Manneback, P., Lufuluabu, U. K., Bert, N., Messaoudi, D., & Guttadauria, A. (2022). RAMi: a new real-time internet of medical things architecture for elderly patient monitoring. *Information*, *13*(9), 423. https://www.mdpi.com/2078-2489/13/9/423

Ganguly, S. (2020). *Healthcare startup MedIoTek's IoMT device helps prescreen COVID-19 patients*. YourStory.com. https://yourstory.com/2020/04/coronavirus-healthtech-startup-mediotek-iomt-device-covid-19

Gao, Y., Al-Sarawi, S. F., & Abbott, D. (2020). Physical unclonable functions. *Nature Electronics*, *3*(2), 81–91.

Hireche, R., Mansouri, H., & Pathan, A.-S. K. (2022). Security and privacy management in internet of medical things (IoMT): a synthesis. *Journal of Cybersecurity and Privacy*, *2*(3), 640–661. https://www.mdpi.com/2624-800X/2/3/33

Inami, A., Kan, T., & Onoe, H. (2022, 9–13 Jan. 2022). Ingestible Wireless Capsule Sensor Made from Edible Materials for Gut Bacteria Monitoring. In *Proceedings of the IEEE 35th International Conference on Micro Electro Mechanical Systems (MEMS)*, Tokyo, Japan.

Irfan, M., & Ahmad, N. (2018). Internet of Medical Things: Architectural Model, Motivational Factors and Impediments. In *Proceedings of the 15th Learning and Technology Conference (L&T)*, Jeddah, Saudi Arabia.

Javaid, M., Haleem, A., Pratap Singh, R., Suman, R., & Rab, S. (2022). Significance of machine learning in healthcare: Features, pillars and applications. *International Journal of Intelligent Networks*, *3*, 58–73. https://doi.org/https://doi.org/10.1016/j.ijin.2022.05.002

Jia, X., Feng, Q., Fan, T., & Lei, Q. (2012). RFID Technology and Its Applications in Internet of Things (IoT).In *Proceedings of the 2nd International Conference on Consumer Electronics, Communications and Networks (CECNet)*, Yichang, China.

Kavitha, B. C., Reddy, K. V., Udaya Kumar, J., Sivakrishna, K., Sankaran, K. S., & Rani, G. V. (2021). IOT Based Remote Health Monitoring System. In *Proceedings of the International Conference on Computational Performance Evaluation (ComPE)*, Shillong, India.

Kotha, H. D., & Gupta, V. M. (2018). IoT application: a survey. *International Journal of Engineering and Technology*, *7*(2.7), 891–896.

Mandal, S. (2020). *Internet of Things (IoT) - Part 1 (Introduction)*. c#Corner. https://www.c-sharpcorner.com/uploadfile/f88748/internet-of-things-iot-an-introduction/

Molteni, M. (2018). Ingestible Sensors Electronically Monitor Your Guts. In *Wired*, Cambridge, MA, USA. https://www.wired.com/story/this-digital-pill-prototype-uses-bacteria-to-sense-stomach-bleeding/

Moosavi, S. R., Gia, T. N., Nigussie, E., Rahmani, A. M., Virtanen, S., Tenhunen, H., & Isoaho, J. (2016). End-to-end security scheme for mobility enabled healthcare Internet of Things. *Future Generation Computer Systems*, *64*, 108–124. https://doi.org/https://doi.org/10.1016/j.future.2016.02.020

Mourse, A. A. A., El-Bahnasawy, N. A., Elsisi, A., & El-Sayed, A. (2021). Development of Fog Computing Based Patient Behavior Monitoring System. In *Proceedings of the International Conference on Electronic Engineering (ICEEM)*, Menouf, Egypt.

NekoPoi. (2020). Asthma Management. https://www.iotforall.com/use-case/asthma-management

Pelekoudas-Oikonomou, F., Zachos, G., Papaioannou, M., de Ree, M., Ribeiro, J. C., Mantas, G., & Rodriguez, J. (2022). Blockchain-based security mechanisms for IoMT Edge networks in IoMT-based healthcare monitoring systems. *Sensors*, *22*(7), 2449.

Razdan, S., & Sharma, S. (2022). Internet of medical things (IoMT): overview, emerging technologies, and case studies. *IETE Technical Review*, *39*(4), 775–788. https://doi.org/10.1080/02564602.2021.1927863

Research, G. V. (2019). IoT in Healthcare Market Worth $534.3 Billion By 2025|CAGR: 19.9%. In *Grand View Research*. San Francisco, CA. https://www.grandviewresearch.com/press-release/global-iot-in-healthcare-market

Saldamli, G., Upadhyay, C., Jadhav, D., Shrishrimal, R., Patil, B., & Tawalbeh, L. A. (2022). Improved gossip protocol for blockchain applications. *Cluster Computing*, *25*(3), 1915–1926.

Sampathkumar, A., Tesfayohani, M., Shandilya, S. K., Goyal, S., Shaukat Jamal, S., Shukla, P. K., Bedi, P., & Albeedan, M. (2022). Internet of Medical Things (IoMT) and reflective belief design-based big data analytics with convolution neural network-metaheuristic optimization procedure (CNN–MOP). *Computational Intelligence and Neuroscience*, *2022*, 2898061.

Sarmiento-Rojas, J., Aya-Parra, P. A., Sayo, J. M. P., Torres, L. F. C., & Ferreira, O. L. C. (2021). Validation of an IoMT-based monitoring system for pressure ulcer prevention in a hospital environment: a pilot. In *Proceedings of the IEEE 2nd International Congress of Biomedical Engineering and Bioengineering (CI-IB&BI)*, Bogota D.C., Colombia.

Sharma, D., Nawab, A. Z. B., & Alam, M. (2021). Integrating M-health with IoMT to counter COVID-19. In K. Raza (Ed.), In *Computational Intelligence Methods in COVID-19: Surveillance, Prevention, Prediction and Diagnosis* (pp. 373–396). Springer, Singapore. https://doi.org/10.1007/978-981-15-8534-0_20

Srivastava, J., Routray, S., Ahmad, S., & Waris, M. M. (2022). Internet of Medical Things (IoMT)-based smart healthcare system: trends and progress. *Computational Intelligence and Neuroscience*, *2022*, 7218113. https://doi.org/10.1155/2022/7218113

Sun, L., Jiang, X., Ren, H., & Guo, Y. (2020). Edge-cloud computing and artificial intelligence in Internet of Medical Things: architecture, technology and application. *IEEE Access*, *8*, 101079–101092. https://doi.org/10.1109/ACCESS.2020.2997831

Tiwari, D., Prasad, D., Guleria, K., & Ghosh, P. (2021). IoT based smart healthcare monitoring systems: a review. In *Proceedings of the 6th International Conference on Signal Processing, Computing and Control (ISPCC)*, Solan, India.

Vishnu, S., Ramson, S. R. J., & Jegan, R. (2020). Internet of Medical Things (IoMT) – an overview. In *Proceedings of the 5th International Conference on Devices, Circuits and Systems (ICDCS)*, Coimbatore, India.

Yan, W. (2018). Toward better management for asthma: from smart inhalers to injections to wearables, researchers are finding new ways to improve asthma treatment. *IEEE Pulse*, *9*(1), 28–33. https://doi.org/10.1109/MPUL.2017.2772398

2 Diabetic Healthcare Data Analytics and Application

Dev Kapadia, Pratyush Mohanty,
Anupama Namburu, and Aravapalli Rama Satish
JNU

2.1 INTRODUCTION

The study of healthcare data known as medical analytics is used to make the health-care delivery system better by extracting information from patient histories, physician records, lab and diagnostics test results, etc. Analytics plays a pivotal role in converting data into valuable insights; enabling swift communication and collaboration with patients, thereby ultimately elevating the quality of patient care and simultaneously reducing healthcare costs.

A few years ago, a physical examination in the hospital was required to diagnose a disease. Technological advancements have really improved the healthcare system over the years. It has allowed us to check pulse rate, and monitor electrocardiogram (ECG/EKG) and blood oxygen levels using smartwatches. IoT increases the importance of mobile health services because they are crucial for monitoring and managing patients with chronic illnesses like cardiovascular disease and diabetes [1, 2]. The increase in the number of diabetic patients around the world implies a rise in the usage of continuous glucose monitoring (CGM) devices, which have since superseded other forms of continuous monitoring. They give immediate glucose level information [3].

Regarding diabetes, various analysis and prediction models have been created and put into use by several researchers utilising various data mining approaches. In Ref. [4], the authors used a classification technique with the Naive Bayes and Decision Tree algorithms and the Weka tool to find out patterns from the diabetes datasets [3]. These innovative breakthroughs are essential for significantly improving the quality of life for those afflicted from sickness. Three key metrics, namely, immediate glucose, glucose trends, and direction information, can be obtained by utilising a tiny sensor [5] and are helpful in treatment planning of the diabetic patients.

A novel wearable CGM system employs a cloud-based DL algorithm to forecast blood glucose levels utilising previous glucose levels. Moreover, it has been demonstrated that recurrent neural networks (RNNs) are able to recognise temporal auto-correlation features in data, while restricted Boltzmann machines (RBMs) are able to recognise complex distributions in the data [5].

In this chapter, different architectures available for the diabetic healthcare are discussed in Section 2.2; different devices, sensors and applications used are discussed in Section 2.3; and the data processing and data mining techniques for diabetic healthcare

DOI: 10.1201/9781003359951-3

are discussed in Section 2.4. The complications and challenges related to healthcare analytics are presented in Section 2.5. The conclusions are presented in Section 2.6.

2.2   ARCHITECTURES FOR DIABETIC HEALTHCARE

2.2.1   IoT ARCHITECTURE FOR HEALTHCARE

The architecture of an HIoT framework is shown in the (Figure 2.1). (reproduced from [6] under Creative Commons License).

A basic HIoT system contains mainly three components, namely, publisher, broker, and subscriber [7].

Publisher: The publisher is an integral function of sensors and other medical equipment that can act independently or in unison to record the patient's statistics.

Broker: The acquired data must be processed and stored in the cloud by the broker. Finally, the subscriber uses a phone, laptop, tablet, etc. to access and visualise the data as part of the ongoing patient information monitoring [8]. The use of cloud computing in healthcare lowers costs while boosting business effectiveness. Cloud computing makes it much simpler and safer to share medical records while also streamlining back-end activities and facilitating the creation and upkeep of telehealth applications.

The e-healthcare cloud design envisions that every patient will have a profile with his or her fundamental information. The doctors who sign up will also have profiles. If a patient does not already have an account, the registered doctor will create one for them before beginning therapy. Everyone will eventually have an account as a result. The doctor can access the patient's medical history thanks to the patient identification

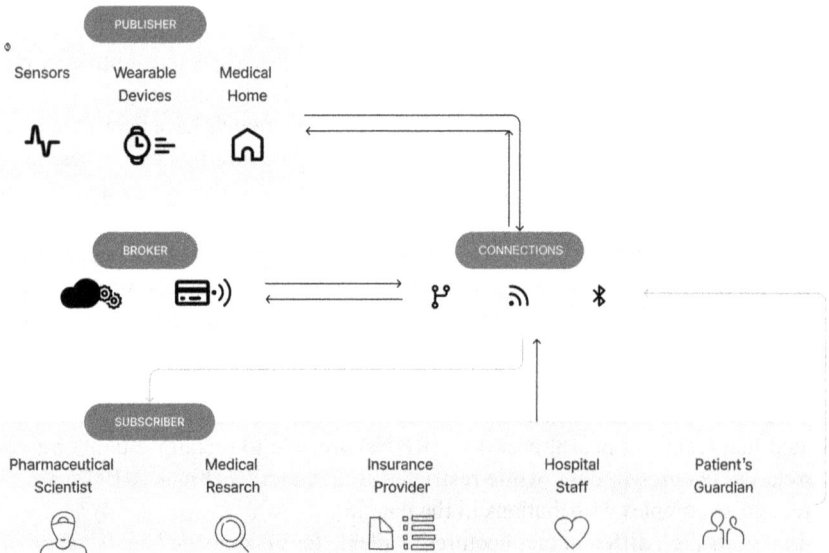

FIGURE 2.1   Components of HIoT system.

number. The client tier provides a rich GUI (UI) for users (doctors, patients) to interface with the systems, while the logic tier executes rules for applications and delivers UI for admins (i.e., government departments, hospitals) [9]. Patients must locate a qualified physician, schedule an appointment, and obtain medical care [10]. Doctors who have enrolled will have their data, including their doctor code, posted. The patient will then schedule an appointment from there. The doctors who are enrolled with the government will log in and obtain their own dashboard [11]. The dashboard includes the doctor's basic information, including their education, affiliation, specialisation, and availability [12].

Subscribers: Monitoring patients and accessing the data from their locations are taken by subscribers. They incorporate the framework components into a computing gird where it serves the purpose on IoT network.

2.2.2 FOG COMPUTING AND EDGE COMPUTING ARCHITECTURE FOR HEALTHCARE

In order to function as a portable fog computing centre and execute some basic data analysis as well as send the monitored data to the cloud, wearable and portable diabetes devices are increasingly offering Bluetooth communication into a tablet or a smartphone [13]. Using a fog computing node, such as a mobile smart device (such a smartphone) or a desktop hub, router, or gateway system, wireless portable diabetes devices can transmit data to the cloud. These tools include closed-loop artificial pancreas systems, insulin pumps, continuous glucose monitors, insulin pens, and insulin pens containing insulin (which are comprised of a continuous glucose monitor and an insulin pump working together) [14].

Sensors, computing, cloud computing, and cloud technologies for wireless diabetic devices are composed of a series. Three different kinds of computing nodes are depicted in the diagram (Figure 2.2): (1) hubs, routers, gateways, smart devices,

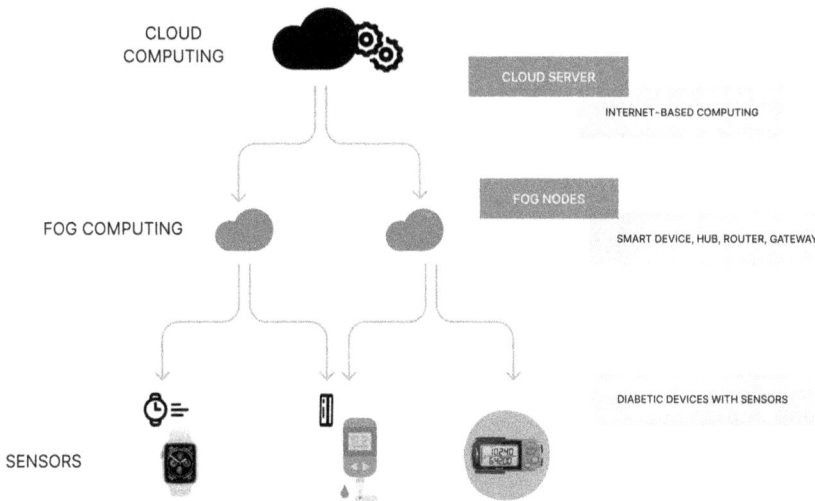

FIGURE 2.2 Tree structure of cloud computing.

and insulin pens are examples of cloud computing nodes. (2) Sensor-equipped mellitus devices, such as glucose levels monitors, smart insulin pens, continuous glucose monitors, and insulin pens, are another category. (3) Cloud servers are Internet-based computing nodes.

2.2.3 Machine Learning Architecture for Healthcare

The layers of the architecture are described (Figure 2.3) in the following: "Data Source," "Data Fusion," "Pre-processing," "Application," and "Fusion." Here is the algorithm:

1. Begin.
2. Enter the data.
3. Employ the data fusion technique.
4. Preprocess the data using multiple techniques.
5. Data division using K-fold cross-validation.
6. Diabetes and healthy individuals' classification using SVM and ANN.
7. SVM and ANN fusion.
8. Analyses architecture performance using a different evaluation matrix.
9. Conclusion.

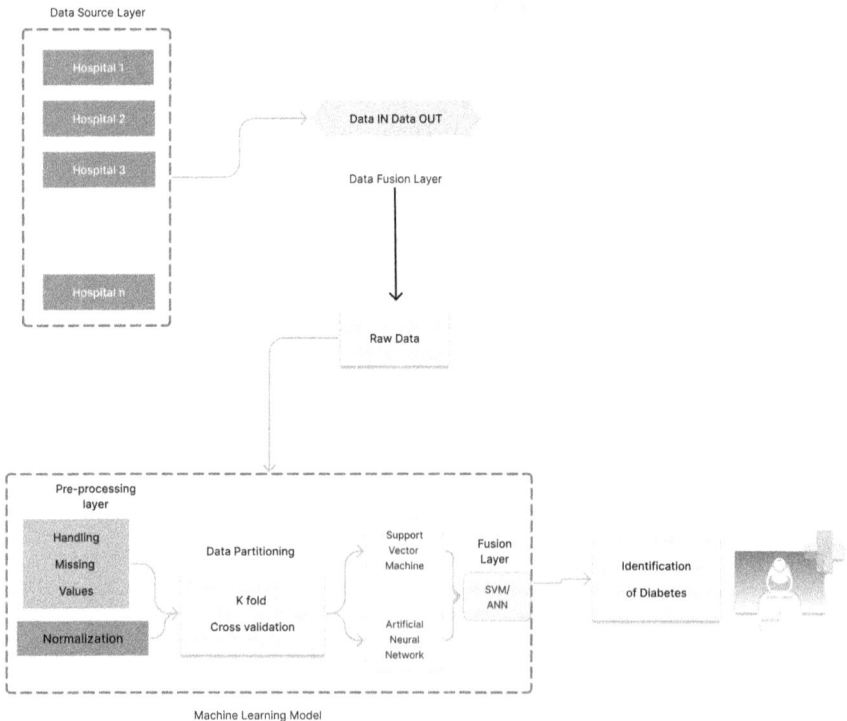

FIGURE 2.3 Machine Learning Architecture to identify types of diabetes.

The most fundamental aspect of the fusion process, input and output data, takes the input data layer and cleans it to better meet the requirements of the production of algorithms for machine learning [15, 16].

2.2.4 BIG DATA ARCHITECTURES FOR HEALTHCARE

Sensing Information: Healthcare monitoring solutions use a wide range of sensors. As they assess a variety of medical indications like body temperature, blood pressure, respiration rate, heart rate, and cardiovascular condition, those devices are crucial for keeping tabs on a patient's health. Patients' living areas could be packed with gadgets like pressure sensors, microphones, and surveillance cameras in order to assure effective health monitoring. As a result, the volume of data created by health-monitoring systems tends to increase dramatically, necessitating the use of advanced methods during the processing stage [17].

Biomedical Imaging: When it comes to identifying diseases and providing healthcare, biomedical scanning is seen to be a potent tool. However, analysing these kinds of images is difficult since they contain noisy data that must be removed to allow doctors to make precise decisions [17].

Social Network Analysis: In order to do a network analysis, you must collect information from online social networks. This stage extracts information that may bring numerous benefits in prediction analysis, such as identifying infectious diseases. In general, social media network data are characterised by uncertainty, making it risky to use them to develop prediction models.

Data filtering: In the presence of enormous amounts of data, data filtering is accomplished by eliminating information that is not pertinent to health monitoring according to a predetermined criterion.

Data cleaning involves many processes including normalisation, noise reduction, and missing data management [17].

The number of traits and situations in the healthcare monitoring has substantially expanded as a result of the increase in health data collection technology in recent years. Selection and extraction of the most crucial features becomes crucial when working with such enormous amounts of data [18] and the detailed process is depicted in (Figure 2.4).

2.2.5 ARCHITECTURE BASED ON DEEP LEARNING

Deep learning architectures, in terms of applications in diabetes diagnosis, are primarily used for classifications. An example will be explored further in application of various deep learning architecture and convolutional neural networks in order to classify images of the retina in order to diagnose the presence of diabetes retinopathy.

In such a case (Figure 2.5), an image dataset is passed through several layers of activation functions as specified by the model (DESNET 12) that contain subsequent Max Pooling layers and layers of functional layers like ReLU in order to improve classification. In another case [19], other deep learning models like ResNET-34 are used on similar retina image dataset along with image data augmentation to improve variety and increase features in dataset. Deep learning algorithms particularly excel in these cases as they rely on the forte of object/feature detection.

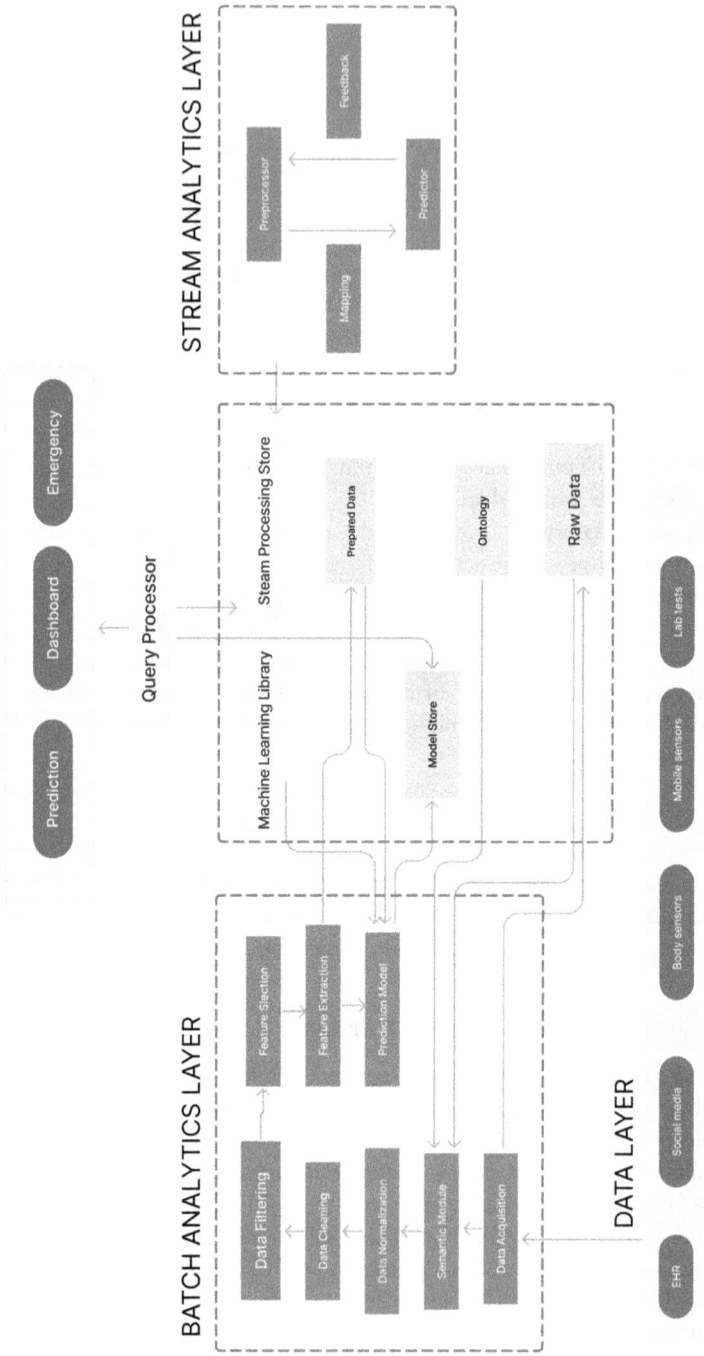

FIGURE 2.4　Big data Architecture of Layers.

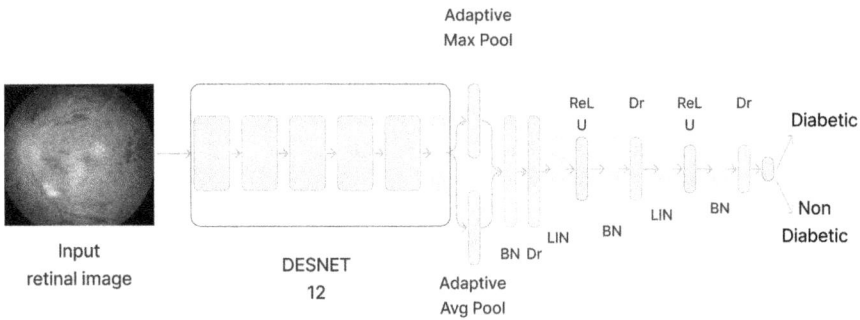

FIGURE 2.5 DESNET 12 architecture.

2.3 SENSOR, DEVICES AND APPLICATIONS FOR DIABETIC HEALTHCARE

2.3.1 CONTINUOUS GLUCOSE METERS

The invention of continuous glucometers has resulted in a shift in the management of diabetes. With a sampling frequency typically between once per minute and once every 5 minutes, these devices enable continuous measurement of blood glucose present in the subcutaneous interstitial fluid, and they display the blood glucose results using correction algorithms. The patient can continue to monitor his blood sugar in this way. The Abbot Freestyle Libre1, one of the most well-known meters, is shown in Figure 2.1. It shows the patient's glycaemia on request via a near-field communications monitoring system. The name of this function is Flash Glucose Monitoring [20].

2.3.2 WEARABLE DEVICES

With research focusing on physiological/pathological factors that caused biosensing and active delivery of therapeutic agents in an on-demand manner, the field of intelligent wearable medical devices has quickly advanced to meet the increasing demand for personal healthcare and overcome the limitations of passive therapy [21].

Wearable biosensing devices (WBDs) have been created to detect a wide range of physiological parameters for biodetection and monitoring, including mechanical deformations, electrocardiogram (ECG) signals, and body temperature [22–24].

The days of pricking fingertips for a drop of blood to test glucose at home are quickly disappearing. In the near future, the elementary home testing devices to measure sugar levels could well be replaced by wearables that can track glucose/sugar in the body all through the day, without the need to prick the fingertip with the needle (lancet) for blood. Service providers have already started offering wearables, known as CGM devices that track glucose levels in the body 24/7. The devices have been developed to empower diabetics to have the ability to monitor their sugar levels throughout the day. These are compact medical systems that come with small sensors (like a patch) that can be attached to the arm, and they take glucose readings in the fluids that surround cells in the body. The sensors come with a limited life, and they have to be replaced after the end of a week or a couple of weeks. Remote health-monitoring applications are a component of the end-to-end intervention system in the majority of the aforementioned intervention categories. Applications for remote monitoring of health connect the user to healthcare professionals and resource managers using sensors and a home hub [25].

2.3.3 IoT and Sensor Devices

"Internet of Things" is an umbrella term that covers various topics related to the augmentation of the Web and the Internet into the physical world, through the massive implementation of spatially devices with embedded identification, sensing, and/or actuation performance characteristics, to enable a whole new category of applications and services [26, 27]. Device-to-Device, Device-to-Cloud, Device-to-Gateway, and Back-End-Data-Sharing are the four common forms of communication listed by the Internet Architecture Board. These models highlight how adaptable IoT devices are when it comes to connecting and giving users value [28].

In Ref. [29], the authors have examined wearable sensor-based systems for tracking and predicting health. The majority of patients in the clinics were ambulatory and could be observed by wearing sensors. The suggested design uses sensors to gather physiological inputs from the patient's body and transmits them to a nearby central node, which can be a smartphone or any smart device equipped with a GUI, alarm signal, logical database, and sensor control unit. By keeping track of the data, the central node transmits the signal or alarm to the ambulance and medical facility. The functioning of a wireless temperature and heartbeat monitoring system using sensors were described in IoT-based healthcare monitoring systems in Ref. [1]. This systems shown in (Figures 2.6 and 2.7) can assist a patient in an emergency situation by monitoring their body temperature and heartbeat, and is controlled by a microcontroller.

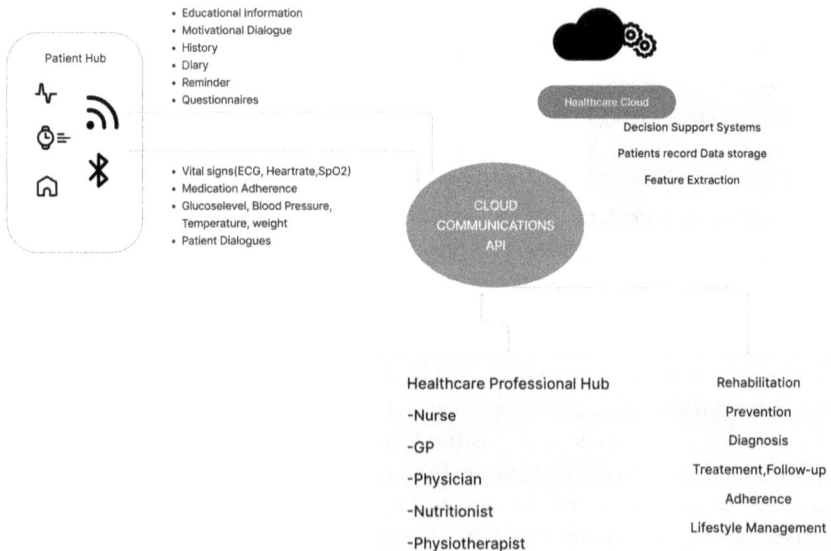

FIGURE 2.6 Various cloud communications API.

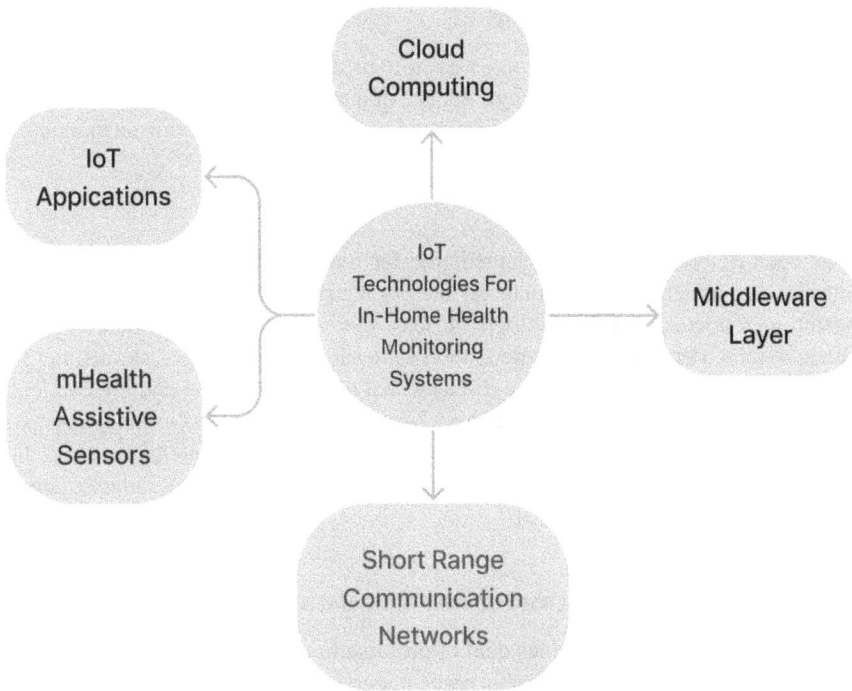

FIGURE 2.7 Various IoT Technologies for in-Home health monitoring systems.

Two key areas of interest in bioelectronic medicine research for the immediate management of diabetes have emerged recently: (1) sampling from the peripheral nerves to obtain metabolic information, and (2) manipulating their electrical activity to reduce swings in glycaemic index. Bioelectronic medicine is a brand-new, very effective therapeutic strategy that has emerged as a result of recent advancements in transdisciplinary research. Recent studies on the former aim to discriminate between hypoglycaemia conditions and neural readings by using the nervous system as a glucometer [30]. Although mainly applicable to type 2 diabetes, control of insulin sensitivity with neuromodulation could also be applied to persons with type 1 diabetes. After eating, some people need exogenous insulin injections to reduce their blood glucose levels [30]. For specific patient populations with illnesses that allow for the implementation of self-care strategies, disease management involves a framework of synchronised healthcare interventions and communications. Disease management enables patients by allowing them to work with other medical professionals to control their condition and prevent complications. The development of tools and technologies for the treatment of diabetes and the application of AI for effective data handling are currently being researched and are promising for the near future. The development of cellphones and wearable technology influences and enhances the cohesive integrity for the benefit of diabetic patients [31]. IMIS offers a platform for communication and agent-based services for users (doctors, nurses, pharmacists, and patients) at home such that diabetic patients can acquire and share data when they need it from any location [32]. Healthcare recipients and all care providers are the two categories in IMIS who have access to information from the shared platforms.

2.3.4 Decision Support Systems

The healthcare professionals are assisted in their decision-making by this decision support system. It is possible through the use of a real-time diabetes alert and monitoring system. The M2DM [33] application is a prime example of this technology.

2.3.5 Telemedicine

It has been demonstrated that telemedicine for diabetic patients improves glucose control, drug adherence, and financial factors. Current diabetes guidelines advise consulting a doctor or a diabetologist at least every 3 months to evaluate HbA1c, change diabetes treatment as needed, and best treat cardiovascular risk factors [34]. Programmes for managing specific diseases that use telemedicine and are explicitly offered by health insurance providers have been launched (e.g., in Germany). As part of their contracts with cost bearers, more than 2,000 diabetic patients received diabetes telehealthcare, and the outcomes in terms of HbA1c reductions and user/patient acceptance were encouraging [35].

2.3.6 Advanced Telediabetology Methods Based on New Technologies

Telediabetology has a tremendous deal of promise to remove obstacles and increase access to high-quality diabetes care in rural, underserved parts of developing nations. We were able to screen for diabetes and its complications as well as bridge the gap between patients and diabetes care providers by using a mobile van equipped with the right equipment, trained technicians and satellite technology. It has particular significance for India given its vast geographic spread and predominately rural population where diabetes care is currently not accessible or affordable. We also guaranteed that diabetes health services were accessible in a significant number of villages in southern India and raised awareness through widespread diabetes education about the significance of rigorous glycaemic control, diabetic complications and their prevention [36].

2.3.7 Mobile Applications

Mobile applications that can address some of these problems have been developed since paper-based diabetes journals are frequently imprecise and perhaps inaccurate (due to poor readability, missing data, or misinterpretation). The number of mobile applications for managing diabetes increased to over 1,500 in 2017, making them the most prevalent disease-related app category [37]. Typically, health information obtained via mobile applications can also be transformed into transferrable information that can be given to healthcare practitioners with the user's permission [38].

2.3.8 Smart Pens

Smart pens have been produced especially for paediatric care, patients with cognitive impairment, and individuals who fail to adhere to their therapy well. These gadgets wirelessly send information on the quantity and time of insulin applications to phone applications and immediately record such parameters [39]. Smart pens make it simpler for users to recall when and how much insulin they last administered by providing a more accurate and clear view of their dosage practices.

2.3.9 DECISION SUPPORT SYSTEMS FOR INSULIN THERAPY

Given that insulin medication can occasionally be challenging to manage and modify in outpatient settings, standardised decision support tools for residential care may be useful for addressing some of these barriers. Specifically, among populations at risk, such as senior citizens who need home healthcare services, or those who are residents of nursing homes, such systems aid in achieving sufficient glycaemic control. This strategy could eliminate the requirement for diabetes-related unexpected patient visits to healthcare professionals. One of the systems, which combines a basal algorithm with a basal plus insulin algorithm, was put to the test on type 2 diabetic patients receiving residential nursing care in a proof-of-concept study [40].

2.4   DIABETIC DATA PROCESSING AND MINING

Data mining: The methodology and processes employed in order to gather useful information from large datasets and data repositories is called data mining. The aim here is to find patterns and subsequently utilise the found patterns to enable decision-making. There are several techniques employed which enable the knowledge discovery process which includes Artificial Intelligence, Clustering, Regression, Classification, Deep Learning Methods and Neural Networks, Decision Trees, and Association Rules, among others [41].

There exist several works, e.g., [42], where classification and prediction ensure the detection of diabetes early. The extraction of useful and viable information from diabetes data using various data mining techniques enables patients to control their glycaemic values [43].

2.4.1 USE OF MACHINE LEARNING ALGORITHMS

Machine learning algorithms are the frontier of such detection methodologies since pattern recognition using various models in the field enables results of varying accuracies, and tuning learning parameters can help increase accuracies. For example, in Ref. [44], we can observe that most used models for data mining-based predictive analysis utilising models like K-Nearest Neighbours, Decision Tree, and Random Forest, among others, on clinical datasets on diabetes mellitus yielded a resulting accuracy greater than 90%.

One thing to keep track is the nature of dataset being used since it affects the features being extracted. Many studies are conducted on the Pima Indians Dataset (like [45]); subjects of this dataset are females who are of minimum of 21 years of age and of the Pima heritage (The Pima People—Native American). The general workflow and pipeline followed by a predictive model goes as shown in Figure 2.8.

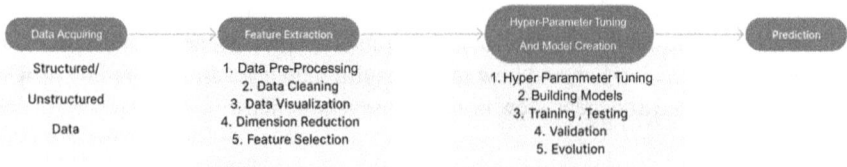

FIGURE 2.8 Essential Learning Process required to develop a predictive model.

2.4.2 FEATURE REDUCTION

Feature reduction is a crucial part in the data mining/analytic workflow since it helps in removing the noise in data and the extra data that may have crept up. Such removal of less important features has three primary goals: reduction of the over-fitting problem, increasing prediction accuracy and reduction of computation time. In works like Ref. [42], there is the application of Multi-Layer Perception (MLP) with novel feature selection utilising various optimisations like Adaptive Particle Swam Optimisation (APSO) to determine early onset of diabetes. Here, these optimisations helped in reducing the number of inputs, and these helped with higher accuracy when compared to methods like Linear Regression, KNN or SVM.

Most machine learning methods and frameworks strive for increasing accuracy by reducing or discarding features. In these circumstances, it can be observed that we are able to achieve the maximum accuracy of the ML models within 95% [45] when the models had been employed on the Pima Indians Dataset (PID). However, it is to note that the PID be explained as a restrictive dataset since it has a mere eight features, 768 records and information only on female patients.

In another instance, diabetes was detected utilising the PID dataset and using SVM with Radial Basis and Linear Kernel, KNN, and ANN with scores of 0.90 and 0.92 for SVM-linear and k-NN, respectively [46].

Such machine learning algorithms and CNN could be employed also to predict conditions which are similar or are a direct result of diabetes, for example, diabetes retinopathy, which can be caused by diabetes mellitus. Early detection of diabetes retinopathy can be done by training deep learning models applied to the images to photographs of the same, as done in Ref. [19] where on obtaining a certain image of the retina, features like the Disc and Macula are detected, which are followed by augmentation of the images, application of the ResNET-34 for training and the subsequent validation and grading. Convolutional neural networks are particularly useful for image classification/detection, and on utilisation with architectures like VGG16 [47], we can variably evaluate their viability in the space of deep learning and machine learning models.

Application of CNN techniques along with more advanced hybrid CNN–Singular Value Decomposition (CNN-SVD) to extract features and reduce them from 256 to 100 leads to decreasing model complexity and helps improve the performance of the model. This shows that tweaking and applying feature reduction in the dataset can sometimes help it outperform when compared to cases of more complex models which can be taxing when it comes to the time taken to execute such models.

2.5 DIABETIC COMPLICATIONS AND ITS MONITORING

Diabetes often comes with a host of complications. Since insulin production is one of the key regulators of chemical equilibrium in our body, its lack or absence can lead to other hosts of problems or symptoms. Diabetes is a widespread disease; in 2015, about 415 million individuals were suffering from the illness (International Diabetes Federation) and that number can go as high as 640 million by 2040 [48]. It has also been estimated that half of the people suffering from it are unaware of it. This makes them more prone when it comes to its complications.

The role of machine learning comes here, with its main function of classifying datasets that plays a major role in the prediction of such complications. It particularly helps in the diagnosis of diabetes or its complications, for instance, prediction of retinopathy based on classifying images. Medical test data are of utmost importance since such tests are present in abundance as most diabetic patients or otherwise would have gone through a series of tests in their adult life.

Several complications experienced by patients include irregular metabolic events such as hyperlipidaemia, hypertension, obesity and other physical symptoms like neuropathy and nephropathy. We can categorise such into thematic areas: microvascular complications, macrovascular complications and other miscellaneous complications [48].

Cardiovascular and neurological problems are also one of the more severe complications in diabetic patients. In Ref. [49], Machine Learning algorithms like Convolutional Neural Networks, Random Under-sampling Boost and Support Vector Machine were used to screen cardiovascular autonomic neuropathy in individuals suffering from diabetes after feature dataset of HRV (heart rate variability) under various scenarios, and these models obtained accuracy levels of 87%–90% on CNN to 98.3% on SVM models.

Similarly, basing the model on features like body mass index and percussive entropy, and subsequently applying Logistic Regression [50] predict future peripheral neuropathy in people suffering from type II diabetes. Diabetic kidney disease (DKD), aka diabetic nephropathy, can be predicted using Big Data analysis and AI models [51], in which feature extraction can include natural language processing and time series pattern recognition using deep learning.

Running a feature prediction model enables us to understand feature importance and key symptomatic indicators or test values to look out for. As viewed in Ref. [52], in the case of feature importance of diabetes classifiers, we see that fields such as Waist Size, Age, Sodium Intake, Blood-Urea Nitrogen, and Diastolic and Systolic Pressures have higher importance than fields like Household Income and Ethnicity. Similarly, while applying the same to the cardiovascular disease classifier, fields like Age, Systolic BP, LDL Cholesterol, and Chest pain rank high on the feature importance [52].

In a similar vein, complications like diabetic foot are characterised by symptoms like infection followed by neuropathy and ischaemia. Here decreasing onset and prevention of symptoms to prevent diabetic foot is of utmost importance. Individuals who are at the highest risk include people with conditions like peripheral vascular disease, neuropathy and foot deformities, and/or the presence of callus. Classification of diabetes foot can be done primarily based on the analysis of the following symptoms:

1. The Neuropathic Foot, wherein neuropathy is the main cause.
2. The Neuroischemic Foot, whose main cause is vascular disease that is occlusive [53].

A severe complication of diabetes mellitus is a degenerative disease that affects the retina of the eye called diabetes retinopathy, which led to blindness or visual impairment in 20.6 million people in 2015 and 3.2 million people in 2020 globally [19]. As mentioned earlier, novel models enable great image processing and classification using CNN-based models like ResNET [19]. Image-based techniques can be used for classification of the severity of the retinopathy, which is characterised by the type of lesions that develop as part of this complication, and it can be categorised into haemorrhages (HM), microaneurysms (MA) and hard/soft exudates (EX) [54].

Since these lesions are essentially round dots or spots, various machine learning models can be run on retinal fundus images obtained during the screening phase of the retina from spatial domain optical coherence tomography (SD–OCT) scans [55].

Neuropathy is essentially degeneration of nerves, and in individuals suffering from diabetes, it manifests in the form of motor, sensory and autonomic parts of the nervous system [56]. Its symptoms predominantly present themselves in limbs; for instance, damage to the internal nerves of the intrinsic foot can lead to imbalance between extension and flexion regions of the affected foot. This can lead to foot deformities that display as pressure points and bony regions which in severe cases can cause skin tear and ulcers [53].

Such a case of skin break or ulcer can particularly get infected and lead to greater consequences as the infection deepens. Neuropathy can be particularly devastating from the fact that such an infection can spread to other compartments of the foot due to anatomical peculiarities, and the individual's lack of pain from the infected area and continual ambulation can accelerate the spread. This is compounded by the fact that there are many more soft tissues in the foot like muscle sheaths, tendons, and fascia that are more susceptible to the infection. So, the problem of neuropathy is amplified by the presence of hyperglycaemia and ischaemia [57].

2.5.1 CHALLENGES IN DIABETIC HEALTH MONITORING

As we have seen across the length of the text, we are mostly using classifiers in order to find whether a complication or the disease exists or not, which is the main information we require. The goal here is to make sure the models being constructed aid in the same while keeping into account how well features influence the dataset. The challenge that rises here is two-fold:

1. Train the model on a diverse dataset that accounts for various scenarios of populous getting diabetes and/or complications along with it.
2. Generate a model with high training as well as validation accuracy while ensuring that the model accounts for relevant features.

In terms of datasets, many of the works we see use the Pima Indian Dataset [45], which is easier to work with; however, as mentioned earlier, it contains data on females which doesn't lend itself to diversity especially when it comes to other attributes/features and can create inherent bias in our classifications. We also see various approaches to these challenges, for example: use of CNN–Singular Value Decomposition [58], use of the random under-sampling boosting (RUSBOOST) [42], and other use cases of optimisers to improve the model like the Grey Wolf Optimiser when applied in integration to a Multi-Layer Perceptron (GWO-MLP).

Similarly the use cases of advanced classifiers and novel approaches with Convolution Neural Networks, in contrast to more simplistic machine learning approaches, usually result in better outcomes as it highlights the importance of updating and improving characteristics of the model and trains it in order to decrease error while optimising it. Utilising these predictive models as an alternative to medical diagnosis is still in question. However, in several cases, early screening of data obtained from patients through such models can be helpful since a machine learning model can pick up on patterns humans may not be able to diagnose.

Diabetes complications that arise in patients can now be addressed by several new antidiabetic drugs—particularly for type 2 diabetes (e.g., receptor agonists, insulin analogues, sodium–glucose co-transporter-2 [SGLT2]). However, accessibility to such medicines is also a point of contention. Insulin and drug pricing is extremely capitalised and competitive, which leaves the end consumers as the losers, particularly patients in low-income and middle-income countries. In Ref. [59], an attempt is made to estimate price targets in order to pursue negotiations for inclusion in national formularies given the addition of these novel agents to WHO's Essential Medicines List. Financial burden to the patients and their supporters is a real setback to any healthcare infrastructure, and a major part of this burden can be attributed to complication that becomes chronic as a result of leaving them untreated, particularly cardiovascular system, eye-, feet- and kidney-related diabetic complications.

The reason such a systemic problem exists can be explained by the fact that wherever access to institutions like insurance are suboptimal, patients need to pay out of their own pocket for both the underlying cause (diabetes in this case) and later its complications. Many studies have shown that managing type 2 diabetes in its early onset through medication and treatment can delay, if not can prevent, its complications [60]. Nowadays, with the leaps in medical research, we have more than seven classes of antidiabetic medications, in addition to insulin, for type 2 diabetes. Many newer medications that are antidiabetic have positive effects on renal and cardiovascular outcomes, which are over and above their primary use in ameliorating hyperglycaemia [61]. Such medicines include the aforementioned SGLT2 and glucagon-like peptide 1 (GLP-1).

2.6 CONCLUSION

The purpose of this chapter is to enlighten and present base projects that are founded on research. It will present the main line of research in machine learning and deep learning techniques. It gives an idea about IoT-related types of equipment for health monitoring and prognosis. The importance of obtaining a diverse dataset pertaining to diabetes or its complications cannot be understated enough. Understanding and deriving efficient models while ensuring certain fields are given more weightage, and reducing unnecessary fields through feature reduction is necessary to develop an optimised model. We have seen various algorithms and models of machine learning, and deep learning is used to develop prediction models: from approaches like Random Forest, CNN-SVD, and MLP to deep neural networks to process images in the case of diabetes retinopathy. We have also discussed how machine learning models with greater accuracy and optimisation may help in the early detection of cases of diabetes, neuropathy and retinopathy along with challenges faced in deploying such models to the real world.

REFERENCES

[1] Yuehong, Y. I. N., Zeng, Y., Chen, X., & Fan, Y. (2016). The internet of things in healthcare: an overview. *Journal of Industrial Information Integration*, 1, 3–13.

[2] Guariguata, L., Whiting, D. R., Hambleton, I., Beagley, J., Linnenkamp, U., & Shaw, J. E. (2014). Global estimates of diabetes prevalence for 2013 and projections for 2035. *Diabetes Research and Clinical Practice*, 103(2), 137–149.

[3] Rghioui, A., Naja, A., Mauri, J. L., & Oumnad, A. (2021). An IoT based diabetic patient monitoring system using machine learning and node MCU. In *Journal of Physics: Conference Series* (Vol. 1743, No. 1, p. 012035). IOP Publishing.

[4] Eswari, T., Sampath, P., & Lavanya, S. J. P. C. S. (2015). Predictive methodology for diabetic data analysis in big data. *Procedia Computer Science*, 50, 203–208.

[5] Nasser, A. R., Hasan, A. M., Humaidi, A. J., Alkhayyat, A., Alzubaidi, L., Fadhel, M. A., & Duan, Y. (2021). Iot and cloud computing in health-care: a new wearable device and cloud-based deep learning algorithm for monitoring of diabetes. *Electronics*, 10(21), 2719.

[6] Dang, L. M., Piran, M. J., Han, D., Min, K., & Moon, H. (2019). A survey on internet of things and cloud computing for healthcare. *Electronics*, 8(7), 768.

[7] Oryema, B., Kim, H. S., Li, W., & Park, J. T. (2017, January). Design and implementation of an interoperable messaging system for IoT healthcare services. In *2017 14th IEEE annual consumer communications & networking conference (CCNC)* (pp. 45–52). IEEE.

[8] Pradhan, B., Bhattacharyya, S., & Pal, K. (2021). IoT-based applications in healthcare devices. *Journal of Healthcare Engineering*, 2021, 1–18.

[9] Biswas, S., Akhter, T., Kaiser, M., Mamun, S., & others. (2014). Cloud based healthcare application architecture and electronic medical record mining: an integrated approach to improve healthcare system. In *2014 17th International Conference on Computer and Information Technology (ICCIT)* (pp. 286–291). IEEE.

[10] Pandey, R., Paprzycki, M., Srivastava, N., Bhalla, S., & Wasielewska-Michniewska, K. (2021). *Semantic IoT: Theory and Applications: Interoperability, Provenance and Beyond*. Springer International Publishing.

[11] US Department of Health and Human Services. (2008). The nationwide privacy and security framework for electronic exchange of individually identifiable health information. Office of the National Coordinator for Health Information Technology.

[12] Reichman, A. (2011). *File Storage Costs Less in the Cloud than in-House*. Forrester Research, Cambridge, MA.

[13] Bluetooth. (2016). Medical & health: bluetooth is changing the face of healthcare. Available at: E.bluetooth.com/what-is-bluetooth-technology/where-to-find-it/medical-health. Accessed March 3, 2017.

[14] Comstock, J. (2016). CES 2016: running list of health and wellness devices. Available at: https://www.mobihealthnews.com/content/ces-2016-running-list-health-and-wellness-devices. Accessed March 3, 2017.

[15] Samreen, S. (2021). Memory-efficient, accurate and early diagnosis of diabetes through a machine learning pipeline employing crow search-based feature engineering and a stacking ensemble. *IEEE Access*, *9*, 134335–134354.

[16] Nadeem, M. W., Goh, H. G., Ponnusamy, V., Andonovic, I., Khan, M. A., & Hussain, M. (2021). A fusion-based machine learning approach for the prediction of the onset of diabetes. *Healthcare (Basel, Switzerland)*, *9*(10), 1393. https://doi.org/10.3390/healthcare9101393

[17] El aboudi, N., & Benhlima, L. (2018). Big data management for healthcare systems: architecture, requirements, and implementation. *Advances in Bioinformatics*, *2018*, 4059018. https://doi.org/10.1155/2018/4059018

[18] Jovic, A., Brkić, K., & Bogunovic, N. (2015). A review of feature selection methods with applications. In *2015 38th International Convention on Information and Communication Technology, Electronics and Microelectronics (MIPRO)* (pp. 1200–1205). IEEE. https://doi.org/10.1109/MIPRO.2015.7160458

[19] Oh, K., Kang, H. M., Leem, D., Lee, H., Seo, K. Y., & Yoon, S. (2021). Early detection of diabetic retinopathy based on deep learning and ultra-wide-field fundus images. *Scientific Reports*, *11*(1), 1897. https://doi.org/10.1038/s41598-021-81539-3

[20] Rodriguez, I. (2021). On the management of type 1 diabetes mellitus with IoT devices and ML techniques. *arXiv preprint arXiv:2101.02409.*

[21] Lee, M. Y., Lee, H. R., Park, C. H., Han, S. G., & Oh, J. H. (2018). Organic transistor-based chemical sensors for wearable bioelectronics. *Accounts of Chemical Research*, *51*(11), 2829–2838.

[22] Mao, Y., Shen, M., Liu, B., Xing, L., Chen, S., & Xue, X. (2019). Self-powered piezoelectric-biosensing textiles for the physiological monitoring and time-motion analysis of individual sports. *Sensors*, *19*(15), 3310.

[23] Wang, X., Gu, Y., Xiong, Z., Cui, Z., & Zhang, T. (2014). Silk-molded flexible, ultrasensitive, and highly stable electronic skin for monitoring human physiological signals. Advanced *Materials*, *26*(9), 1336–1342.

[24] Webb, R. C., Bonifas, A. P., Behnaz, A., Zhang, Y., Yu, K. J., Cheng, H., & Rogers, J. A. (2013). Ultrathin conformal devices for precise and continuous thermal characterization of human skin. Nature *Materials*, *12*(10), 938–944.

[25] Tan, M., Xu, Y., Gao, Z., Yuan, T., Liu, Q., Yang, R., & Peng, L. (2022). Recent advances in intelligent wearable medical devices integrating biosensing and drug delivery. *Advanced Materials*, *34*(27), 2108491.

[26] Miorandi, D., Sicari, S., De Pellegrini, F., & Chlamtac, I. (2012). Internet of things: Vision, applications and research challenges. *Ad Hoc Networks*, *10*(7), 1497–1516.

[27] Saranya, M., Preethi, R., Rupasri, M., & Veena, S. (2018). A survey on health monitoring system by using IOT. *International Journal for Research in Applied Science & Engineering Technology*, *6*, 778–782.

[28] Pandey, R., Pandey, A., Maurya, P., & Singh, G. D., 2022. Prenatal healthcare framework using IoMT data analytics. In *The Internet of Medical Things (IoMT) and Telemedicine Frameworks and Applications* (p. 76). IGI Global.

[29] Pantelopoulos, A., & Bourbakis, N. G. (2009). A survey on wearable sensor-based systems for health monitoring and prognosis. *IEEE Transactions on Systems, Man, and Cybernetics, Part C (Applications and Reviews)*, *40*(1), 1–12.

[30] Güemes Gonzalez, A., Etienne-Cummings, R., & Georgiou, P. (2020). Closed-loop bioelectronic medicine for diabetes management. *Bioelectronic Medicine*, *6*, 11. https://doi.org/10.1186/s42234-020-00046-4

[31] Behera, A. (2021). Use of artificial intelligence for management and identification of complications in diabetes. *Clinical Diabetology*, *10*(2), 221–225.

[32] Hernando, M. E., García, G., Gómez, E. J., & del Pozo, F. (2004). Intelligent alarms integrated in a multi-agent architecture for diabetes management. *Transactions of the Institute of Measurement and Control*, *26*(3), 185–200.

[33] Hernando, M. E. (2003) Multi-agent architecture for the provision of intelligent tele-medicine services in diabetes management. In *The Workshop on Intelligent and Adaptive Systems in Medicine*. Prague, Czech.

[34] Care, D. (2020). Standards of medical care in diabetes-2020. *Diabetes Care*, *43*, S1–S224.

[35] Salzsieder, E., Heinke, P., Thomas, A., Puchert, A., Augstein, P., Kohnert, K. D., & Vogt, L. (2020). Wirksamkeit und Akzeptanz telemedizinisch unterstützter Gesundheitsberatungsdienstleistungen in der diabetologischen Versorgungspraxis. *Diabetes Stoffwechsel und Herz*, *29*, 277–293.

[36] Mohan, V., Prathiba, V., & Pradeepa, R. (2014). Tele-diabetology to screen for diabetes and associated complications in rural India: the chunampet rural diabetes prevention project model. *Journal of Diabetes Science and Technology*, *8*(2), 256–261. https://doi.org/10.1177/1932296814525029

[37] Doupis, J., Festas, G., Tsilivigos, C., Efthymiou, V., & Kokkinos, A. (2020). Smartphone-based technology in diabetes management. *Diabetes Therapy*, *11*, 607–619.

[38] Efat, M. I. A., Rahman, S., & Rahman, T. (2020). IoT based smart health monitoring system for diabetes patients using neural network. In *Cyber Security and Computer Science: Second EAI International Conference, ICONCS 2020, Dhaka, Bangladesh, February 15-16, 2020, Proceedings 2* (pp. 593–606). Springer International Publishing.

[39] Klonoff, D. C., & Kerr, D. (2018). Smart pens will improve insulin therapy. *Journal of Diabetes Science and Technology*, *12*(3), 551–553.

[40] Libiseller, A., Kopanz, J., Lichtenegger, K. M., Mader, J. K., Truskaller, T., Lackner, B., & Donsa, K. (2020). Study protocol for assessing the user acceptance, safety and efficacy of a tablet-based workflow and decision support system with incorporated basal insulin algorithm for glycaemic management in participants with type 2 diabetes receiving home health care: a single-centre, open-label, uncontrolled proof-of-concept study. *Contemporary Clinical Trials Communications*, *19*, 100620.

[41] Bharati, M., & Ramageri, B. (2010). Data mining techniques and applications. *Indian Journal of Computer Science and Engineering*, *1*(4), 301–305.

[42] Le, T., Vo, T., Pham, T., & Dao, S. (2021). A novel wrapper-based feature selection for early diabetes prediction enhanced with a metaheuristic. *IEEE Access*, *9*, 7869–7884.

[43] Khan, F. A., Zeb, K., Al-Rakhami, M., Derhab, A., & Bukhari, S. A. C. (2021). Detection and prediction of diabetes using data mining: a comprehensive review. *IEEE Access*, *9*, 43711–43735.

[44] Shanmugavalli, M., & Sivakumar, K. (2022). Data mining based predictive analysis of diabetic diagnosis in health care: overview. *Mathematical Statistician and Engineering Applications*, *71*(4), 572–588.

[45] Lakhwani, K., Bhargava, S., Hiran, K., Bundele, M., & Somwanshi, D. (2020). Prediction of the onset of diabetes using artificial neural network and PIMA Indians diabetes dataset. In *2020 5th IEEE International Conference on Recent Advances and Innovations in Engineering (ICRAIE)* (pp. 1–6). IEEE.

[46] Kaur, H., & Kumari, V. (2022). Predictive modelling and analytics for diabetes using a machine learning approach. *Applied Computing and Informatics*, *18*(1/2), 90–100. https://doi.org/10.1016/j.aci.2018.12.004

[47] Pandey, R., Sahai, A., & Kashyap, H. (2022). Chapter 13- Implementing convolutional neural network model for prediction in medical imaging. In R. Pandey, S. K. Khatri, N. kumar Singh, & P. Verma (Eds.), *Artificial Intelligence and Machine Learning for EDGE Computing* (pp. 189–206). https://doi.org/10.1016/B978-0-12-824054-0.00024-1

[48] Papatheodorou, K., Banach, M., Bekiari, E., Rizzo, M., & Edmonds, M. (2018). Complications of diabetes 2017. *Journal of Diabetes Research*, *2018*, 3086167. https://doi.org/10.1155/2018/3086167

[49] Alkhodari, M., Rashid, M., Mukit, M., Ahmed, K., Mostafa, R., Parveen, S., & Khandoker, A. (2021). Screening cardiovascular autonomic neuropathy in diabetic patients with microvascular complications using machine learning: a 24-hour heart rate variability study. *IEEE Access*, *9*, 119171–119187.

[50] Xiao, M.-X., Lu, C.-H., Ta, N., Wei, H.-C., Haryadi, B., & Wu, H.-T. (2021). Machine learning prediction of future peripheral neuropathy in type 2 diabetics with percussion entropy and body mass indices. *Biocybernetics and Biomedical Engineering*, *41*(3), 1140–1149. https://doi.org/10.1016/j.bbe.2021.08.001

[51] Makino, M., Yoshimoto, R., Ono, M. et al. (2019).Artificial intelligence predicts the progression of diabetic kidney disease using big data machine learning. *Scientific Reports*, *9*, 11862. https://doi.org/10.1038/s41598-019-48263-5

[52] Dinh, A., Miertschin, S., Young, A., & Mohanty, S. D. (2019). A data-driven approach to predicting diabetes and cardiovascular disease with machine learning. *BMC Medical Informatics and Decision Making*, *19*(1), 211. https://doi.org/10.1186/s12911-019-0918-5

[53] Pendsey, S. P. (2010). Understanding diabetic foot. *International Journal of Diabetes in Developing Countries*, *30*(2), 75–79. https://doi.org/10.4103/0973-3930.62596

[54] Burns, S. A., Elsner, A. E., Sapoznik, K. A., Warner, R. L., & Gast, T. J. (2019). Adaptive optics imaging of the human retina. *Progress in Retinal and Eye Research*, *68*, 1–30. https://doi.org/10.1016/j.preteyeres.2018.08.002

[55] Atwany, M., Sahyoun, A., & Yaqub, M. (2022). Deep learning techniques for diabetic retinopathy classification: a survey. *IEEE Access*, *10*, 28642–28655.

[56] Bowering, C. K. (2001). Diabetic foot ulcers. Pathophysiology, assessment, and therapy. *Canadian Family Physician Medecin de Famille Canadien*, *47*, 1007–1016.

[57] Pendsey, S. (2003). *Diabetic Foot: A Clinical Atlas*. Jaypee Brothers Medical Publishers.

[58] Nahiduzzaman, M., Islam, M., Islam, S., Goni, M., Anower, M., & Kwak, K. S. (2021). Hybrid CNN-SVD based prominent feature extraction and selection for grading diabetic retinopathy using extreme learning machine algorithm. *IEEE Access*, *9*, 152261–152274.

[59] Basu, S., Brown, C., Beran, D., Flood, D., Seigle, J., Manne-Goehler, J., Mezhrahid, J., Yudkin, J. S., Davies, J., Lipska, K. J., Sibai, A. M., Houehanou, C., Labadarios, D., Farzadfar, F., Aryal, K. K., Marcus, M-E., Mayige, M. T., Theilmann, M., Geldsetzer, P., & Domainico, P. (2021). Expanding access to newer medicines for people with type 2 diabetes mellitus in low- and middle-income countries: a microsimulation and price target analysis. *The Lancet Diabetes and Endocrinology*, *9*(12), 825–836. https://doi.org/10.1016/S2213-8587(21)00240-0

[60] Chasseloup, F., Bourdeau, I., Tabarin, A., Regazzo, D., Dumontet, C., Ladurelle, N., Tosca, L., Amazit, L., Proust, A., Scharfmann, R., Mignot, T., Fiore, F., Tsagarakis, S., Vassiliadi, D., Maiter, D., Young, J., Lecoq, A.L., Demeocq, V., Salenave, S., & Kamenicky, P. (2021). Loss of KDM1A in GIP-dependent primary bilateral macronodular adrenal hyperplasia with Cushing's syndrome: a multicentre, retrospective, cohort study. *The Lancet Diabetes & Endocrinology*, *9*(12), 813–824.

[61] Cushing, H. (1994). The basophil adenomas of the pituitary body and their clinical manifestations (pituitary basophilism). *Obesity Research*, *2*(5), 486–508. https://doi.org/10.1002/j.1550-8528.1994.tb00097.x

3 Blockchain for Handling Medical Data

T. Vigneswari, Farjana Farvin Sahapudeen, and G. Kalaiselvi
Anjalai Ammal Mahalingam Engineering College

3.1 INTRODUCTION

The Internet of Medical Things (IoMT) is a network of devices connected to the Internet to provide enhanced medical service. IoMT [1] is an association of health-related medical devices, software applications and services that have a deep impact on patients' health monitoring, diagnosis and treatment. The network inclusive of medical devices, sensors and edge devices facilitates efficient remote monitoring of patients by automating medical procedures and workflow management. With the use of IoMT, the diagnosis and treatment are done at a rapid pace with unbelievable accuracy, which capacitates medical professionals to enhance the health status of their patients.

The major use cases involved in the healthcare sector are [2] remote health monitoring, health record management, pharmaceutical supply chain management and process of health insurance claim. The major use case is healthcare monitoring and data management. The digital health record of patients is sent to the hospital through a smartphone or a tracking device. The device tracks the cardiac activity and sends the details to a physician at the other end. If any abnormality is found in the cardiac activity by the other end, immediate care is given to the patient at the earlier stage itself. The data are received continuously and stored in a centralized or decentralized storage which is possibly a cloud environment. A massive amount of clinical data are collected, processed and shared among entities like patients, caretakers, and doctors and hospital administration involved in the scenario.

The process may not be as simple as we think as it involves many challenges in making the IoMT realize its full potential. The major challenges posed are as follows [3]:

- Lack of user-centred interoperability.
- Centralized data management which leads to single-point failure.
- Management of a huge volume of data raises issues of data management, acquisition, processing and analysis.
- Data standards to facilitate secure data sharing.
- Issues related to networking such as bandwidth, communication protocols and energy efficiency.
- Scalability.

DOI: 10.1201/9781003359951-4

Among these challenges, security and privacy play pivotal roles in the successful adoption of IoMT systems.

The medical gadgets or devices integrated in the IoMT framework employ either wireless or remote communications where providing security is of the utmost importance [4]. The entire environment can be significantly impacted by both incoming and outgoing data if there is any vulnerability or point of information leakage from the IoMT devices due to inadequate methods of authentication and access control [5]. Thus, security measures must be put into place or deployed at key locations throughout the network or system. If not, attackers capable of employing unique tactics can quickly get through security safeguards and access patient data without proper authorization.

Most of the IoMT systems currently in use deploy centralized security paradigms which are very inappropriate. This vividly indicates a need for distributed sharing and securing mechanisms for ensuring better privacy preservation to the user [6, 7]. Moreover [8], it needs a comprehensive method for data transaction between patients and IoMT healthcare providers.

IoMT healthcare systems necessitate the deployment of proper data integrity features to function properly. It majorly ensures that the data don't get altered during wireless transmission [9] and safeguards medical data from being changed or deleted by unauthorized entities. IoMT frameworks have realized the need of data integrity as it is more crucial in providing healthcare since data represent diagnoses, treatments and health statuses [10]. This characteristic can also be described in this context as the ability to detect unauthorized data alteration or distortion that results in perpetual catastrophe [11]. Thus, proper data integrity precautions must be incorporated to avoid hostile attempts to change delivered data.

Some of the sample attacks due to inferior or a feeble security and integrity framework are given below:

- An attack may alter or manipulate a patient's medication dosage, which could have lethal consequences in their health status [12].
- Medical data are vulnerable to falsification in healthcare systems, which can result in incorrect medication administration and patient prognosis, both of which increase the risk of an allergic reaction [13–15].
- Attackers may transmit inappropriate medical alerts and create large financial losses as a result of security flaws.

Hence, there is an essential need for a complementing technology which empowers IoMT to overcome the challenges and improve utilization of the same more prevalently. Recently, blockchain is emanating as a secured and decentralized platform without involvement of any third party. It offers a number of features like tamperproof, confidentiality, immutability, traceability, data integrity, secrecy, and privacy which make it more ideal for using along with IoMT.

The following section briefs about the fundamentals of blockchain technology preceded by its implementation along with IoMT.

3.2 BLOCKCHAIN—AN OVERVIEW

The blockchain is a distributed ledger [16] built with numerous sequences of blocks, and each block consists of the details about all the transactions happening in the network. Blockchain enables peer-to-peer distributed sharing of data and computation. The core components of blockchain are shown in Figure 3.1. Data that are shared among the entities involved in the process are stored in each block. Whenever a transaction is done, a new block is created. The initial block is known as genesis block and every block (i) consist of a hash of the ($i-1$) block. Consensus is a mechanism by which the blockchain participants agree to add new data to the blockchain. The two major components [17] of each block are header and body. The header of the block consist of a hash value, a timestamp, nonce and a Merkle tree. The header holds the hash of the preceding block; hence, the structure of a particular block depends on the preceding block, thus forming a chain. The timestamp indicates the time during which the block is published. Nonce is an arbitrary value that is changed by the miners often to receive a hash value which aids in solving certain mathematical puzzles. Merkle tree is a data structure used in blockchain to decrease the efforts required to verify the transaction in a block. Merkle trees create a root hash of all the data in a block and use that hash to compile the entire collection of data. By continually hashing data node pairs until only one node remains, the root

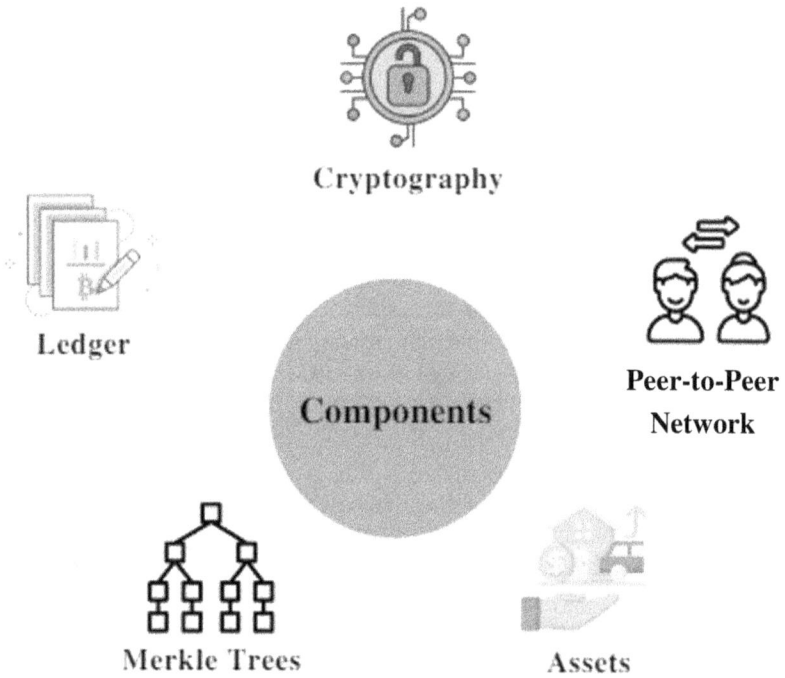

Cryptography

Ledger

Components

Peer-to-Peer Network

Merkle Trees

Assets

FIGURE 3.1 Core components of blockchain.

hash is discovered. The Merkle root is the final child node that remains. Blockchain stores blocks which are micro-level tasks, and it is referred to as a transaction. The transactions are shared among the participants, and the data are validated or approved by them through consensus algorithms which also make them tamper-proof. A similar copy of ledger is maintained among the participating entities to ensure immutability.

Each block has a self-executing piece of code encapsulated in it to augment the various actions involved in the application logic. Smart contracts are tamper-proof as they are interspersed with the blockchain, and only the process owner can modify the smart contracts if necessity arises. They are self-enforcing as the rules are followed ardently and self-verifying due to automated possibilities.

The classification of blockchain is given below [18]:

- Public blockchain is completely decentralized, where access and permission to participate in consensus are provided to all the participants.
- Private blockchain is used to monitor the transactions among different individuals or departments of particular single-enterprise solutions. All the users need approval from other users in the group to join the network and become known members.
- Consortium blockchain is where only privileged users can have public access to a permissioned network.
- Hybrid blockchain is a combination of both public and private frameworks. The ledger is made completely accessible by employing public blockchain, whereas the private blockchain is employed to validate the modifications that are carried out in the ledger during transactions.

Features of blockchain are discussed in Ref. [19] as follows:

- Ledger is shared among all users without third-party intervention enabling decentralization and eliminating intermediaries. This creates a direct impact on the cost involved and efficiency of the application.
- The ledger is resistant to tampering and hence ensures immutability.
- The transaction records are time-stamped, and each participant holds a copy of the same to warrant availability. This assures redundancy and fault-tolerance capability in the blockchain framework.
- A unique address is generated with which interaction is performed among the users without revealing their unique identity.
- Security ensures that an authorized user retrieves data with an assurance of privacy, authentication, integrity and privacy.
- Blockchain has introduced a concept called tokenization that creates new economic opportunities for the production of tradeable tokens with real-world value.

As we have seen the basics of blockchain, now we will proceed to the working of the same with IoMT [20]. The massive amount of data that are generated from the immense number of sensors, edge devices, etc. are stored and managed by utilizing the blockchain without the involvement of a centralized server or cloud infrastructure. Blockchain facilitates various anonymous parties to execute transactions in a network without mutual trust in a decentralized manner which are then stored in ledger. Blockchain maintains the ever-scaling number of records by using these distributed digital ledgers which are tamper-proof as mentioned earlier. The chains of blocks in the ledger are linked to one another cryptographically with the use of private key. Once a data block is stored in a blockchain ledger, changing or eliminating a data block is trivial. All the participants have access to view the transaction, but they are unaware of the actual contact who generated the data block. In light of the fact that IoMT fundamentally requires these qualities, the blockchain's secure, decentralized, and autonomous capabilities make it a perfect paradigm to use alongside it.

3.3   BLOCKCHAIN FRAMEWORK FOR IoMT

The stakeholders included in any IoMT framework [21] are medical experts, patients, hospital administration, insurance providers, etc. Smart contracts are developed to provide access permissions to manage the sensitive medical data among the stakeholders involved in the process. Data integrity, metadata about ownership and concerned permissions can be maintained through deploying appropriate smart contracts. Consensus algorithms like Proof of Stake (PoS), Proof of Work (PoW) and Design Proof of Stake (DPoS) are used in IoMT–blockchain framework through which transactions are validated. It uses hashing and encryption techniques to secure the transactions of sensitive medical information. The most common platforms [22] used in healthcare sectors are Ethereum, hyperledger fabric, hyperledger sawtooth, hyperledger Iroha, Ripple and Quorum. The healthcare records may be stored either in on-chain or off-chain modality in blockchain.

Furthermore, the following reasons necessitate the use of blockchain with IoMT [23, 24]:

- The data in patients' health record are private and sensitive; hence, they should be managed with privacy, confidentiality and security.
- As far as security is concerned, the IoMT architecture integrates devices with constrained resources which cannot meet the requirements of traditional security methods. Apart from this, they follow a centralized approach, but IoMT ensembles networks of distributed nature. Hence, traditional security methods are not advisable for incorporating in IoMT to provide privacy preservation which is an essential feature.
- Numerous medical devices are being added dynamically in the IoMT, and a decentralized paradigm to manage the transactions is needed, which makes the integration of blockchain with IoMT as indispensable.
- Blockchain secures the health records in real time along with providing tamper-proof access to all authorized entities present in the IoMT.

A layered architecture integrating blockchain and IoMT is presented in Ref. [25]. The architecture has four layers. The application layer integrates all the entities, such as doctors, patients, health workers and government authorities, involved in the framework. Management layer monitors usage, management and storage of medical data. Data integrity is ensured in this layer by employing the generic Secure Privacy Conserving Provable Data Possession (SPCPDP) framework. The network layer not only collaborates with networking components and protocols but also accommodates blockchain, which serves as the foundational framework for this architecture. Perception layer collects data from the sensors and medical gadgets attached to the patient. Blockchain-based approaches deployed in IoMT are Ethereum-Based, Hyperledger-Based, Modified Consensus Protocol and Modified Cryptographic Technique.

Some of the familiar IoMT–blockchain applications in healthcare are listed in Table 3.1.

Due to the divergent requirements of both the technologies, integrating them is far from simple and poses a number of adversities during implementation. The following are challenges ahead in integrating the two pioneering technology [32, 33]:

- Blockchain involves cryptography techniques and mining which requires abundant computational resources for processing, and it also consumes more energy wherein the IoMT is a resource-constrained framework which also has energy constraints.
- A massive amount of data are generated by the IoMT applications. To ensure integrity, they must be stored in the distributed ledger of blockchain which may not be affordable by the networked devices in IoMT with less storage capacity.

TABLE 3.1
Healthcare Applications

Company	Solution Provided
Patientory [26]	Managing the health data of patients by providing better access control. Providing the status of the patient to the other entities by creating a network among them
Healthcombix [27]	Providing confidentiality in managing data asset of patients and diagnosing diseases, and decentralized payment networks create robust new healthcare ecosystems
Doc.AI [28]	The blockchain technology is employed to encrypt the health-related data of patients, thus ensuring a secured sharing among the users. Later, these data are given as input to machine learning techniques for prediction of disease and treatment
Medicalchain [29]	Effective medical data sharing among the users along with security by providing proper access control permissions. Telemedicine facility is also provided for interaction between patient and doctor for remote consultation and sharing health records
Exochain [30]	Each data owner is given access to control how the other users utilize their data
Novartis [31]	Confluence of IoT and blockchain is deployed to identify fake medicines and monitor real-time temperature for proper distribution of drugs

- The wearable medical gadgets that are part of IoMT are mobile, and they form a dynamic network, but for blockchain, it is much essential to have a static network.
- The time consumed for creating a block may decrease the response time of the IoMT application which is a time critical mission.
- Traffic overhead is created by blockchain as they generate a continuous sequence of communication, and in contrast, Internet of Things (IoT) devices involved in IoMT have limited bandwidth.
- There are no standard protocols for the interaction between the IoMT, blockchain, cloud and end-user devices using the application. This hinders the adoption of this converged technology to a greater extent.
- In-depth programming knowledge may be required agonistic of the users' knowledge level to abstract, manage the devices involved in IoMT and configure blockchain for the particular IoMT application.
- The available integrated IoMT–blockchain framework does not use open standard technical paradigms forbidding it from extension by others.
- Finally, the scope of application has been limited to data management of the healthcare sector, and furthermore, exploration is needed. They may be extended to monitor the activities related to diagnosis and treatment to provide preventive measures.

3.4 BLOCKCHAIN AND DATA MANAGEMENT

Blockchain has emerged as one of the pioneering solutions for working with an immense amount of sensitive medical data, whose efficacy has been proved in various areas of the medical sector. The medical data collected from the healthcare sector are used for diagnosis of diseases, prediction of pandemic diseases, providing remote treatment and monitoring the progress of recovery of patients.

Hence, there arises a necessity to resolve these issues so that medical data received from various IoMT frameworks are managed in a better way. The blockchain technology allows management, analysis, and security of medical data as shown in Figure 3.2, which could be implemented in IoMT to overcome these challenges and augment the performance of IoMT.

3.4.1 DATA STORAGE IN BLOCKCHAIN

IoMT generates immense amounts of data, and it is not advisable to store them in blockchain as transactions. In order to maintain these large volumes of medical records, there is a need for a distributed storage system with a peer-to-peer structure. A decentralized distributed storage along with persistence is needed to maintain the voluminous data available in the IoMT framework. Data storage in blockchain-based IoMT systems may be classified into two major categories [34].

3.4.1.1 On-Chain Blockchain Storage

These are transactions that take place inside the blockchain. Data storage is done in the decentralized ledger, and it is accessed by users who own a copy of the same. Data are updated in the blockchain during each on-chain transaction. The request for

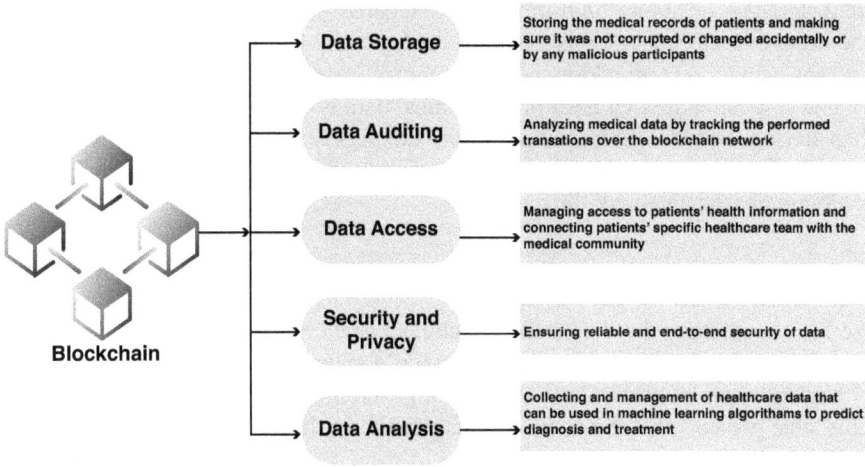

FIGURE 3.2   Data management scenario in IoMT.

accessing the document and providing approval for using the document are deployed in blockchain as transactions. These transactions include details such as time stamp of data access, source of data, metadata, smart contracts for requesting, and approval or rejection of access to data.

The major disadvantage of storing in on-chain blockchain storage is it is more expensive as IoMT generates a huge amount of data which cannot be handled by the blockchain.

3.4.1.2   Off-Chain Blockchain Storage

This type of storage can be managed in two ways. The first one is to identify a legitimate authority that provides the storage facilities and holds the responsibility to manage the medical data; the second option is the management of clinical data by storing them in individual organizations where the patients' data are registered.

In either method, the blockchain maintains a uniform resource identifier to abstract the medical datasets stored in off-chain. Data integrity is provided by using a hashing summary of the original medical data available in the blockchain. The list of users who are provided with access permission to use the data is also available in the blockchain to enable auditing. Asymmetric encryption is used to encrypt the medical data to provide secured storage and transaction.

The major blockchain storage offering systems are Interplanetary File System (IPFS), Sia, Storj and MaidSafe. Among these, IPFS is the most prominent one.

3.4.1.3   Interplanetary File System (IPFS)

A huge volume of medical data can be stored effortlessly by using IPS [35, 36], which is a distributed peer-to-peer storage system. IPFS uses a distributed hash table (DHT) to store files with their content-addressed hash and eliminates redundant files that use log of version control. Additionally, it locally generates a hash of files with a high request volume to ensure quick response during subsequent access. IPFS provides characteristics including high throughput, secured transaction by mapping hashes and accessing the transactions simultaneously among the peers involved in the network by using a content-addressed block storage system.

The content-addressed hash of the clinical data is available in blockchain rather than storing the whole information. Information and corresponding hash is used to retrieve the particular medical record. Version control is also available in IPFS by combining hash value with the record.

3.4.2 DATA ACCESS

We have seen about the data storage techniques involved in IoMT–blockchain, but accessing the data with proper authorization is a major issue.

Various types of medical records [37, 38] used in IoMT are briefed in Table 3.2. This table presents that the control of data may be either with the hospital administration/doctor or with the patients.

In this section, a data access technique provided in Ref. [39] warrants better patient control over their medical data. Personal health record (PHR) is shared among the entities in this process. The entities commonly involved here in accessing the PHR are as follows:

- Regulatory agency which registers patients, doctors and healthcare systems.
- Hospitals which are responsible for transferring the medical record of the patient by using a unique symmetric key which is shared between hospital and the patient concerned.
- Patient who is the owner of a health record which is deployed by using an application that manages the PHR. Patients hold a crucial role in managing the PHR and creating response to the queries, and they have the rights to accept or deny the request.
- The doctor requests an encrypted Personal Health Record (PHR) from the patient and possesses the capability to decrypt it on their own end.

TABLE 3.2
Types of Records Used in IoMT

Record Type	Electronic Medical Record (EMR)	Electronic Health Records (EHRs)	Personal Health Records (PHRs)
Content	Digitized form of the prescription, diagnosis reports, etc. that is available with the hospital/doctor in paper form. It consists of the entire clinical history and treatment provided to the patient during a single consultation.	Collection of clinical history of the patient which has the entire details of diagnosis, treatment, medicines prescribed allergic reactions and results of laboratory tests.	Medical data gathered from gadgets attached with patients, pharmacies, hospitals, etc.
Control	Hospital administration/doctor	Hospital administration/doctor	Patient

- Trusted oracles which perform re-encryption of patients symmetric key to doctor. A reputation system is attached with this.
- Decentralized database is used to store encrypted PHR. The most commonly used decentralized database is IPFS.
- The storage and trusted oracle are paid by the insurance company.
- Ethereum platform which is used to deploy this application as it is public and supports decentralization.

The smart contract deployed in this application is shown in Table 3.3.

The sequential flow of action among the entities to share the PHR to the doctor is shown in Figure 3.3.

TABLE 3.3
Smart Contract Deployment for Data Access

Smart Contract (SC)	Responsibilities
Controller (CSC)	Register all the entities involved in the process.
	Audit the reputation of oracle.
	Grant permission to the doctor for submitting an evaluation about the oracle.
Patient record (PRSC)	Each patient owns a separate PRSC to store the metadata about PHR and request it from the doctor.
	Concedes the patient to accept or deny the access request from the doctor.
	Records are sent to the doctor with the participation of oracle and evaluation of the same is done.

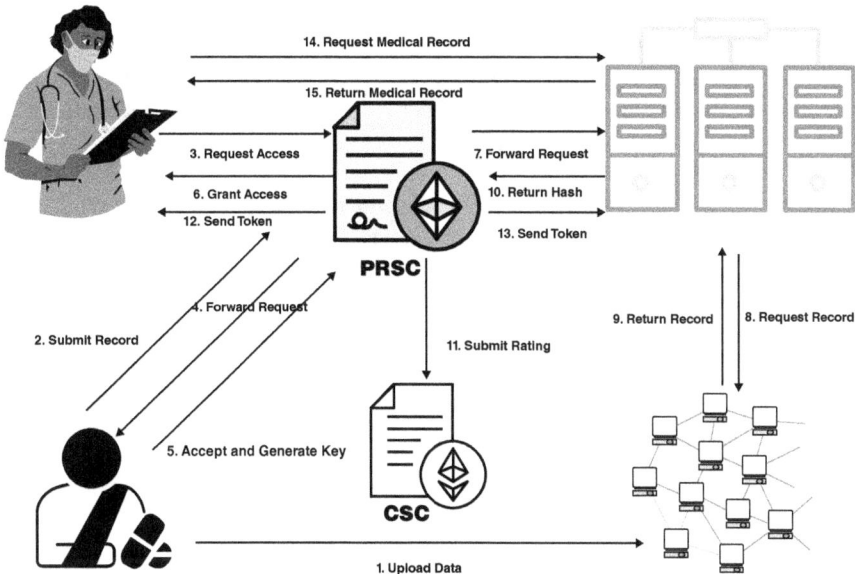

FIGURE 3.3 Data access with patients' control over data.

The patient uploads the PHR to the IPFS, and it is submitted to the Patient Record Smart Contract (PRSC) eventually. All the access requests to the record are directed to the PRSC. When a doctor is in need of a particular patient record, an access request is sent to the PRSC. The request is then forwarded to the patient, and they may either approve or deny it.

3.4.2.1   Case 1: Accepting the Request

- When a request is accepted by the patient, a re-encryption key is generated and communicated to the PRSC. The PRSC in turn provides the doctor with the access permission to view the particular patient's record.
- Concurrently, the request is also forwarded to the re-encryption oracle through which request for medical record is sent to IPFS. The requested medical record and a return hash are sent back to the oracle by IPFS to the oracle. The PRSC evaluates the oracle and submits the ratings to the Controller Smart Contract (CSC) to update the oracle's reputation.
- A token is generated by PRSC to the doctor and IPFS. Upon receiving the token, the doctor sends a request to the re-encryption oracle for accessing the patient's record. The oracle reencrypts the patient's record using the re-encryption key and returns the medical record to the doctor.
- At the other end, the re-encrypted symmetric key is decrypted with a private key owned by the doctor. Then the patient's record is decrypted with the plain text symmetric key. The doctor also submits oracle ratings and updates its reputation.

3.4.2.2   Case 2: Denying the Request

- The doctor submits the request to the PRSC, and the patient is made aware of the new request. The patient's decision about declining the specific request is updated with the PRSC. Then, the decision of denying the request is notified to the doctor.

In another work [40], MedRec is a blockchain-integrated IoMT framework that works with electronic medical record (EMR) which is also a decentralized system to manage health records. The users are guaranteed with effortless access to their medical records along with an immutable usage history as data are shared with hospital administration, doctors, etc. by using the properties of blockchain. The other stakeholders access the data as miners in a blockchain where they access data as mining rewards. The local data storage solutions can be merged with MedRec to leverage interoperability and adaptability.

The smart contracts used in MedRec are shown in Table 3.4.

TABLE 3.4
Smart Contracts in MedRec

Smart Contract (SC)	Responsibility
Registrar contract (RC)	Ethereum address identity of entity is mapped with identification string. Maps identity strings to an address in the blockchain, where the summary contract can be found.
Patient–provider relationship contract (PPR)	Provide identification to the records of the owner by combining access permissions along with its data pointers.
Summary contract (SC)	Provides references to Patient–Provider Relationship (PPR) contracts by depicting past and recent transactions among all the nodes in the network.

3.4.3 DATA PRESERVATION SYSTEM

In the course of medical diagnosis and treatment, enormous amounts of medical data are produced, including prescriptions for medications, lab results, prognoses, suggested treatments, digital images like ECG, scans, etc. Such lifesaving data should be preserved from tampering and also data loss, which may delay the progress of treatment and thus endangering the patient's life.

The following requirements should be met in order to protect the more sensitive medical records:

- Data consistency should be maintained after submission by the user.
- Prevention of data from tampering, forging and deletion.
- Anonymity and encryption.

A data preservation system using the blockchain smart contract is proposed in Ref. [41] which include two steps as given in Figure 3.4.

3.4.3.1 Data Submission

The initial procedure is data submission with the following steps:

Data submission and processing

The authorized users submit the medical record which may be in the form of text such as vital signs, diagnosis or medicines prescribed or multimedia data such as scan report, ECG, X-ray, etc. The multimedia data are stored in a folder which is created in a random manner and store the text data in a temporary database which is followed by generating a hash for medical data for further reference.

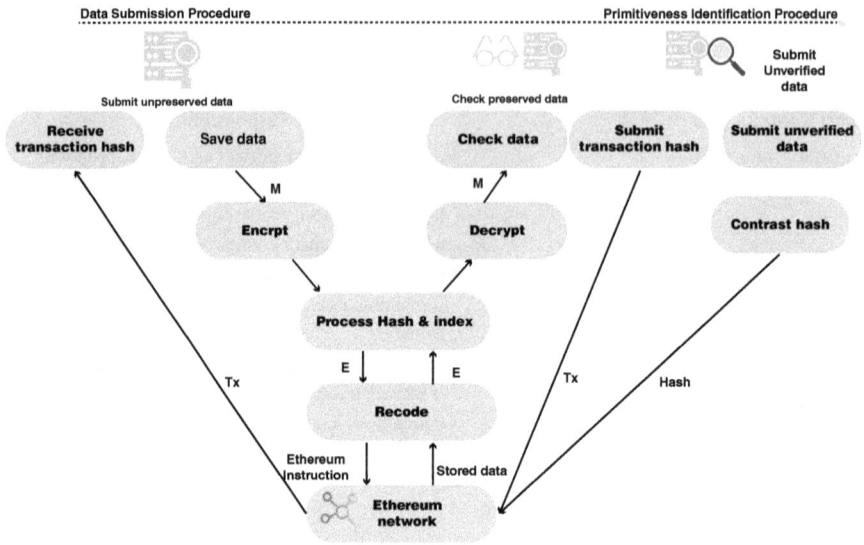

FIGURE 3.4   Data preservation system.

Data validation

The validation of uploaded medical data is done by receiving feedback and generating a confirmation hash.

Data storage

During the storage phase, the text data are stored in the blockchain by encrypting it along with its hash, and in the case of the multimedia file, encrypted location is stored in the blockchain. The multimedia files are split and stored in various locations, and hence, an index is also stored in the blockchain.

Cancelling the preservation

The stored medical data can be deleted anytime by the user.

Once the medical data are uploaded, validated and stored in the blockchain, it is followed by the next course of action which is the procedure for primitiveness identification.

3.4.3.2   Primitiveness Identification

This process involves authentication methods of two types.

Acquisition of preservation content methods allows users to access the data by using the private key generated initially during the preservation of the data. Verifying consistency method allows users to verify whether data available in storage are the same as the data stored. A hash is extracted from the data and compared with the hash already generated during the storage. If both the hash values are similar, then integrity of data is assured or else a warning is issued to the user.

The two smart contracts involved in the privacy protection system are briefed in Table 3.5.

TABLE 3.5
Smart Contract for Data Preservation

Smart Contract (SC)	Responsibilities
Preservation submission (PS)	The correctness of the data uploaded by the user is confirmed and stored in the privacy system officially, which then becomes unchangeable.
Primitiveness verification (PV)	Detects if there is any change in the original data submitted by the user before any other user requests for viewing the record. If any unauthorized changes are found, it is reported back to the origin.

3.4.4 PRIVACY PROTECTION

The health record shared in IoMT records consist of sensitive information about individuals such as personal information (name, identity number, date of birth), details of doctors attending the individual, diagnosis reports and treatment records. It has been shown in Ref. [42] that more than 20 billion patient records can be accessed without any difficulty from the centralized storage which affects the privacy of the data owners. The primitive methods of warranting privacy involve either storing the medical data in a local database along with access control permissions or encrypting the medical data and sharing the key among authorized users. This method shows inefficiency due to the risk involved in tampering the key and modifying the access permission by third-party users. Apart from this, privacy and access control for private data was ensured by utilizing regulations like Health Insurance Portability and Accountability Act (HIPAA) and General Data Protection Regulation (GDPR) [43, 44]. When blockchain is integrated with IoMT, it eliminates such problems and aids in providing proper privacy protection mechanisms by preserving anonymity of the patient detail.

Privacy protection involving blockchain along with techniques like group signature, trust execution environment (TEE), attribute based encryption and sibling intractable function families (SIFF) are discussed in this section.

3.4.4.1 Privacy Protection with Group Signature

An IoMT [45] data privacy protection system that assimilates blockchain, group signature and asymmetric encryption to enhance data sharing among various hospitals and protecting data privacy of patients is detailed in the following section:

- The medical data of the patient are uploaded by the doctor. The patient's name and identity number are combined to create a hash which will be the unique identity of the patient. After this, other medical details along with the time of treatment are uploaded, and each of these transactions is encrypted by using Advanced Encryption Standard (AES) keys.
- A group is then formed by combining several hospitals, and a group manager is also identified among them. The group manager owns two keys: a private key that belongs to him, and a public key that encrypts the AES key.

As the keys are generated randomly, the group manager holds a table which maps the hash values of public and private keys. Then, the encrypted data along with AES key are shared among the group members by the doctor.

- The transaction collection is completed by creating transaction signatures by the group members. The group members sign the hash value of the cipher text and collect the serial number, signature, cipher text and the encrypted AES key.
- The practical byzantine fault-tolerance (PBFT) consensus algorithm is used to validate the transactions, and when all the nodes agree with the transaction details, data are uploaded in the blockchain by the hospital. The PBFT algorithm is selected as it provides low latency, high throughput and better security boundary.
- Once the data are stored, it can be retrieved only among the group members by using the serial number, and the group manager grants access permission.

The transaction consists of a serial number, a cipher text, an encrypted symmetric encryption key and a group signature. The privacy of the data is ensured through the serial numbers which is a hash value that cannot be easily cracked. The location of the hospital is concealed in the group signature which is only revealed to the group manager. The medical data of the patient are anonymized through encryption and split into many complex pieces, and hence, the sensitive data cannot be obtained by others who don't belong to the group.

3.4.4.2 Privacy Protection with Trust Execution Environment (TEE)

This approach [46] uses a combination of blockchain, TEE, and IPFS to provide privacy. Off-chain distributed data storage is deployed, and data are executed in TEE using smart contracts to provide privacy. Medical data are stored by the user in TEE, and smart contract provided by the blockchain is also executed in TEE. The user has to submit a public hash key to retrieve the data which is verified against the list of hash keys available in TEE.

3.4.4.3 Sibling Intractable Function Families (SIFF)

A shared key is created to encrypt and store a particular patient's health record in a blockchain. This key can only be rebuilt by the legitimate parties to access the data before the commencement of treatment. The shared key is created using the SIFF [47], and encrypted data are stored using the hyperledger fabric. A small number of entities are given a shared key by the SIFF, which is then used to encrypt and store in a blockchain the data used in the diagnosis and treatment process.

3.4.4.4 Attribute-Based Encryption (ABE) Privacy Protection

A privacy preservation mechanism for medical data is given in Ref. [48] which uses searchable encryption technique and K-anonymity algorithm. It leverages consortium blockchain and hyperledger fabric, which allow users to search for encrypted medical data records. The smart contract employs attribute-based access control mechanisms to warrant data access by the user with proper attributes. The K-anonymity

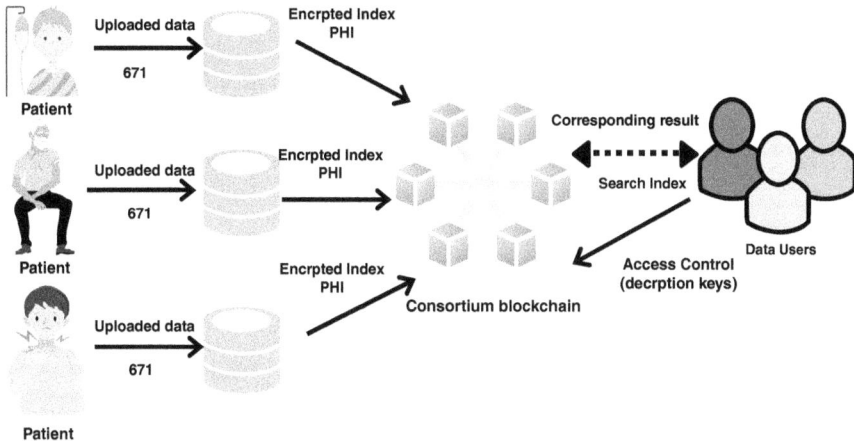

FIGURE 3.5   Architecture of attribute-based privacy system.

and searchable encryption ensure that the medical data are shared without privacy leaking. The workflow of the system is given in Figure 3.5.

MedSBA [49] is another framework using ABE to provide fine-grained access along with privacy protection based on general data protection regulation. The medical data are shared between the involved entities and stored in cloud by using ABE along with blockchain. MedSBA uses both KP-ABE (Key-Policy Attribute-Based Encryption) and CP-ABE (Ciphertext-Policy Attribute-Based Encryption) for providing access control to the patients. This scheme also deploys both permissioned and permissionless versions of practical byzantine fault-tolerance (PBFT) consensus-based private blockchain. While the latter is used to set the information of key and storage places on cloud-storing systems, the former is used to distribute public medical information and the structure for permitted access to medical data.

3.4.4.5   Attribute-Based Signature (ABS) Privacy Protection

An IoMT data sharing scheme discussed in Ref. [50] stores the medical data in the cloud and utilizes blockchain to store the address and metadata. The storage in the cloud and mapping in the blockchain provides better storage and avoids tampering of medical data. During retrieval, part of decryption is done to lessen the computation overhead in blockchain. In this method, the patient provides an access policy for the other users to handle the medical data by encrypting it with ABE. Thus, this method warrants integrity and authentication to the medical data, and also the patient identity is not revealed to preserve privacy. Then, it deploys ABS by which the owner signs the medical data by using a collection of attributes which make the identity anonymous to enhance the privacy preservation further.

The security analysis of the privacy protection system discussed in this section is shown in Table 3.6.

TABLE 3.6

Security Analysis of Various Privacy Protection Schemes

	Group Signature	TEE	SIFF	Attribute-Based Encryption	ABE with ABS
Anonymity	Yes	Yes	–	Yes	Yes
Traceability	Yes	–	–	–	–
Integrity	Yes	Yes	Yes	Yes	Yes
Confidentiality	Yes	Yes	Yes		Yes
Authentication		Yes	Yes	Yes	Yes
Privacy	Yes	Yes	Yes	Yes	Yes
Resistance to attacks	DDoS Modification attack User impersonation attack	–	–	–	Chosen Ciphertext Attack (CCA)

3.5 CONCLUSION

IoMT is the convergence of IoT with the healthcare system to afford efficient health services by exploiting the power of IoT. IoMT provides a potential communication network between the healthcare service providers, clients and the devices used. Some of the issues that curtail the use of IoMT are discussed in the chapter; and thus, IoMT necessitates the development of a ground-breaking technology to provide a platform for masking the instability of IoMT and encouraging widespread adoption of IoMT in healthcare systems. Blockchain is such a technology that utilizes a distributed ledger managed by peer-to-peer networks. Blockchain works without a centralized administrator or data storage management. Data are shared between several nodes, and quality of data is maintained through replication and encryption. In this chapter, an elaborate overview of managing the IoMT data by integrating blockchain technology is presented.

REFERENCES

[1] Al-Turjman, F., Nawaz, M. H., & Ulusar, U. D. (2020). Intelligence in the Internet of Medical Things era: a systematic review of current and future trends. *Computer Communications*, *150*, 644–660.

[2] Ben Fekih, R., & Lahami, M. (2020). Application of blockchain technology in healthcare: a comprehensive study. In *The Impact of Digital Technologies on Public Health in Developed and Developing Countries: 18th International Conference, ICOST 2020, Hammamet, Tunisia, June 24–26, 2020, Proceedings 18* (pp. 268–276). Springer International Publishing.

[3] Farouk, A., Alahmadi, A., Ghose, S., & Mashatan, A. (2020). Blockchain platform for industrial healthcare: vision and future opportunities. *Computer Communications*, *154*, 223–235.

[4] Benslimane, Y., Benahmed, K., & Benslimane, H. (2019). Security mechanisms for 6LoWPAN network in context of internet of things: a survey. In *Renewable Energy for Smart and Sustainable Cities: Artificial Intelligence in Renewable Energetic Systems 2* (pp. 49–69). Springer International Publishing.

[5] Yaacoub, J. P. A., Noura, M., Noura, H. N., Salman, O., Yaacoub, E., Couturier, R., & Chehab, A. (2020). Securing internet of medical things systems: limitations, issues and recommendations. *Future Generation Computer Systems, 105*, 581–606.

[6] Alsubaei, F., Abuhussein, A., & Shiva, S. (2017, October). Security and privacy in the internet of medical things: taxonomy and risk assessment. In *2017* IEEE 42nd *Conference on Local Computer Networks Workshops (LCN Workshops)* (pp. 112–120). IEEE.

[7] Khalid, U., Asim, M., Baker, T., Hung, P. C., Tariq, M. A., & Rafferty, L. (2020). A decentralized lightweight blockchain-based authentication mechanism for IoT systems. *Cluster Computing, 23*(3), 2067–2087.

[8] Sun, W., Cai, Z., Li, Y., Liu, F., Fang, S., & Wang, G. (2018). Security and privacy in the medical internet of things: a review. *Security and Communication Networks, 2018*, 1–9.

[9] Koutras, D., Stergiopoulos, G., Dasaklis, T., Kotzanikolaou, P., Glynos, D., & Douligeris, C. (2020). Security in IoMT communications: a survey. *Sensors, 20*(17), 4828.

[10] Hatzivasilis, G., Soultatos, O., Ioannidis, S., Verikoukis, C., Demetriou, G., & Tsatsoulis, C. (2019, May). Review of security and privacy for the Internet of Medical Things (IoMT). In *2019* 15th *International Conference on Distributed Computing in Sensor Systems (DCOSS)* (pp. 457–464). IEEE.

[11] Sun, Y., Lo, F. P. W., & Lo, B. (2019). Security and privacy for the internet of medical things enabled healthcare systems: a survey. *IEEE Access, 7*, 183339–183355.

[12] Meinert, E., Van Velthoven, M., Brindley, D., Alturkistani, A., Foley, K., Rees, S., & de Pennington, N. (2018). The internet of things in health care in oxford: protocol for proof-of-concept projects. *JMIR Research Protocols, 7*(12), e12077.

[13] Parmar, A. (2012). Hacker shows off vulnerabilities of wireless insulin pumps. *MedCity News*.

[14] Huang, X., & Nazir, S. (2020). Evaluating security of internet of medical things using the analytic network process method. *Security and Communication Networks, 2020,* 1–14.

[15] Solangi, Z. A., Solangi, Y. A., Chandio, S., bin Hamzah, M. S., & Shah, A. (2018, May). The future of data privacy and security concerns in Internet of Things. In *2018* IEEE *International Conference on Innovative Research and Development (ICIRD)* (pp. 1–4). IEEE.

[16] Bigini, G., Freschi, V., & Lattanzi, E. (2020). A review on blockchain for the internet of medical things: definitions, challenges, applications, and vision. *Future Internet, 12*(12), 208.

[17] Iansiti, M., & Lakhani, K. R. (2017). The truth about blockchain. *Harvard Business Review, 95*(1), 118–127.

[18] Ray, P. P., Dash, D., Salah, K., & Kumar, N. (2020). Blockchain for IoT-based healthcare: background, consensus, platforms, and use cases. *IEEE Systems Journal, 15*(1), 85–94.

[19] Dilawar, N., Rizwan, M., Ahmad, F., & Akram, S. (2019). Blockchain: securing internet of medical things (IoMT). *International Journal of Advanced Computer Science and Applications, 10*(1).

[20] Razdan, S., & Sharma, S. (2022). Internet of medical things (IoMT): overview, emerging technologies, and case studies. *IETE Technical Review, 39*(4), 775–788.

[21] Sharma, D., & Sharma, S. K. (2023). The use of blockchain technology in IoT-based healthcare: a concise guide. In *Blockchain Technology Solutions for the Security of Iot-Based Healthcare Systems* (pp. 183–198). Academic Press.

[22] Tith, D., Lee, J. S., Suzuki, H., Wijesundara, W. M. A. B., Taira, N., Obi, T., & Ohyama, N. (2020). Application of blockchain to maintaining patient records in electronic health record for enhanced privacy, scalability, and availability. *Healthcare informatics research, 26*(1), 3–12.

[23] Sun, W., Cai, Z., Li, Y., Liu, F., Fang, S., & Wang, G. (2018). Security and privacy in the medical internet of things: a review. *Security and Communication Networks, 2018*, 1–9.

[24] Seliem, M., & Khalid, E. (2019). BIoMT: blockchain for the internet of medical things. In *2019 IEEE International Black Sea Conference on Communications and Networking (BlackSeaCom)*. IEEE.

[25] Indumathi, J., Shankar, A., Ghalib, M. R., Gitanjali, J., Hua, Q., Wen, Z., & Qi, X. (2020). Block chain based internet of medical things for uninterrupted, ubiquitous, user-friendly, unflappable, unblemished, unlimited health care services (BC IoMT U6 HCS). In *IEEE Access* (Vol. 8, pp. 216856–216872). IEEE.

[26] McFarlane, C., Beer, M., Brown, J., & Prendergast, N. (2017). *Patientory: a healthcare peer-to-peer EMR storage network v1.* Entrust Inc.: Addison, TX, USA.

[27] Yaeger, K., Martini, M., Rasouli, J., & Costa, A. (2019). Emerging blockchain technology solutions for modern healthcare infrastructure. *Journal of Scientific Innovation in Medicine, 2*(1).

[28] Corea, F., & Corea, F. (2019). The convergence of AI and blockchain. *Applied Artificial Intelligence: Where AI Can be Used in Business* (pp. 19–26). Springer.

[29] Armstrong, S. (2018). Bitcoin technology could take a bite out of NHS data problem. *BMJ, 361*, k1996.

[30] Exochain, [Online]. Available: https://exochain.com/.

[31] Scorringe, M. (2019). More than medicine: pharmaceutical industry collaborations with the UK NHS. *Sustainable Entrepreneurship: The Role of Collaboration in the Global Economy* (pp. 111–137).

[32] Ellouze, F., Fersi, G., & Jmaiel, M. (2020). Blockchain for internet of medical things: a technical review. In *The Impact of Digital Technologies on Public Health in Developed and Developing Countries: 18th International Conference, ICOST 2020, Hammamet, Tunisia, June 24–26, 2020, Proceedings 18* (pp. 259–267). Springer International Publishing.

[33] Siyal, A. A., Junejo, A. Z., Zawish, M., Ahmed, K., Khalil, A., & Soursou, G. (2019). Applications of blockchain technology in medicine and healthcare: challenges and future perspectives. *Cryptography, 3*(1), 3.

[34] Xiao, Z., Li, Z., Liu, Y., Feng, L., Zhang, W., Lertwuthikarn, T., & Goh, R. S. M. (2018, December). EMRShare: a cross-organizational medical data sharing and management framework using permissioned blockchain. In *2018 IEEE 24th International Conference on Parallel and Distributed Systems (ICPADS)* (pp. 998–1003). IEEE.

[35] Kumar, S., Bharti, A. K., & Amin, R. (2021). Decentralized secure storage of medical records using blockchain and IPFS: acomparative analysis with future directions. *Security and Privacy, 4*(5), e162.

[36] Sun, J., Yao, X., Wang, S., & Wu, Y. (2020). Blockchain-based secure storage and access scheme for electronic medical records in IPFS. *IEEE Access, 8*, 59389–59401.

[37] EHRs Have Made it Easy for Cardiologists to Treat Their Patients. Accessed: Jul. 8, 2020. [Online]. Available: https://tbrcinfo.blogspot.com/2018/12/ehrs-have-made-it-easy-for.html

[38] Garets, D., & Davis, M. (2006). Electronic medical records vs. electronic health records: yes, there is a difference. *Policy White Paper. Chicago, HIMSS Analytics* (pp. 1–14).

[39] Madine, M. M., Battah, A. A., Yaqoob, I., Salah, K., Jayaraman, R., Al-Hammadi, Y., & Ellahham, S. (2020). Blockchain for giving patients control over their medical records. *IEEE Access, 8*, 193102–193115.

[40] Azaria, A., Ekblaw, A., Vieira, T., & Lippman, A. (2016, August). Medrec: Using block-chain for medical data access and permission management. In *2016 2nd International Conference on Open and Big Data (OBD)* (pp. 25–30). IEEE.

[41] Li, H., Zhu, L., Shen, M., Gao, F., Tao, X., & Liu, S. (2018). Blockchain-based data preservation system for medical data. *Journal of medical systems*, *42*, 1–13.

[42] Vaidya, R. (2019). Cyber security breaches survey 2019. *Department for Digital, Culture, Media and Sport* (p. 66).

[43] Ahamad, S. S., & Khan Pathan, A. S. (2021). A formally verified authentication pro-tocol in secure framework for mobile healthcare during COVID-19-like pandemic. *Connection Science*, *33*(3), 532–554.

[44] Scholl, M. A., Stine, K., Hash, J., Bowen, P., Johnson, L. A., Smith, C. D., & Steinberg, D. (2008). An introductory resource guide for implementing the health insurance por-tability and accountability act (HIPAA) security rule.

[45] Wang, B., & Li, Z. (2021). Healthchain: a privacy protection system for medical data based on blockchain. *Future Internet*, *13*(10), 247.

[46] Yahmed, F., & Abid, M. (2020). Trust execution environment and multi-party computa-tion for blockchain E-health systems. In The Impact of Digital Technologies *on* Public Health in Developed and Developing Countries*:* 18th International Conference, ICOST 2020, Hammamet, Tunisia, June 24–26, 2020, Proceedings 18 (pp. 277–286). Springer International Publishing.

[47] Tian, H., He, J., & Ding, Y. (2019). Medical data management on blockchain with pri-vacy. *Journal of Medical Systems*, *43*, 1–6.

[48] Chen, Y., Meng, L., Zhou, H., & Xue, G. (2021). A blockchain-based medical data shar-ing mechanism with attribute-based access control and privacy protection. *Wireless Communications and Mobile Computing*, *2021*, 1–12.

[49] Pournaghi, S. M., Bayat, M., & Farjami, Y. (2020). MedSBA: a novel and secure scheme to share medical data based on blockchain technology and attribute-based encryption. *Journal of Ambient Intelligence and Humanized Computing*, *11*, 4613–4641.

[50] Yang, X., Li, T., Pei, X., Wen, L., & Wang, C. (2020). Medical data sharing scheme based on attribute cryptosystem and blockchain technology. *IEEE Access*, *8*, 45468–45476.

4 Cloud Computing for Complex IoMT Data

Akrati Sharma and Priti Maheshwary
Rabindranath Tagore University

4.1 CLOUD INTRODUCTION

Nowadays, things are getting smarter day by day. Starting from mobile phones to home appliances, technology is involved everywhere in order to make things smarter. Several technologies play an important role to make things intelligent. Cloud computing and the Internet of Things are among them, and together, both technologies do a terrific job. Let us understand both technologies in detail.

Cloud computing is not just typical traditional network computing, but it also comprises a collection of multiple resources like storage, server, desktop, databases, and the list goes on and on. Whatever is physically available in traditional systems can be virtualized and used in cloud systems. The given resources are assigned to the other users as per need, on a metering basis. A user can request any type and any amount of resources from its cloud service provider (CSP). The resources can be customized as per the requirement of the user, and resources like storage space, RAM, operating systems, and other software can be given to the user.

As mentioned in Figure 4.1, the cloud infrastructure has several services. Suppose that a person demands for a Linux operating system to perform some task from CSP, and another person demands Windows 10 operating system to perform some other task from a similar CSP.

Cloud computing encompasses several benefits, and some of the major ones are as follows:

- On-demand self-service
- Resource pooling
- Location independency
- Broad network access
- Pay as per demand
- 24×7 availability

The digital revolution in the medical world started way earlier, and it has grown gradually but covered the pace with the speedy expansion of science and technology [1]. Now medical issues and its cure have become a necessity for people, but the traditional medical model has so many demands like meeting a physician, exclusive management of disease, and limited medical information. However, with the introduction

DOI: 10.1201/9781003359951-5

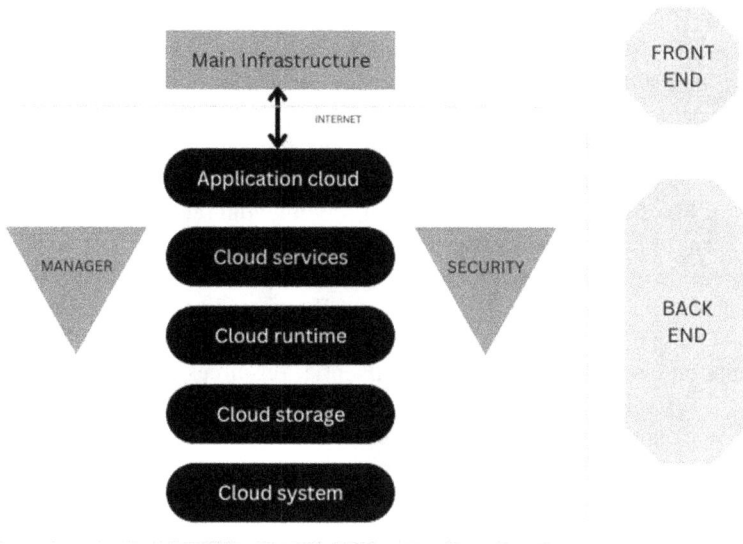

FIGURE 4.1   Cloud infrastructure.

of the notion of the Internet of Things (IoT) [2] in the year 1999, the usage of IoT in various chores has been included in all phases [3, 4]. As much as the medical world is notified, the Internet of Medical Things (IoMT) is focused on the quintessence of IoT technique in the medical field and also the essentials of the medical digital sensors and positioning technology [5], which is additionally put to the medical applications in amalgamation using mobile nodes and interaction with additional devices to apprehend the proper communication between patients and doctors.

The IoMT aids more or less all facets of the medical field like storing the critical information of patients, monitoring symptoms, suggestions, and formation of drugs. The wireless medical sensor [6] is the essential component of IoMT, and it performs operations like pressure sensing, electrocardiogram sensing, etc. Such sensors are vitally important in gaining crucial information about patients. Currently, sensors have been extensively used in various medical operation rooms, emergency rooms, and intensive care units (ICUs) to monitor and display the main conditions of patients. Benefits of wearable devices are as follows:

- Increase the interaction between doctor and patient.
- 24 × 7 availability of patient records and information.
- Easy to store patient's medical history.
- No location dependency between both parties.

Advancement in technology plays a vital role in the enhancement of the applications of the digital medical world like drug designing, protein structure prediction, and gene expression prediction. These are some crucial tasks that require accurate result; a minor mistake can cost someone's life. These new technological advancements help minimize human intervention, increase the involvement of technology

[7], and reduce the chances of error in all such applications. Several technological aspects were taken into account to design such an accurate decision support system that will help take decisions with precision. The accuracy of the result majorly depends upon the experiences collected from the historical data, hence, the rapid and real-time processing, scrutiny, and ultimate interpretation of collected medical data are crucial, as they have a direct bearing on the health and lives of patients. Since medical data comprise the confidential records of patients, it is also significant to reinforce the security of profound data and crucial patient data. IoT techniques can support medical centers and help them to satisfy the demand for fast-serving medical treatment to patients. Every medical organization is an individual entity; hence, resource sharing is always part of the concern. The IoMT constructed on cloud computing offers influential resources and significantly lessens medical costs. It realizes the distribution of medical material using the cloud platform, to expand the competence and excellence of medical facilities. Conversely, relying on cloud computing can consume extensive network transmission resources and introduce significant delays, that will impose a potential risk to the security and integrity of patient's medical data.

- Edge computing can be one of the solutions to the above issues, and it will decrease the dependence on a remote-centralized server or a distributed local server, and resolves the complications that reside in cloud computing via various applications of resources on edge devices. This will make medical sectors more responsive using IT networks so that patients can avail medical services whenever required on an early basis. Edge computing alone does not suffice to resolve the problem, but it becomes more efficient when combined with the cloud. Cloud computing focuses on global issues, while edge computing focuses on local ones. The balanced use of edge and cloud computing will improve the application of medical treatments. To motivate this collaboration, we need to focus on some points:
- The proper usage of IoMT architecture maximizes its effectiveness in the medical world. Furthermore, it assesses current research applications within the IoMT and explores the unavoidable direction of future progress in this field. Considering the hasty growth of medical data and the issues in its storage, the three-tier architecture of cloud computing for IoMT can work significantly and introduce the methods involved in the application of cloud computing to IoMT.
- In contrast with traditional IoMT and medical cloud IoT, architectural methods of edge computing have many the advantages. Besides, the traditional technology of computing is slower and vulnerable, and therefore, cloud along with edge should be used to provide quicker and more effective computing services. On the one hand, the cloud can store data more effectively; on the other hand, the edge will ensure the security and privacy of sensitive data stored over the cloud. Hence, together as a package, these two technologies can work effectively for data processing.

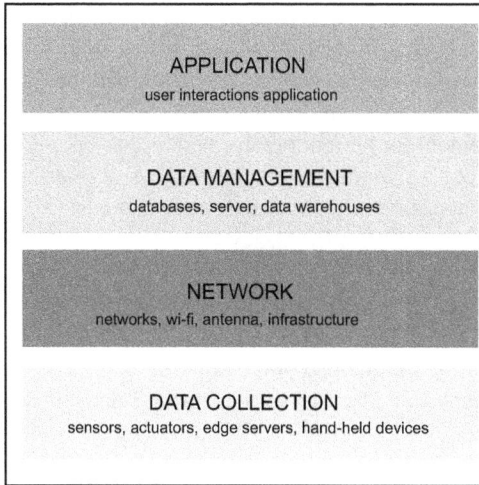

FIGURE 4.2   IoMT architecture.

4.1.1 IoMT System Architecture

Figure 4.2 demonstrates the various IoMT layers. The IoMT architecture functions mostly using four layers [8], as described below:

1. **Data Collection Layer**: It performs different methods of data recognition and gathering. It collects data from patients through various wearable medical devices, like sensors, with data sensing properties.
2. **The network layer**: It allocates the composed medical data from the data collection layer using a guided or unguided media to transport to the third layer which is the data management layer. This layer unites all the collected medical data and then exchanges the information with the above layer.
3. **The data management layer**: This layer uses mid-level applications required by the IoMT devices. It also ensures interoperability between the diverse layers of the architecture. There are some services performed by the data management layer: data gathering, proper data storage, preprocessing of data, etc. Furthermore, the application layer facilitates intelligent interaction between a user and the IoMT system. The user can effortlessly communicate and accomplish medical tasks and show medical data.
4. **The application layer**: It maintains the intellectual collaboration between a user and the IoMT system. The user can communicate and accomplish medical tasks with no trouble and display the stored medical data whenever required.

4.1.2   Technologies in IoMT

- **Radio Frequency Identification (RFID)** is the trending and base technology of IoMT. It consumes the radio frequency technique to recognize the selected target to understand the relevant data and the designated target without physical exchange. Such a method is known as full duplex data communication. The pros of using this technique are as follows:

- It can recognize the data present at remote locations, and hence, it completely gives the location dependency.
- No requirement for human involvement.
- Data collection also minimizes the chances of interference.
- **RFID Components** mainly possess three fragments:
  - Electronic tag for radio frequency
  - Data reader
  - Data manager for proper data management

Electronic tag is one of the major components of an RFID system; it is used for reading radio frequency. It also stores appropriate data of a specific object and interconnects the reader and the writer. RFID works quite efficiently in the identification of high-velocity mobile object, and it also recognizes multi-target object identification. These efficiencies of RFID are highly required in several medical fields like remedial procedures, medical data management, tracking of unwanted medical items, patients recognition, etc. It helps diagnose many symptoms of various diseases in patients by tracking patients' heart rates, recording electrocardiogram (ECG), calculating blood pressure, measuring SpO_2 (oxygen level) and so on. Based on collected samples, medical staff can diagnose the patient's ailment and go for the medication. Also, the data collected through RFID are quite precise as the accuracy rate of RFID is 100%, which ensures data precision and accuracy rate always on the notch.

There are several applications of RFID, which can prominently be used to make it worthwhile. The ink-jet-made RFID tag antenna [9] available in the market is quite popular because of its cost-effectiveness and proper observation of patient's medication dose, and this also doesn't require any human intervention and thus saves time and assets.

4.1.3 WIRELESS SENSOR NETWORK (WSN)

It assimilates sensor technology [10] to perform some major tasks like data gathering, data handling, and data diffusion. This can be done using various inbuilt technologies in sensor networks like data distribution methods, data handling techniques, and data transmission techniques. The wireless network system is made up of an ample number of small sensor nodes, and these nodes are responsible for fast computing competencies and wireless communication. They are easily deployable in the area where one needs to monitor various parameters, and they are also well orchestrated, and hence, no human intervention is required. It is well suited to collect all real-time information. As an example, in the monitoring of diverse environmental conditions, firstly raw data is prepared and transferred by the user using unguided communication channel. Sensor networks in wireless mode are like a boon in the medical field because of its dependability, promptness, security features, and real-time monitoring. The application of WSN is so vast that it can easily be fitted anywhere, and it helps monitor various medical equipment and analyze patients' anatomy based on various parameters like health concerns, disease diagnosis, mental concerns, and so on.

There are several sensors available in the market. Graphene is one of them and is used to improve the attainment of patients' physiological constraints.

As shown in Figure 4.3, various applications of graphene sensors make it a prominent wearable medical sensor across the market of sensor devices. Graphene [11, 12] has awesome features which make it highly demanding in the market. It has highly sensitive sensors, which can sense minute details like fluctuations in blood pressure of patients, motion of patients, the body perspiration details, and patients' ECG, SpO_2, etc. Graphene uses nano biosensors, and these sensors are stretchy and flexible and capable to gather a physiological record of the patient in the wearable monitor. These sensors are interacting with the brain of the person wearing it and create a new brain–machine interface.

WSN has a wide range of applications, but still, there are security concerns in use of WSN. In Ref. [13], targeting the security concerns of WSN, a low-energy ingestion network security technique centered on the WSN smart grid observing application is presented. To decipher the energy preservation of WSN, the Balanced–α Weighted Shortest Path ($B–\alpha$ WSP) routing algorithm is presented in Ref. [14], resulting in energy preservation by effectually dropping the influence of elevated load diffusion on node actions, while Ref. [15] presents an optimal clustering algorithm based on compression sensing and principal component analysis to diminish the usage of energy in the network. In short, WSN technology can present great involvement in the field of medical science.

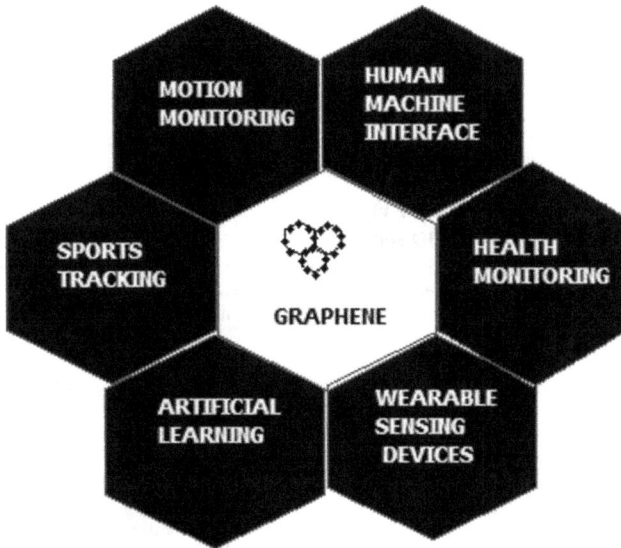

FIGURE 4.3   Graphene applications.

4.1.4 MIDDLEWARE

Middleware is a vital pillar in IoT [16]. Middleware is positioned between the reader-writer and the back-end application system, which plays an arbitrator's role. Middleware can be used in numerous applications that support distributed computing which can be achieved by creating a common platform and making data communication possible between various application platforms. The main task of middleware is to collect data using sensing devices and then store them. After that, some real-time tasks are performed on that data like editing data, processing data like cleaning and binning, and after that sending the processed data to the RFID readers. IoMT middleware accepts typical protocol and interface expertise, which can create various middleware for various types of medical application needs, like e-medical record info communication middleware, medical staff administration middleware, medical equipment supervision middleware, etc. Every single middleware should be made based upon the needs and protocols of IoMT application services to receive data in a standard format.

The large-scale usage of IoMT requires a huge amount of dense sensor nodes and typical commercial methods because the data collected by these sensor nodes are of various kinds. In Ref. [17], a proposed method involves the utilization of a semantic middleware solution that combines technology with semantic elements, leading to a significant enhancement in interoperability at a large scale.

4.1.5 THE COMPARATIVE EXPERIMENTAL STUDY BETWEEN IoMT-BASED SENSORS

RFID and WSN are two famous methods for monitoring and measuring the primary body vitals of patients. Let's have a look at these two well-known technologies (Table 4.1).

TABLE 4.1

Comparison between RFID and WSN

Feature	RFID Sensors	WSN Sensors
Communication range	Short (up to a few meters)	Medium to long (up to hundreds of meters)
Data transmission	One way (reader to tag)	Two way
Battery life	Long (can last for years)	Short (usually a few months to a year)
Deployment	Easy to install and maintain	Complex installation and requires maintenance
Cost	Low per-unit cost	High per-unit cost
Data security	Low security	High security
Interference	Prone to interference	Less prone to interference
Scalability	Limited scalability	Highly scalable
Data collection	Limited data collection capabilities	Advanced data collection capabilities
Mobility	Limited mobility	High mobility
Accuracy	High accuracy	Moderate to high accuracy

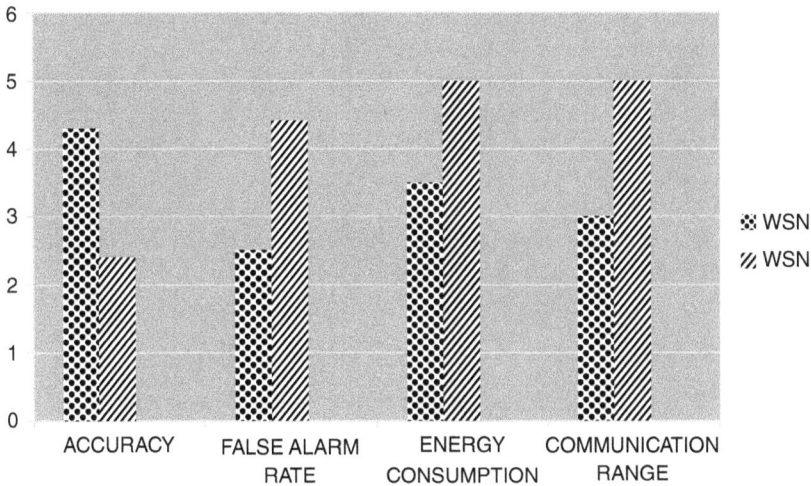

FIGURE 4.4 Comparison chart of WSN and RFID.

4.1.6 IoMT Implementations

The IoMT understands the needs of the medicalstaff so well. It cuts off the huge cost for the diagnosis and treatment of diseases, and it also assures the betterment of human health. As medical data are quite crucial, their maintenance attracts huge attention, and IoMT has satisfied this need as well. There are several applications of IoMT which majorly involve IDS, i.e., identity recognition systems and biometric identification. Biometric identification [18] involves various aspects like face recognition, fingerprint recognition, and retina recognition. The most favored approach is fingerprint recognition, mainly because of its outstanding recognition accuracy in utilizing finger veins. These applications stand out as the core of IoMT. IoMT, much like portable devices such as smartphones, tablets, and laptops, plays a pivotal role. These e-medicines made huge contributions to the betterment of resource utilization. Real-time observations of disease diagnosis [19] can help figure out the patient's physical state and take assessments on analysis and treatment information. If the patient is not able to visit the doctor, there are several technologies that can connect patients to the doctor virtually.

Without the support of wearable devices, achieving real-time monitoring of medical data is nearly impossible in practice. Devices such as smart wristwatch, health wristbands, and rings act as a bridge between patients and doctors to make better coordination between them. Although all these technologies work great together, there are still chances for improvements, which may necessitate interconnection between the devices.

In Ref. [20], a hybrid lean–agile technique is presented that inspires the involvement of wearable devices in the medical field. It also presented various methods that encourage better designing of wearable devices. It should be designed in such a manner so that the large area of the body should be in contact with the sensors. Moreover, the designing methodology emphasizes that the comfort level of the user of those wearable devices should be in higher range.

The deep neural network algorithms in the field of medical science explores the features, applications, and functionality of deep learning. It further compares deep learning and machine learning by critically evaluating the various algorithms in deep learning, thus analyzing the suitability of the convolutional neural network (CNN) algorithm in use for image detection. Deep learning can be utilized by creating a CNN [21] model on chest X-rays. It closely compares the accuracy of the self-created CNN model against the VGG16 CNN model.

4.2   APPLICATION OF IoMT ALONG WITH CLOUD

The increment of various medical centers results in the generation of huge medical data. Such kind of data are quite sensitive and crucial by nature as it consists of important information about patients. The most critical task is to keep this data safe and secure; otherwise, it may lead to data proliferation. According to IBM Global Technology Services, "Data proliferation refers to the prodigious amount of data, structured and unstructured, that businesses and governments continue to generate at an unprecedented rate and the usability problems that result from attempting to store and manage that data. While originally about problems associated with paper documentation, data proliferation has become a major problem in primary and secondary data storage on computers".

Hence, it's a challenge to solve the issue of data proliferation, and cloud computing can be one of the prominent solutions to this problem. Cloud computing in IoMT emphasizes the proper usage of resources so that none of the resources are going to be wasted.

Organization of enormous medical data is a tedious task that requires appropriate data loading, managing, and processing. Here, patient confidentiality is not just a concern, but it also involves so many issues simultaneously. Like impactful disease mechanisms, management of decision support systems and the outflow of extremely delicate data will lead to serious impacts on humans and society. Traditional IoMT cannot ensure the reliability and security of such enormous data. IoMT along with the advantages of the cloud can achieve the goal; together, these two promising technologies give high scalability, dependency and security to the data.

The IoMT has the potential to revolutionize the field of healthcare by providing real-time patient monitoring and remote care. However, the vast amount of data generated by IoMT devices require powerful storage and processing capabilities, which can be difficult for healthcare providers to handle on their own. Cloud computing provides an ideal solution to these challenges by offering scalable and cost-effective data storage and processing capabilities. In this chapter, we will explore the application of cloud computing for IoMT data and the benefits and challenges associated with this technology.

One of the key advantages of cloud computing is its ability to handle large volumes of data generated by IoMT devices. These devices can collect data such as heart rate, blood pressure, and oxygen saturation continuously and generate a massive amount of information that needs to be processed, analyzed and stored. By using cloud computing, healthcare providers can store and process this data in a centralized location that can be accessed remotely, thus reducing the need for expensive and complex IT infrastructure.

Cloud computing also enables real-time processing of IoMT data, which is critical for providing timely interventions and improving patient outcomes. Real-time analysis of data can detect early warning signs of health problems and help healthcare providers make timely and informed decisions. Additionally, cloud computing can facilitate data sharing and collaboration among healthcare professionals, leading to more coordinated and effective care.

Another advantage of cloud computing is the ability to perform advanced analytics on IoMT data using machine learning and artificial intelligence algorithms. These analytics can provide valuable insights into patient health, which would be impossible by using traditional methods. For example, by analyzing patterns in IoMT data, machine learning algorithms can detect early signs of disease, predict patient outcomes, and identify potential risk factors. These insights can help healthcare providers develop personalized treatment plans and improve the overall quality of care.

However, there are also challenges associated with using cloud computing for IoMT data. One of the biggest challenges is data security and privacy. Healthcare data are highly sensitive and need to be protected from unauthorized access. Cloud service providers need to implement robust security measures to ensure confidentiality, integrity and availability of data. Additionally, healthcare providers need to ensure compliance with all relevant regulatory requirements, such as HIPAA in the United States and GDPR in Europe.

Another challenge is data sovereignty. Healthcare data are subject to different regulatory requirements in different countries, and some countries may require that data be stored locally. Cloud service providers need to ensure compliance with all relevant regulatory requirements and provide transparency around where the data are being stored and processed. Additionally, healthcare providers need to ensure that they have the appropriate legal agreements in place with cloud service providers to protect the privacy and security of their data.

Despite these challenges, the benefits of cloud computing for IoMT data are clear. By leveraging cloud computing, healthcare providers can improve the quality of care while reducing costs and increasing efficiency. The scalability, flexibility and real-time processing capabilities of cloud computing make it an ideal solution for handling the massive amounts of data generated by IoMT devices. It is likely that we will see increasing adoption of cloud computing in healthcare in the coming years as more providers recognize the potential benefits of this technology.

Here are some real-life examples of the cloud using IoMT:

- **Philips healthcare**: Philips Healthcare is using Microsoft's Azure cloud computing platform to power its HealthSuite Digital Platform, which is designed to enable remote monitoring of patients with chronic conditions. The platform collects data from a variety of IoMT devices, including blood pressure monitors, weight scales, and glucose meters, and uses machine learning algorithms to analyze the data and provide personalized care recommendations.

• **Medtronic**: Medtronic, a medical device manufacturer, is using the Amazon Web Services (AWS) cloud computing platform to store and process data from its remote patient monitoring devices. The company's CareLink platform collects data from implantable cardiac devices and sends it to the cloud for analysis. This allows healthcare providers to remotely monitor patients and detect potential health problems before they become serious.

• **GE Healthcare**: GE Healthcare is using the AWS cloud computing platform to power its Edison platform, which is designed to enable real-time analysis of medical data. The platform collects data from a variety of sources, including IoMT devices, and uses machine learning algorithms to analyze the data and provide insights into patient health. This platform can be used to identify potential health problems, monitor patients remotely, and develop personalized treatment plans.

• **Philips VitalMinds**: Philips VitalMinds is a cloud-based platform that uses machine learning algorithms to analyze data from sensors attached to patient beds. The platform can detect early signs of delirium in elderly patients, enabling healthcare providers to intervene early and improve patient outcomes. The platform is designed to be highly scalable, allowing it to handle large volumes of data from multiple hospitals.

These are just a few examples of how cloud computing is being used to handle IoMT data in the healthcare industry. With the rapid growth of the IoMT and the increasing adoption of cloud computing, we can expect to see many more innovative applications of this technology in the future.

As shown in Figure 4.5, there is a cloud data center that can store the data. These data centers are established as per the zone of a particular area. Every small

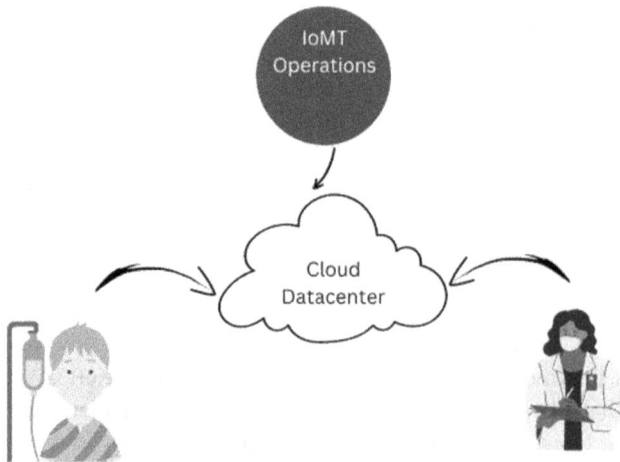

FIGURE 4.5   IoMT along with cloud services.

area has user bases that can store the data locally, and the main copy is stored in data centers. The location of these data centers is unknown as it provides location independence. The availability of data centers and user bases depends upon the number of users present in that area. For a large number of users, more data centers are required to avoid the problem of data center overloading. Users can store their data in these data centers, and the user can be a patient or a doctor or any staff. Patients can use the data center to store the data like their medical history, medical reports, disease symptoms, and so on. Doctors use the data center to read a patient's medical record and diagnose the patient using his historical record and symptoms. To record all these data, various IoMT services are used. Mainly IoT-enabled wearable devices are quite helpful to collect all these data, and they are responsible for data monitoring, resource monitoring, system monitoring, user data management, and so on. IoMT services leveraging cloud computing proficiency involve the transmission of medical data from the user's location to a remote cloud, where the data is processed, and then the results are sent back to the user's endpoint after the entire procedure. However, if the enormous quantity of data produced by the increasing medical data are shifted to the cloud, it will result high pressure to the cloud. This may cause great energy ingestion and vast delay due to the overloading of the cloud. The essential strategy for enhancing the power of IoMT lies in the efficient enhancement of cloud computing capabilities and the utilization of distributed computing resources.

4.3 CONCLUSION

Cloud computing offers so many services to consumers; hence, it has become more popular in recent days. Cloud computing has become the solution to many issues, and IoMT is a field that benefits from the same. The IoMT is a new field of IoT which includes all medical aspects along with intelligent use of devices. Several technologies that work along with the cloud make IoMT better; for example, RFID and WSN. The key ingredient of IoMT is wearable devices, which provide location independency to the user. Without their physical presence, patients can communicate with doctors, and these wearables collect various information about the patient and make them available to the doctor for further examination. Cloud helps store the data from both ends and keep it secure and safe for a long time.

REFERENCES

[1] Ning, Z., Xia, F., Ullah, N., Kong, X., & Hu, X. (2017). Vehicular social networks: enabling smart mobility. *IEEE Communications Magazine*, *55*(5), 16–55.
[2] Al-Fuqaha, A., Guizani, M., Mohammadi, M., Aledhari, M., & Ayyash, M. (2015). Internet of Things: a survey on enabling technologies, protocols, and applications. *IEEE Communications Surveys & Tutorials*, *17*(4), 2347–2376.
[3] Wang, W., Yu, S., Bekele, T. M., Kong, X., & Xia, F. (2017). Scientific collaboration patterns vary with scholars' academic ages. *Scientometrics*, *112*(1), 329–343.

[4] Wang, X., Ning, Z., Hu, X., Wang, L., Guo, L., Hu, B., & Wu, X. (2019). Future communications and energy management in the Internet of Vehicles: toward intelligent energy-harvesting. *IEEE Wireless Communications, 26*(6), 87–93.

[5] Shit, R. C., Sharma, S., Puthal, D., & Zomaya, A. Y. (2018). Location of things (LoT): a review and taxonomy of sensors localization in IoT infrastructure. *IEEE Communications Surveys & Tutorials, 20*(3), 2028–2061.

[6] Liu, X., Liu, A., Qiu, T., Dai, B., Wang, T., & Yang, L. (2020). Restoring connectivity of damaged sensor networks for long-term survival in hostile environments. *IEEE Internet of Things Journal, 7*(2), 1205–1215.

[7] Yu, J., Liu, J., Zhang, R., Chen, L., Gong, W., & Zhang, S. (2019). Multi-seed group labeling in RFID systems. *IEEE Transactions on Mobile Computing, Early Access*. doi:10.1109/TMC.2019.2934445

[8] Hemanth, J. A. D. J., & George, A. (2021). *Internet of Medical Things: Remote Healthcare Systems and Applications*. Springer.

[9] Sharif, A., Ouyang, J., Yan, Y., Raza, A., Imran, M. A., & Abbasi, Q. H. (2019). Low-cost inkjet-printed RFID tag antenna design for remote healthcare applications. *IEEE Journal of Electromagnetics, RF, and Microwaves in Medicine and Biology, 3*(4), 261–268.

[10] Wei, W., & Qi, Y. (2011). Information potential fields navigation in wireless ad hoc sensor networks. *Sensors, 11*(5), 4794–4807.

[11] Pang, Y., Yang, Z., Yang, Y., Wu, X., Yang, Y., & Ren, T.-L. (2019). Graphene based wearable sensors for healthcare. In *Proceedings of the International Conference on IC Design and Technology (ICICDT)* (pp. 1–4).

[12] Rachakonda, L., Sundaravadivel, P., Mohanty, S. P., Kougianos, E., & Ganapathiraju, M. (2018). A smart sensor in the IoMT for stress level detection. In *Proceedings of the IEEE International Symposium on Smart Electronic Systems (iSES)* (pp. 141–145).

[13] Dhunna, G. S., & Al-Anbagi, I. (2019). A low power WSNs attack detection and isolation mechanism for critical smart grid applications. *IEEE Sensors Journal, 19*(13), 5315–5324.

[14] Yadav, S., Kumar, V., Dhok, S. B., & Jayakody, D. N. K. (2019). Energy-efficient design of MI communication-based 3-D non-conventional WSNs. *IEEE Systems Journal*, Advance online publication. https://doi.org/10.1109/JSYST.2019.2918184

[15] da Cruz, M. A. A., Rodrigues, J. J. P. C., Al-Muhtadi, J., Korotaev, V. V., & de Albuquerque, V. H. C. (2018). A reference model for Internet of Things middleware. *IEEE Internet of Things Journal, 5*(2), 871–883.

[16] Zgheib, R., Conchon, E., & Bastide, R. (2019). Semantic middleware architectures for IoT healthcare applications. In *Enhanced Living Environments* (pp. 263–294). Springer.

[17] Xin, Y., Kong, L., Liu, Z., Wang, C., Zhu, H., Gao, M., Zhao, C., & Xu, X. (2018). Multimodal feature-level fusion for biometrics identification system on IoMT platform. *IEEE Access, 6*, 21418–21426.

[18] Banou, S., Swaminathan, M., Reus Muns, G., Duong, D., Kulsoom, F., Savazzi, P., Vizziello, A., & Chowdhury, K. R. (2019). *Beamforming galvanic coupling signals for IoMT implant-to-relay communication. IEEE Sensors Journal, 19*(19), 8487–8501.

[19] Ning, Z., Zhang, K., Wang, X., Guo, L., Hu, X., Huang, J., Hu, B., & Kwok, R. Y. K. (2020). Intelligent edge computing in Internet of Vehicles: a joint computation offloading and caching solution. *IEEE Transactions on Intelligent Transportation Systems, 2020*, 1–14. https://doi.org/10.1109/TITS.2020.2997832

[20] Glazkova, N., Fortin, C., & Podladchikova, T. (2019). Application of lean-agile approach for medical wearable device development. In *Proceedings of the 14th Annual Conference on Systems of Systems Engineering (SoSE)* (pp. 75–80).

[21] Pandey, R., & Pande, M. (2021). Provenance data models and assertions: a demonstrative approach. In M. Pande (Ed.), *Semantic IoT: Theory and Applications: Interoperability, Provenance and Beyond* (pp. 103–129).

5 The Potential of IoMT Devices in Early Detection of Suicidal Ideation

Archana Sahai and Farooqui UK
Amity University

5.1 INTRODUCTION

An individual's mental well-being is one of the most valuable aspects to consider, especially in the post-pandemic era. Due to recent recessions, global geopolitical instability and rising rates of addiction and obsessions, the public's mental health has been deteriorating and the number of cases related to mental health has been on the rise. Global Data's epidemiological forecasts estimate that major depressive disorder (MDD) affects almost 40% more people than previously estimated across the US, France, Germany, Italy, Spain, the UK, and Japan. Global MDD prevalence is expected to increase further after the pandemic, from 53.4 million people in 2019 to 55.4 million in 2029 [1].

There is also a deficiency of mental health and addiction care providers, which has left many patients struggling hard to access care and resources [2]. This situation is more critical in the Indian scenario as still people don't report mental health and still lots of superstitious faiths about handling mental health exist. Society is still not educated about mental health-related issues. The mental health of a person is measured by a high grade of affective disorder which results in major depression and different anxiety disorders. Many conditions are recognized as mental disorders including anxiety disorder, depressive disorder, mood disorder, personality disorder, suicidal tendencies, and apparent suicide.

Suicide due to mental health issues is one of the major problems that the Indian government is facing. If we consider the overall global scenario that about 8,00,000 people commit suicide worldwide every year, then 17% of them are residents of India. The male-to-female suicide ratio has been about 2:1 in India. On average, the total number of suicides in India per day is 300. According to the suicide reports in India and National Crime Records Bureau (NCRB), the total number of suicides in India as per 2014 statistics is 1,09,456 [3]. The NCRB released data on suicidal deaths in India in August 2022, and the figures were startling. A total of 1,64,033 suicides were reported in the country in 2021, which is an increase of 7.2% in comparison to the previous year in terms of total numbers. In terms of the rate of suicide, India reported

DOI: 10.1201/9781003359951-6

a rate of 12 (per lakh population), and this rate reflects a 6.2% increase during 2021 as compared to 2020. The number reported is the highest ever recorded in the country since the inception of reporting suicides by the NCRB in 1967 [4]. Mental health issues in most cases result in suicide. So, it needs to be monitored and treated.

As the proportion of people living with mental health conditions continues to grow, this will drive innovation in the mental health space and will lead to the increased uptake of novel digitally assisted care and diagnosis methods. In the future, there will be a diverse selection of personalized, digital tools available in the mental health space for consumers to choose from.

5.2   INTERNET OF THINGS (IoT)

The era of the Internet of Things (IoT) has arrived, where smart, connected technologies are being embedded in everyday objects such as cars, toothbrushes, washing machines, and physical infrastructure on a massive scale. The use of IoT devices for remote monitoring, diagnosis, and treatment is viewed as an important way to improve and expand individualized medical care and assist with lowering costs, especially for bipolar disorders and other mental illnesses [5]. While there is no standard definition, the IoT is describes as "the extension of network connectivity and computing capability to objects, devices, sensors and items not ordinarily considered to be computers" (Internet Society 2015) [6]. IoT devices can be thought of as physical devices with embedded technology that can sense, generate, and store data, and sometimes respond to commands via actuators that can modify the physical world. Increasingly, IoT devices will be installed in the home for medical purposes as selected by patients or as recommended by physicians. Today, a diverse range of IoT devices are found in homes, retail businesses, public spaces, hospitals and healthcare facilities, vehicles, and utility infrastructure, and they are directly worn by consumers. Virtually every consumer electronics device is now sold as a connected IoT device (NIST 2019). The scale of the IoT is unprecedented, with estimates of 30 billion connected devices by 2020 (Nordrum 2016) [7], and half the total global Internet traffic will be machine-to-machine connections by 2022 (Cisco 2019) [8].

IoT tools such as mobile health (mHealth), which capture patient data, are expanding rapidly, with publicly available mobile apps becoming ubiquitous, alongside a large portion of the population in the US and Europe now owning smartphones [9]. Mobile apps are particularly suited to treating mental illnesses such as depression and anxiety, where stigma and a lack of self-belief act as barriers to treatment and engagement. For example, almost 8,000 medical, lifestyle, or health and fitness apps are listed on the Apple and Google app stores under the non-specific terms 'mental health' or 'depression'.

5.3   INTERNET OF MEDICAL THINGS (IoMT)

IoMT (Internet of Medical Things) is a term used to describe a collection of medical devices and applications that are connected to the Internet and can transmit data between them and to healthcare providers. These devices are often wearable or implantable and are designed to gather and transmit health-related data, such as vital signs, medication usage, and activity levels.

IoMT devices can be used for a variety of applications in healthcare, such as remote monitoring of chronic conditions, real-time tracking of vital signs, medication management, and behavioral health monitoring. The data collected from IoMT devices can be used to inform healthcare decision-making, improve patient outcomes, and provide personalized treatments.

IoMT devices typically use wireless and Internet-enabled technologies to collect and transmit data. They often incorporate sensors, such as accelerometers and heart rate monitors, to gather data about patient health and wellness. These data are then transmitted to a cloud-based platform where it can be stored and analyzed.

The use of IoMT devices has the potential to revolutionize healthcare by providing real-time, personalized data that can help healthcare providers make more informed decisions about patient care. However, as with any technology, there are also concerns about privacy, security, and accuracy of the data collected. As such, it is important for healthcare providers to use IoMT devices responsibly and in compliance with applicable regulations and laws. The IoMT refers to the interconnected system of medical devices, sensors, and software applications that allow healthcare providers to collect and analyze real-time data from patients. IoMT devices can be particularly useful for monitoring mental health, as they provide a way to collect objective data on a patient's emotional state and behavior over time.

5.4   IoMT DEVICES AND MENTAL HEALTH

IoMT devices offer a promising avenue for mental health monitoring, as they provide a way to collect objective data in real time. However, it's important to note that privacy and security concerns must be carefully considered when using these devices, as they can potentially expose sensitive health information. It's also important to work with a healthcare provider who can interpret the data collected by these devices and provide appropriate treatment recommendations.

The use of IoMT devices for mental health has the potential to help people in several ways. Here are a few examples:

1. **Early detection and prevention**: IoMT devices can be used to monitor vital signs, such as heart rate, respiratory rate, and sleep patterns. This data can be used to detect early signs of mental health problems, such as depression, anxiety, and sleep disorders, allowing for early intervention and prevention.
2. **Remote monitoring and support**: IoMT devices can be used to remotely monitor patients' mental health status, enabling healthcare providers to provide support and treatment even if the patient is not physically present. This can be especially helpful for patients who live in rural areas, where access to mental healthcare may be limited.
3. **Personalized treatment**: IoMT devices can be used to collect data on a patient's mental health status, enabling healthcare providers to develop personalized treatment plans based on the patient's specific needs and symptoms.

4. **Improved self-awareness**: IoMT devices can be used to track a patient's mood, sleep, and physical activity levels, helping the patient to become more self-aware and identify potential triggers or patterns that may be affecting their mental health.

5. **Reduced stigma**: The use of IoMT devices for mental health can help reduce the stigma associated with mental health problems by making it easier for patients to monitor and manage their mental health in a discreet and non-intrusive way.

The use of the IoMT for mental health diagnosis related to suicidal tendencies is an emerging field of research. There is no definitive set of medical parameters that can accurately diagnose suicidal tendencies. However, some studies have suggested the following parameters may be useful in predicting suicidal tendencies:

Neurotransmitter imbalances: Imbalances in neurotransmitters such as serotonin, dopamine, and norepinephrine have been linked to depression and suicidal tendencies.

Hormonal imbalances: Hormonal imbalances, including changes in cortisol levels and thyroid hormones, have also been linked to depression and suicidal tendencies.

Sleep disturbances: Sleep disturbances, including insomnia and hypersomnia, have been linked to an increased risk of suicidal ideation.

Inflammation: Inflammation in the body has been associated with depression and suicidal tendencies.

Cardiovascular measures: Heart rate variability, blood pressure, and other cardiovascular measures have been found to be associated with suicidal tendencies.

Cognitive factors: Certain cognitive factors, such as hopelessness, low self-esteem, and negative thinking patterns, have also been linked to suicidal tendencies.

It is important to note that suicidal tendencies are a complex and multifactorial phenomenon, and the above parameters may not be sufficient to diagnose or predict suicidal tendencies on their own. A comprehensive evaluation by a mental health professional is necessary to accurately diagnose and treat suicidal tendencies.

5.5 PARAMETERS MEASURED BY IoMT DEVICES

Adoption of IoMT-enabled healthcare devices became a more common custom primarily during COVID-19 that emphasized more on the use of remote health-monitoring devices like smart inhalers, oxygen saturation monitors, blood glucose monitors including smart pens, smart wearable devices like smart watches, smart implants, smart toothbrushes, sleep trackers, and loneliness detectors [10]. Various IoMT devices can be used to measure the parameters that may be associated with suicidal tendencies. Some examples of IoMT devices that can measure these parameters include the following:

Neurotransmitter measuring devices: Various devices can measure neurotransmitter levels in the body. For example, electroencephalogram (EEG) devices can measure brainwave activity, which may be associated with neurotransmitter imbalances.

Hormone measuring devices: Devices such as blood glucose monitoring and continuous glucose monitoring (CGM) systems can measure hormonal imbalances, including changes in cortisol levels.

Sleep tracking devices: Devices such as wearable activity trackers and smart-watches can track sleep patterns and provide information on sleep disturbances.

Inflammatory markers measuring devices: Devices such as C-reactive protein (CRP) monitors can measure inflammation in the body.

Cardiovascular monitoring devices: Devices such as heart rate monitors, blood pressure monitors, and pulse oximeters can be used to monitor cardiovascular measures.

Cognitive assessment tools: Various IoMT devices can be used to assess cog-nitive factors associated with suicidal tendencies. For example, digital cognitive assessment tools can be used to measure cognitive functioning, including attention, memory, and processing speed (Figure 5.1).

It is important to note that not all these devices are widely available or have been validated for use in predicting or diagnosing suicidal tendencies. A healthcare pro-fessional with expertise in mental health and IoMT may be able to provide more information on which devices are appropriate for specific situations.

IoMT devices can potentially help in predicting suicide rates by continuously monitoring and collecting data on various physiological and behavioral parameters of an individual that may be indicative of a suicidal state. For example, IoMT devices such as smartwatches or fitness trackers can collect data on heart rate variability, sleep patterns, and activity levels, while wearable sensors can monitor changes in sweat, temperature, and other physiological parameters.

The collected data can be analyzed using machine learning algorithms to identify patterns and anomalies that may indicate a potential risk for suicidal ideation. For instance, changes in sleep patterns and activity levels, coupled with low heart rate variability, may indicate depression or other mental health issues, which can increase the risk of suicide.

As for the number of datasets available for predicting suicidal tendencies in the Indian scenario, there are several publicly available datasets that researchers can use to develop predictive models. One such dataset is the NCRB dataset, which provides

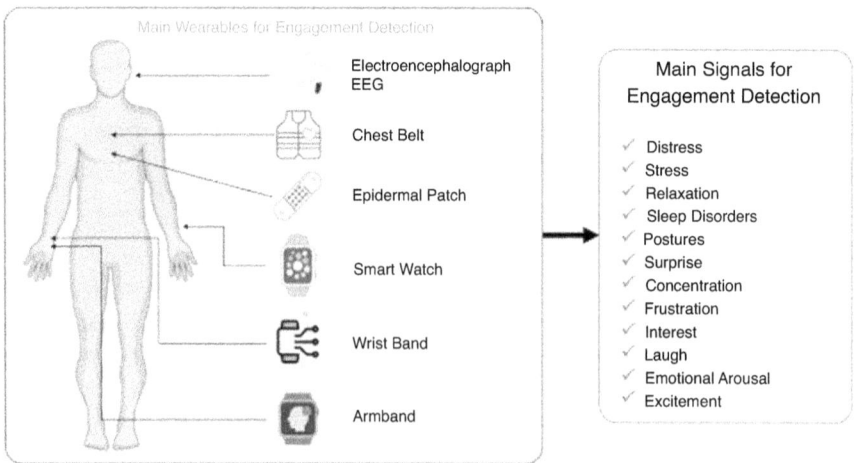

FIGURE 5.1   Main wearables for engagement detection [11].

information on suicide rates in India, including demographic information, causes, and methods of suicide.

However, it is essential to note that suicide is a complex and multi-dimensional issue, and relying solely on predictive models based on physiological and behavioral data may not be sufficient to prevent suicide. It is crucial to take a holistic approach that includes social, cultural, and environmental factors in suicide prevention.

In terms of getting in touch with someone suffering from suicidal ideation, it is crucial to seek professional help from mental health experts, such as psychologists and psychiatrists. The National Institute of Mental Health and Neurosciences (NIMHANS) in Bangalore is one such institution that provides mental health services in India. Additionally, several helplines, such as the Suicide Prevention India Foundation (SPIF) and the AASRA helpline, are available to provide support and counseling to those in need.

5.6   IoMT DEVICES IN PREDICTING SUICIDE RATES

IoMT devices can potentially help in predicting suicide rates based on physiological and behavioral data.

1. **Smartwatches**: Smartwatches with health-monitoring features can collect data on heart rate, sleep patterns, and activity levels, which can provide insights into an individual's emotional and mental states (Figure 5.2).
2. **Wearable sensors**: Wearable sensors can be attached to the body to collect data on physiological parameters such as skin conductance, body temperature, and respiration rate. These parameters may be indicative of an individual's emotional state, and changes in these parameters can potentially help in predicting suicide risk (Figures 5.3 and 5.4).

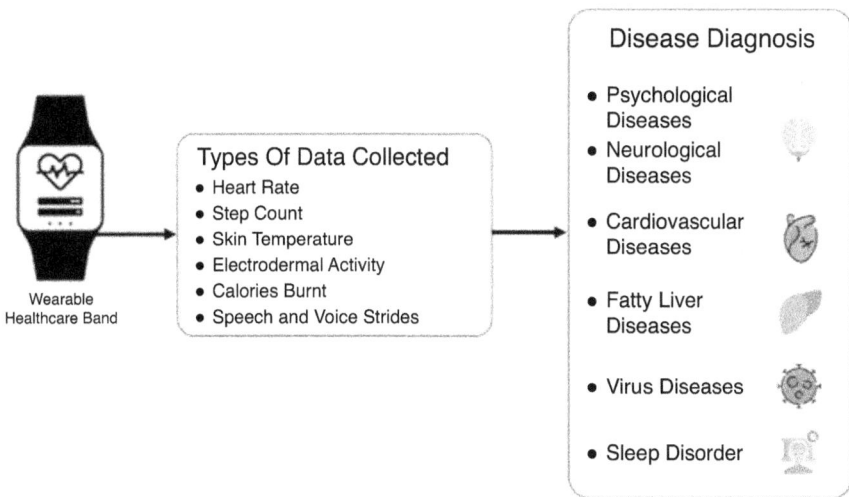

FIGURE 5.2   Types of data collected by wearable healthcare bands [12].

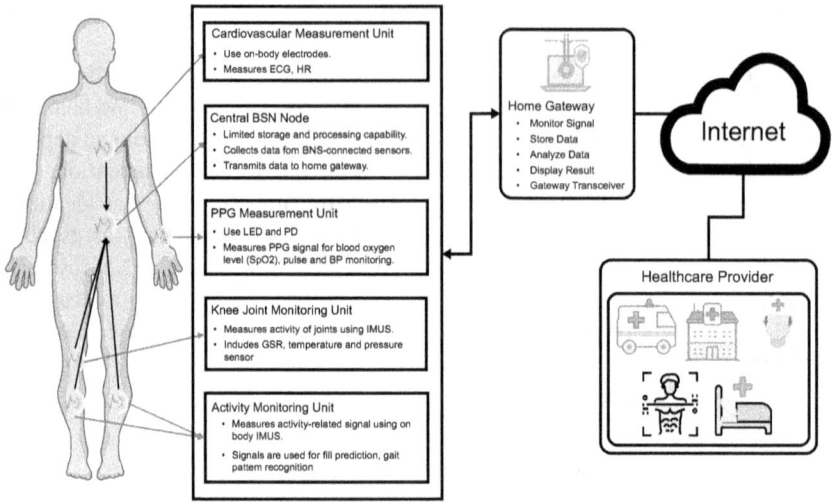

FIGURE 5.3   Wearable sensors to collect data on physiological parameters [13].

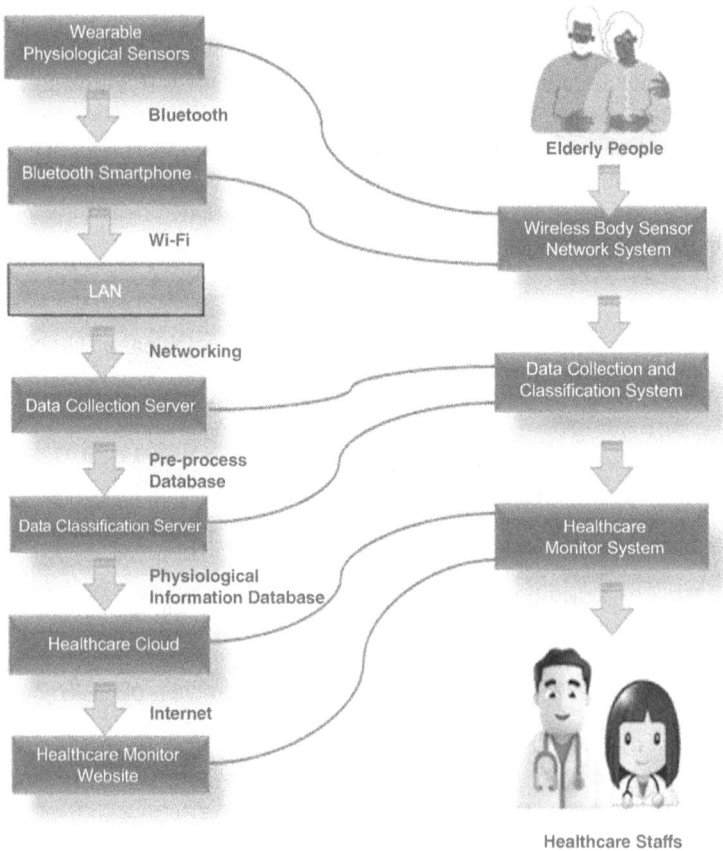

FIGURE 5.4   Transfer of physiological parameters to healthcare staffs [14].

3. **Mobile apps**: Mobile apps can collect data on an individual's behavior, including social media usage, text messages, and phone calls. Analysis of this data can help identify changes in behavior that may be indicative of a suicidal state (Figure 5.5).

4. **Medical devices**: Medical devices, such as electroencephalography (EEG) and functional magnetic resonance imaging (fMRI), can collect data on brain activity and may provide insights into an individual's mental state (Figure 5.6).

1. **Smart home devices**: Devices like smart speakers and security cameras can be used to monitor a patient's behavior in their home. For example, they can detect changes in movement patterns or identify potential safety risks, such as leaving the stove on (Figures 5.7 and 5.8).

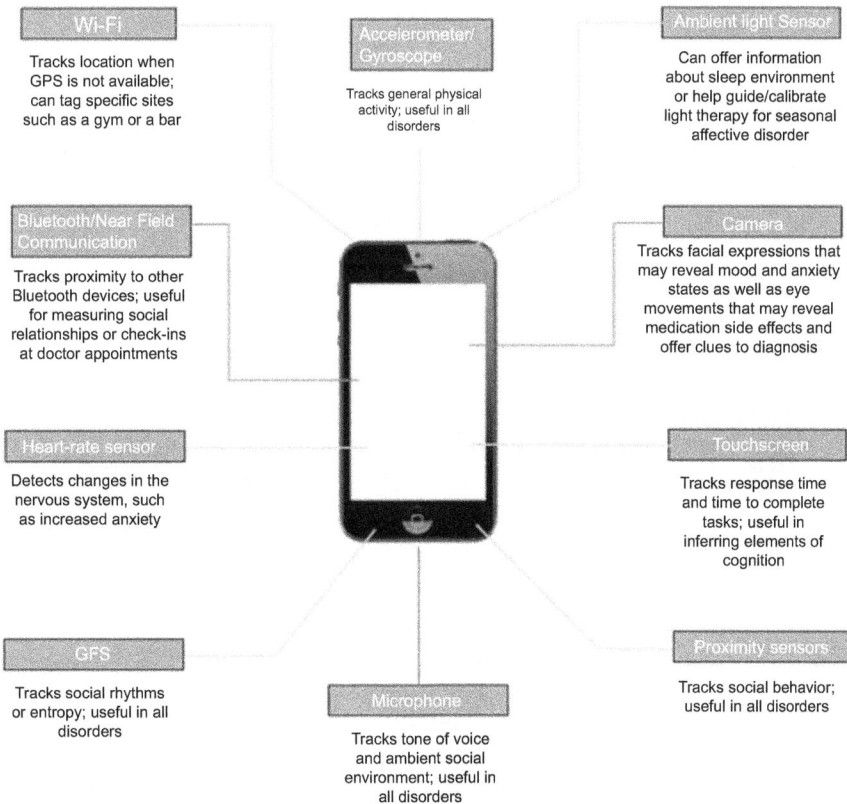

Wi-Fi
Tracks location when GPS is not available; can tag specific sites such as a gym or a bar

Accelerometer/ Gyroscope
Tracks general physical activity; useful in all disorders

Ambient light Sensor
Can offer information about sleep environment or help guide/calibrate light therapy for seasonal affective disorder

Bluetooth/Near Field Communication
Tracks proximity to other Bluetooth devices; useful for measuring social relationships or check-ins at doctor appointments

Camera
Tracks facial expressions that may reveal mood and anxiety states as well as eye movements that may reveal medication side effects and offer clues to diagnosis

Heart-rate sensor
Detects changes in the nervous system, such as increased anxiety

Touchscreen
Tracks response time and time to complete tasks; useful in inferring elements of cognition

GFS
Tracks social rhythms or entropy; useful in all disorders

Microphone
Tracks tone of voice and ambient social environment; useful in all disorders

Proximity sensors
Tracks social behavior; useful in all disorders

FIGURE 5.5 Diagnosis possibilities by smartphone [15].

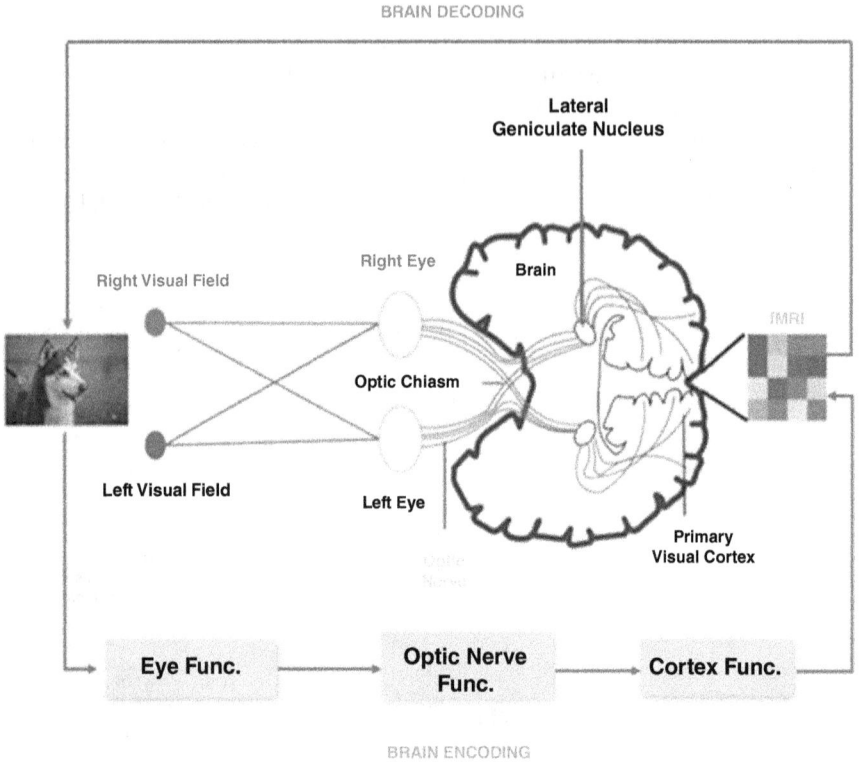

FIGURE 5.6   Brain decoding [16].

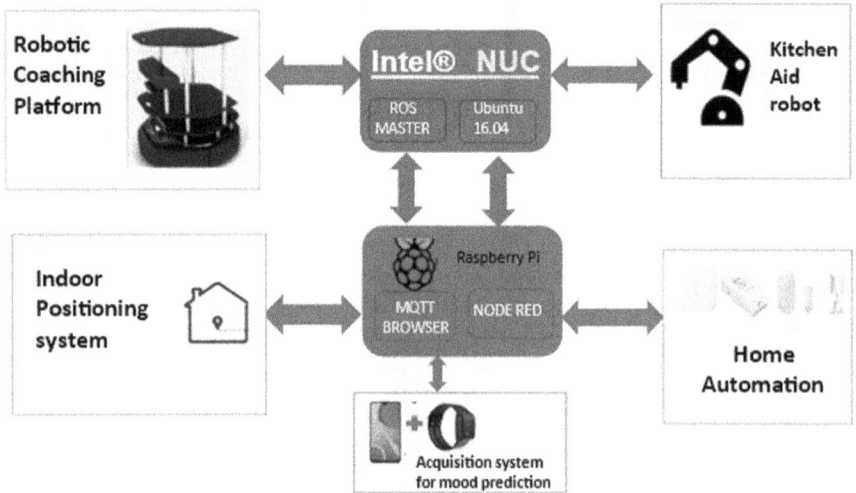

FIGURE 5.7   Smart home devices to monitor patient's behavior at home [17].

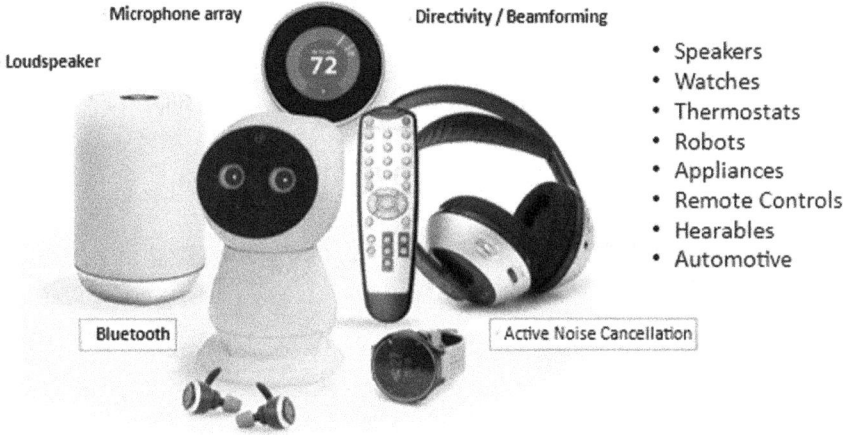

FIGURE 5.8 Smart home devices [18].

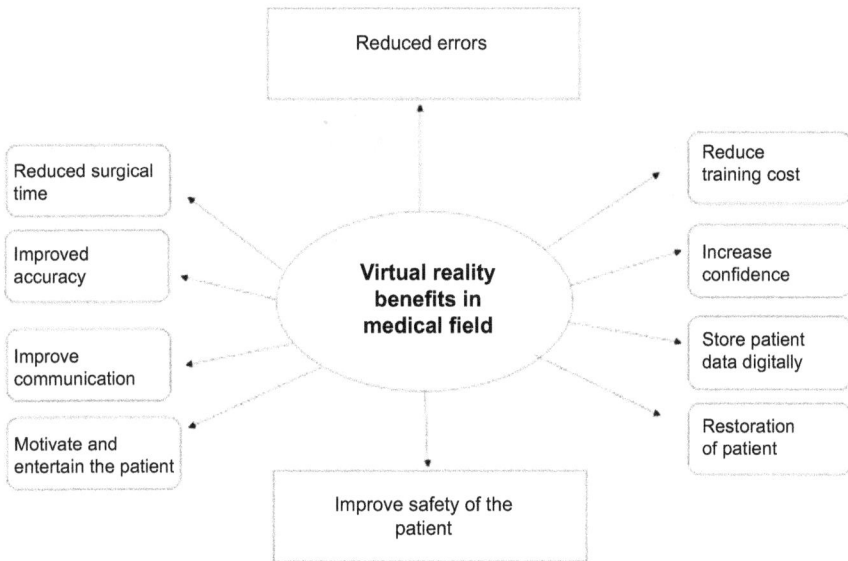

FIGURE 5.9 Improving patient's emotional state by virtual reality systems [19].

2. **Virtual reality systems**: Virtual reality technology can be used to create immersive environments that can help patients manage anxiety, phobias, or PTSD. By tracking a patient's physiological responses in real time, these systems can provide insights into the patient's emotional state (Figures 5.9 and 5.10).

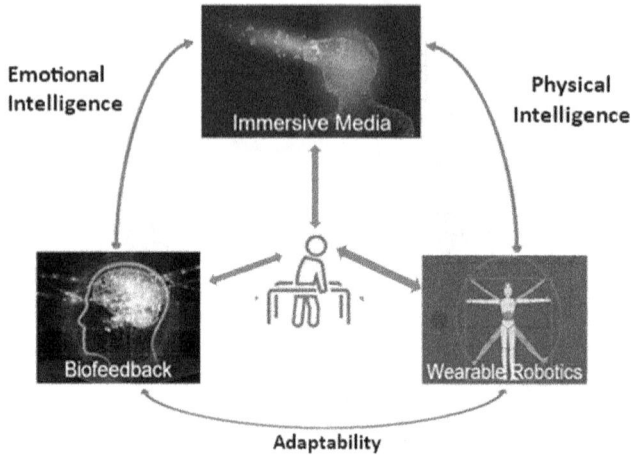

FIGURE 5.10   Monitoring emotional and physical intelligence using virtual reality systems [20].

5.7   IoMT DEVICES SPECIFICALLY FOR MENTAL HEALTH IN INDIA

While there are limited data available on the use of IoMT devices specifically for mental health in India, there have been some studies and reports that provide insights into the potential benefits of IoMT devices in mental healthcare.

A study conducted by the Indian Institute of Technology (IIT) Hyderabad in 2019 explored the use of IoMT devices for monitoring and predicting mental health conditions. The study focused on using wearable devices to monitor heart rate variability, which can be an indicator of stress and anxiety. The researchers found that the wearable devices were able to accurately predict stress levels in study participants, demonstrating the potential of IoMT devices for mental health monitoring.

Another study conducted by researchers at the Indian Institute of Science (IISc) Bangalore in 2020 investigated the use of IoMT devices for monitoring sleep patterns in patients with depression. The study found that the IoMT devices were able to detect changes in sleep patterns that correlated with improvements in depressive symptoms, suggesting that these devices could be useful in monitoring treatment progress [21].

Additionally, a report by the Confederation of Indian Industry (CII) and PwC India in 2019 highlighted the potential of IoMT devices in improving access to mental healthcare in India, particularly in rural areas where access to mental healthcare is limited. The report suggested that IoMT devices could be used to provide remote monitoring and support for patients, as well as to collect data for analysis and research [22].

One recent study that provides data analytics on the use of IoMT devices for mental health in India is a pilot study published in the *Journal of Medical Systems* in 2021. The study focused on using a smartphone application to track mood, sleep, and physical activity in patients with depression. The study found that the application was able to collect data on these parameters and provide insights into the patient's mental health status.

Another study published in the *Journal of Medical Internet Research* in 2019 investigated the use of wearable devices for monitoring stress levels in Indian software professionals. The study found that wearable devices were able to accurately detect changes in stress levels over time and could be used as a tool for stress management.

A systematic review and meta-analysis published in the journal *BMC Psychiatry* in 2019 examined the effectiveness of digital interventions for mental health in low- and middle-income countries, including India. The review found that digital interventions, including those using IoMT devices, can be effective in reducing symptoms of depression and anxiety.

The effectiveness of IoMT devices in predicting and preventing mental health problems, including suicidal ideation, has been the subject of several research studies. A study published in the *Journal of Medical Internet Research* found that IoMT devices that monitor heart rate variability could predict depression and anxiety symptoms in patients with cardiovascular disease. The study found that changes in heart rate variability were significantly associated with changes in depression and anxiety symptoms [23].

While these studies and reports provide promising insights into the potential benefits of IoMT devices for mental healthcare in India, more research is needed to fully assess the effectiveness of these devices in diagnosing, treating, and predicting mental health conditions. It is also important to consider the ethical and privacy implications of using IoMT devices in mental healthcare and to ensure that they are used in a way that respects patient autonomy and confidentiality.

While there is still much research to be done on the effectiveness of IoMT devices for mental health, there is promising evidence to suggest that these devices can help people with mental health-related problems by improving early detection, remote monitoring and support, personalized treatment, self-awareness, and reducing stigma.

5.8 APPLICATION OF IoMT DEVICES IN INDIA

There are hospitals and healthcare organizations in India that use IoMT devices for monitoring mental health-related issues. IoMT devices have gained significant attention in recent years due to their potential to revolutionize healthcare by enabling continuous monitoring of health indicators. One area of interest is the use of IoMT devices to predict mental health outcomes. In India, where mental health has traditionally been stigmatized and under-resourced, such technology could be particularly valuable. *Artificial Intelligence and Machine Learning for Predictive and Analytical Rendering in Edge Computing* focuses on the role of AI and machine learning as they impact the growing number of devices and applications in diversified domains of industry, especially in medical diagnostics [24].

With the help of IoMT technologies, self-care and early diagnosis have become influential services in strengthening the healthcare ecosystem, especially for those which utilize remote monitoring systems [25]. The literature search on the topic yielded a limited number of studies specific to India, but there are some studies conducted in other countries that are relevant to the Indian context.

One study titled "Monitoring the Mental Health of Patients Using Wearable Devices" was conducted in Iran and published in the *Journal of Medical Systems* in 2020. The study evaluated the accuracy of wearable devices in predicting mental health outcomes in a sample of 15 patients with depression. The researchers found that wearable devices were able to accurately predict the level of depression in patients. However, the study was limited by its small sample size and the fact that it focused on depression only [26].

Another study titled "Continuous Monitoring of Depression Symptoms Using Wearable and Mobile Devices" was conducted in the US and published in the *Journal of Medical Internet Research* in 2016. The study evaluated the feasibility and acceptability of using wearable and mobile devices to monitor depression symptoms in 16 patients with MDD. The researchers found that the devices were acceptable and feasible to use and had the potential to provide real-time monitoring of depression symptoms.

In addition, a study titled "The Use of Technology in the Assessment and Treatment of Mental Health Problems in Children and Adolescents – A Review" published in the *Journal of Child Psychology and Psychiatry* in 2017 looked at the potential of technology-based interventions in improving mental health outcomes in children and adolescents. Although the study did not focus specifically on IoMT devices, it highlighted the potential of technology-based interventions in improving mental health outcomes [27].

Overall, while there is limited research specific to India, the studies conducted in other countries suggest that IoMT devices have the potential to accurately predict mental health outcomes and provide real-time monitoring of symptoms. However, further research is needed to evaluate the feasibility and acceptability of these devices in the Indian context, as well as their potential impact on mental health outcomes in this population [28].

For example, the NIMHANS in Bangalore, India, has incorporated IoMT devices in their mental healthcare delivery. They use wearable devices to monitor and track the symptoms of patients with mental illnesses such as depression and anxiety. The data collected from these devices are analyzed using AI algorithms to provide personalized treatment plans and interventions [29].

Additionally, other hospitals and mental health clinics in India are also starting to incorporate IoMT devices in their practices. For instance, the Apollo Hospitals Group has established a virtual mental health clinic that uses IoMT devices to remotely monitor the mental health of their patients.

As the use of IoMT devices continues to grow in the healthcare sector, we can expect to see more hospitals and healthcare organizations in India and around the world adopting this technology for mental health monitoring and treatment.

It's important to note that patient data privacy is a critical concern in the use of IoMT devices, and healthcare organizations must take appropriate measures to safeguard patient data confidentiality and comply with relevant regulations and laws.

5.9 CONCLUSION

The use of IoMT devices in mental healthcare has been proposed as a potential solution to reduce suicide and mental health cases in India. IoMT devices can be used to monitor various parameters related to mental health such as heart rate, sleep patterns, and mood fluctuations, which can be helpful in identifying potential mental health issues before they escalate into serious problems.

While there is some research on the use of IoMT devices for mental health monitoring, there is limited data available on the effectiveness of these devices in reducing suicide and mental health cases in India specifically. However, some studies have shown promising results in using IoMT devices to monitor and treat mental health conditions.

For example, a 2018 study published in the *Journal of Medical Systems* showed that IoMT devices can be used to improve treatment adherence and outcomes for patients with depression. Another study published in the *Journal of Medical Internet Research* in 2020 found that the use of IoMT devices to monitor physical activity and sleep patterns can help reduce symptoms of depression in older adults.

While these studies do not specifically address suicide prevention, they suggest that IoMT devices have the potential to improve mental health outcomes. However, more research is needed to determine the effectiveness of IoMT devices in suicide prevention and to identify the most effective approaches to using these devices in mental healthcare.

It is important to note that IoMT devices should not be seen as a replacement for traditional mental health treatments such as therapy and medication. Rather, they can be used as complementary tools to provide additional support for patients and help healthcare providers monitor their progress. It's important to note that predicting suicide risk is a complex process, and these devices should not be used in isolation. The data collected by these devices should be analyzed by trained professionals who can provide appropriate support and intervention.

REFERENCES

[1] Kessler, R. C., & Bromet, E. J. (2013). The epidemiology of depression across cultures. *Annual Review of Public Health*, *34*, 119–138. https://doi.org/10.1146/annurev-publhealth-031912-114409. PMID: 23514317; PMCID: PMC4100461.

[2] Gupta, D., Bhatia, M. P. S., & Kumar, A. (2021). Real-Time Mental Health Analytics Using IoMT and Social Media Datasets: Research and Challenges (May 10, 2021). *Proceedings of the International Conference on Innovative Computing & Communication (ICICC)*.

[3] Ministry of Health and Family Welfare, Government of India, National Crime Records Bureau. Available from: https://ncrb.gov.in/. Last accessed on 2022 Sep 10

[4] Singh, O. P. (2022). Startling suicide statistics in India: time for urgent action. *Indian Journal of Psychiatry*, *64*(5):431–432. https://doi.org/10.4103/indianjpsychiatry.indianjpsychiatry_665_22. PMID: 36458077; PMCID: PMC9707658.

[5] Monteith, S., Glenn, T., Geddes, J., Severus, E., Whybrow, P. C., & Bauer, M. (2021). Internet of things issues related to psychiatry. *International Journal of Bipolar Disorders*, *9*(1), 1–9.

[6] https://www.internetsociety.org/wp-content/uploads/2017/08/ISOC-IoT-Overview-20151221-en.pdf

[7] Nordrum, A. (2016). The internet of fewer things [News]. *IEEE Spectrum*, *53*, 12–13. 10.1109/MSPEC.2016.7572524

[8] https://www.cisco.com/c/dam/en_us/about/annual-report/cisco-annual-report-2019.pdf

[9] Pandey, R., Paprzycki, M., Srivastava, N., Bhalla, S., & Wasielewska-Michniewska, K. (2021). *Semantic IoT: Theory and Applications*. Springer Nature.

[10] Dwivedi, R., Mehrotra, D., & Chandra, S. (2022). Potential of internet of medical things (IoMT) applications in building a smart healthcare system: a systematic review. *Journal of Oral Biology and Craniofacial Research*, *12*(2):302–318. https://doi.org/10.1016/j.jobcr.2021.11.010. PMID: 34926140; PMCID: PMC8664731.

[11] Bustos-López, M., Cruz-Ramírez, N., Guerra-Hernández, A., Sánchez-Morales, L. N., Cruz-Ramos, N. A., & Alor-Hernández, G. (2022) Wearables for engagement detection in learning environments: a review. *Biosensors*, *12*, 509. https://doi.org/10.3390/bios12070509

[12] Chakrabarti, S., Biswas, N., Jones, L. D., Kesari, S., & Ashili, S. (2022). Smart consumer wearables as digital diagnostic tools: a review. *Diagnostics*, *12*(9), 2110. https://doi.org/10.3390/diagnostics12092110

[13] Majumder, S., Mondal, T., & Deen, M. J. (2017). Wearable sensors for remote health monitoring. *Sensors*, *17*(1), 130. https://doi.org/10.3390/s17010130

[14] Bustos-López, M., Cruz-Ramírez, N., Guerra-Hernández, A., Sánchez-Morales, L. N., Cruz-Ramos, N. A., & Alor-Hernández, G. (2022). Wearables for engagement detection in learning environments: a review. *Biosensors*, *12*, 509. https://doi.org/10.3390/bios12070509

[15] https://trustiser.net/2017/09/30/can-we-trust-digital-psychiatry/

[16] Du, B., Cheng, X., Duan, Y., & Ning, H. (2022). fMRI brain decoding and its applications in brain–computer interface: a survey. *Brain Science*, *12*, 228. https://doi.org/10.3390/brainsci12020228

[17] Barber, R., Ortiz, F. J., Garrido, S., Calatrava-Nicolás, F. M., Mora, A., Prados, A., Vera-Repullo, J. A., Roca-González, J., Méndez, I., & Mozos, Ó. M. (2022). A multirobot system in an assisted home environment to support the elderly in their daily lives. *Sensors*, *22*, 7983. https://doi.org/10.3390/s22207983

[18] https://www.listeninc.com/applications/smart-speakers-robots-and-iot-enabled-devices/

[19] Mohd, J., Abid, H. (2020). Virtual reality applications toward medical field. *Clinical Epidemiology and Global Health*, 8, 600–605.

[20] Virtual Reality and Human Behaviour. *Frontiers in Virtual Reality*, 12 November 2020, Sec., Volume 1–2020. https://doi.org/10.3389/frvir.2020.585993

[21] Aledavood, T., Torous, J., Triana Hoyos, A. M., Naslund, J. A., Onnela, J. P., & Keshavan, M. (2019). Smartphone-based tracking of sleep in depression, anxiety, and psychotic disorders. *Current Psychiatry Reports*, *21*(7), 49. https://doi.org/10.1007/s11920-019-1043-y. PMID: 31161412; PMCID: PMC6546650.

[22] Dash, S. P. (2020). The impact of IoT in healthcare: global technological change & the roadmap to a networked architecture in India. *Journal of the Indian Institute of Science*, *100*(4), 773–785. https://doi.org/10.1007/s41745-020-00208-y. Epub 2020 Nov 3. PMID: 33162693; PMCID: PMC7606063.

[23] Wilkowska, A., Rynkiewicz, A., Wdowczyk, J., Landowski, J., & Cubała, W. J. (2019). Heart rate variability and incidence of depression during the first six months following first myocardial infarction. *Neuropsychiatric Disease and Treatment*, *15*, 1951–1956. https://doi.org/10.2147/NDT.S212528

[24] Pandey, R., Sahai, A., & Kashyap, H. (2022). Implementing convolutional neural network model for prediction in medical imaging. In *Artificial Intelligence and Machine Learning for EDGE Computing* (pp. 189–206). Academic Press.

[25] Pandey, R., Pandey, A., Maurya, P., & Singh, G. D. (2023). Prenatal healthcare frame-work using IoMT data analytics. In *The Internet of Medical Things (IoMT) and Telemedicine Frameworks and Applications* (pp. 76–104). IGI Global.

[26] Long, N., et al. (2022). A scoping review on monitoring mental health using smart wear-able devices. *Mathematical Biosciences and Engineering, 19*, 7899–7919.

[27] Nannan, L., Yongxiang, L., Lianhua, P., Ping, X., & Ping, M. (2022). A scoping review on monitoring mental health using smart wearable devices. *Mathematical Biosciences and Engineering, 19*(8): 7899–7919. https://doi.org/10.3934/mbe.2022369

[28] Naslund, J. A., Aschbrenner, K. A., Araya, R., Marsch, L. A., Unützer, J., Patel, V., & Bartels, S. J. (2017). Digital technology for treating and preventing mental disorders in low-income and middle-income countries: a narrative review of the literature. *Lancet Psychiatry, 4*(6), 486–500. https://doi.org/10.1016/S2215-0366(17)30096-2. Epub 2017 Apr 19. PMID: 28433615; PMCID: PMC5523650.

[29] Pandey, R., et al., editors.(2023). *The Internet of Medical Things (IoMT) and Telemedicine Frameworks and Applications*. IGI Global. https://doi.org/10.4018/978-1-6684-3533-5

Section 2

Machine Learning for
Medical Things

6 Artificial Intelligence and Internet of Medical Things in the Diagnosis and Prediction of Disease

Narendra Kumar Sharma and Shahnaz Fatima
Amity University

Alok Singh Chauhan
Galgotias University

6.1 INTRODUCTION

The healthcare industry is currently the world's most popular Internet of Things (IoT) application domain. Numerous clinical and medical applications could be improved, and this possibility is increasing daily. Examples include elderly concern, remote healthcare monitoring, numerous constant illnesses, and additional health programs. As a result, a variety of medicinal gadgets, imaging technology, diagnostics, and sensors are viewed as smart devices that have the potential to be extremely important in the IoMT [1].

Traditional medical treatment is insufficient for serious conditions. The statistics show that heart attacks are a major factor in many illnesses and fatalities. IoMT allows for gadget-to-gadget communication with real-time data collection, radically transforming the availability, dependability, and cost of healthcare data in the future [2] (Figure 6.1).

The quality of life has significantly improved across many dimensions, a credit to information technology advancements. The field of medicine is not an exception to this constant change. When it comes to embracing the medical profession and science with this trend, the Internet, automation, AI, and telemedicine have all played significant roles in science. However, it is frequently disputed that when it comes to decisive thoughts, no intelligent system can match the intuition of a skilled doctor [3].

6.2 IoMT

A type of IoT equipment recognized as the Internet of Medical Things (IoMT) consists of medical devices that are networked together to track patient care. Devices that use IoMT, sometimes referred to as healthcare IoT, combine computerization,

DOI: 10.1201/9781003359951-8

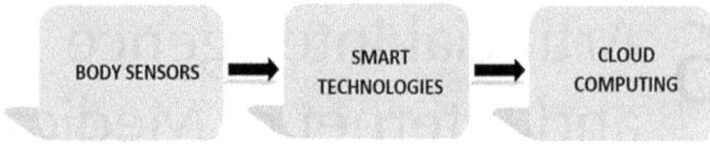

FIGURE 6.1 Semantic view of IoT.

interfacial sensors, and artificial intelligence (AI) built on machine learning (ML) to deliver healthcare monitoring without the need for individual participation. IoMT technology establishes a connection between patients and doctors during the use of medical devices, allowing for distant access for the gathering, dispensation, and communication of health data through a protected system. IoMT equipment offers wireless health parameter monitoring, which lowers unwanted hospitalization and, as a result, the associated medical costs. The point-of-care systems utilized in clinics and hospitals, as well as wearable and at-home real-time health monitoring appliances, are all contained in the IoMT medical technology marketplace [4].

IoMT encounters a number of barriers and difficulties, such as a deficiency in security systems, problems with isolation safeguards, and a requirement for proper education and appreciation of IoMT. Multiple IoT designs have been described in numerous studies due to heterogeneity and the various application aims. A 5-layer formation is frequently utilized to mimic an IoMT-based monitoring system in the medicinal industry; to receive the results of IoMT devices, there is a requirement for an intermediate that preserves and delivers truthful information, and in this role, ML plays a significant part in the IoMT devices.

6.2.1 IoMT AND MACHINE LEARNING

AI is a subdivision of ML that permits computers to grow and learn without being clearly programmed. ML algorithms might be used to evaluate massive quantity of data, or "big data," from electronic health records in order to avoid and analyze disease. Wearable medicinal devices are utilized to constantly examine and save a person's health information in the cloud.

IoMT uses various ML techniques to predict and comprehend changes in healthcare based on various standards [5, 6]. By continuously learning from a person's own data, ML techniques can enhance the framework for monitoring their daily activities and assist them in receiving healthcare at the appropriate moment.

The IoMT is gradually starting to connect doctors and patients through medical services. In order to monitor a patient's health, several methods are utilized, including ultrasounds, blood pressure readings, ECGs, EEGs, and glucose receptor tests; moreover, follow-up doctor visits are crucial. Many hospitals are utilizing intelligent beds that recognize patient movement and adjust the bed's angle and position automatically [7].

6.2.2 IMPORTANCE OF IoMT IN HEALTHCARE

The number of patients is increasing daily, and traditional healthcare institutions are unable to keep up. The IoMT is designed to help solve problems and overcome hurdles so that the healthcare domain's precision, dependability, and efficacy can be improved over time, boosting the health sector's effectiveness [8].

Digital technology advancements like AI, 3D printing, robots, and nanotechnology, among others, are causing the healthcare industry to change before our very eyes. There are numerous potentials to enhance therapeutic results, lower the rate of human error, track data over time, etc. using healthcare digitalization. Many health-associated businesses, including the formation of novel healthcare systems, the storage of patient data and records, and the dealing of various ailments, significantly rely on AI techniques, ranging from ML to deep learning (DL). The most effective approaches for diagnosing diseases also rely on AI. The employment of AI in healthcare opens up previously unimaginable possibilities to recover patient and medical-facility consequences, cut expenses, etc.

6.3 ARTIFICIAL INTELLIGENCE

AI builds programs and algorithms to formulate machines smart and efficient at carrying out tasks that often call for expert human intellect. ML, DL, traditional neural networks, fuzzy logic, and speech recognition are only a few of the subsets of AI that have diverse potential and efficiency that can increase the recital of contemporary medical sciences. These smart systems make clinical diagnostics, medical imaging, and decision-making easier for humans to handle. The IoMT, a next-generation bioanalytical tool that assimilates software applications and network-connected biomedical devices for improving human health, emerges in the same time period.

A number of patient care alternatives and intelligent health systems can be improved with the help of AI. For the analysis of diseases, the expansion of new treatments, and the identification of patient risk factors, AI techniques, such as ML and DL, are extensively used in the healthcare sector. Utilizing AI technology, diseases can be properly detected based on a multiplicity of medical data sources, such as ultrasound, magnetic resonance imaging, mammography, genomics, and computed tomography scans (Figure 6.2).

6.3.1 ARTIFICIAL INTELLIGENCE AND MACHINE LEARNING

AI that could potentially resemble human intelligence is being built for the developing field of instrument learning. ML can be used in the medical industry. Medical professionals have long used ML approaches, primarily to identify illnesses and recommend therapists and treatments. It cannot take the place of human doctors, but it can provide better solutions to healthcare problems and aid doctors through the decision-making process. As opposed to that, DL algorithms are used in the medical industry to gather enormous volumes of data and information needed by patients.

ML is an associate area of computational intelligence that concentrates on the progress of techniques that let computer programs improve their performance using previously collected data. Since the beginning of "Perceptron", novel mathematical models of the brain's functioning mechanisms have been continuously investigated.

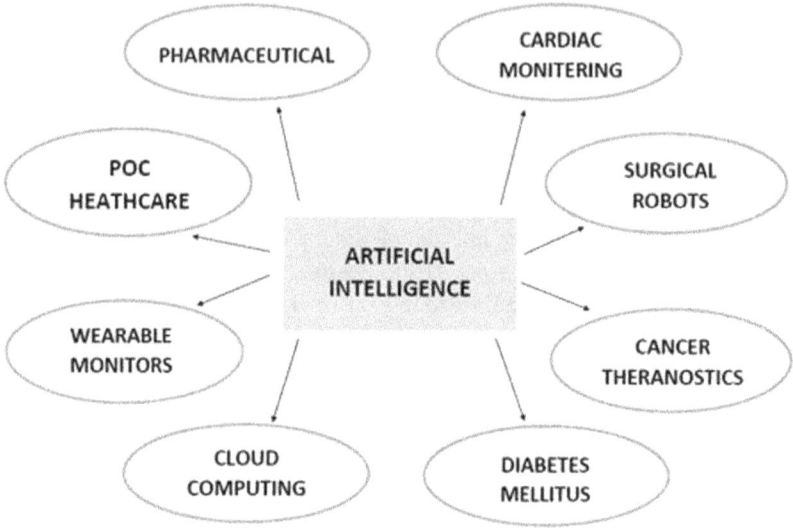

FIGURE 6.2 AI-based approaches in healthcare.

6.4 AI/ML FRAMEWORK FOR DISEASE DETECTION AND MODELING

Disease diagnosis and prediction are very important parts of making an effective treatment policy. Doctors can analyze and predict the illness condition of any patient through an AI- or ML-based diagnosis system. The illustrated framework shows the working process of disease detection and modeling. In that framework, raw data have been collected first, followed by the matrix extraction, normalization, and selection processes. Multiple ML algorithms can be applied to the collected input data based on the illness condition of the patient. Here, some standard AI/ML algorithms like support vector machines, decision trees, and artificial neural networks have been employed to predict the health metrics of patients as collected in the input dataset. We can use the model selection according to our needs to find the best possible result.

Using the web of IoT devices, a doctor can determine and observe a range of metrics from their patients at their own locations, such as their home or place of employment. Early diagnosis and treatment can help a patient avoid hospitalization or even a medical visit, which will cut the cost of healthcare dramatically. IoT devices are used in healthcare in a variety of ways, including wearable health and fitness trackers, biosensors, clinical equipment for monitoring vital signs, and other gadgets or clinical instruments. These IoT devices produce a lot of data about health. If we can integrate these data with other readily accessible healthcare data, such as Electronic health records (EMRs) or Public health records (PHRs), we may forecast a patient's health condition and how it will change from a subclinical to a pathological state [9] (Figure 6.3).

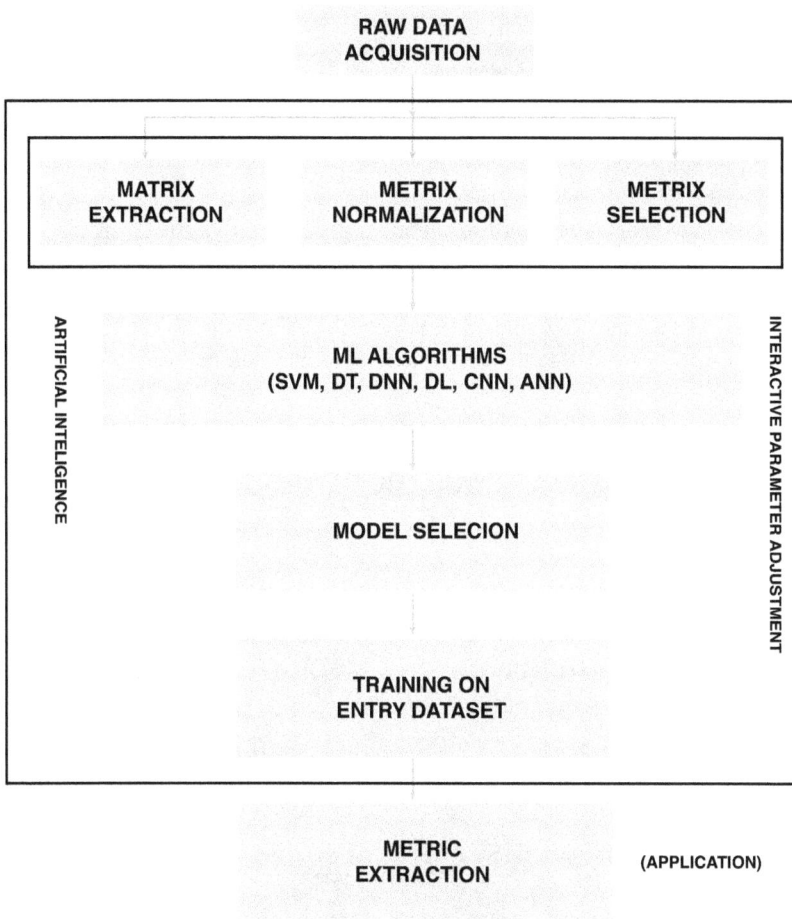

FIGURE 6.3 AI/ML framework for disease diagnosis and prediction [10].

6.4.1 ARTIFICIAL INTELLIGENCE AND DISEASE DIAGNOSTIC

Applications of AI systems include the establishment of treatment protocols, patient monitoring, medication research, personalized drugs in the healthcare industry, and outburst forecasting in universal health.

IoMT devices with AI assistance can continuously monitor patients' health. The use of intelligent robots, virtual assistants, and smart homes may be advantageous to the elderly and disabled. Combining data from IoMT sensors with data from the health system can assist track and prevent pandemic illnesses. When disasters strike, intelligent systems can help authorities provide patients with the support they need and act rationally and quickly [11].

6.4.2 MACHINE LEARNING AND DISEASE DIAGNOSIS

A software program that calculates and extrapolates relevant data, and establishes the characteristics of the corresponding pattern is referred to as ML. The healthcare industry is home to a wealth of knowledge. Electronic medical records that include organized or amorphous data are included.

To solve the IoMT, three types of ML algorithms can be used: supervised, unsupervised, and semi-supervised algorithms. Patient information is crucial in the medical field, and it is also used to anticipate and categorize various diseases. To achieve the best results, prepossessing receives the data from IoMT devices. The training and testing datasets can be split into 75% and 25%, respectively. The output must also be authenticated, and without validating the data, the findings may still be striking (Figure 6.4).

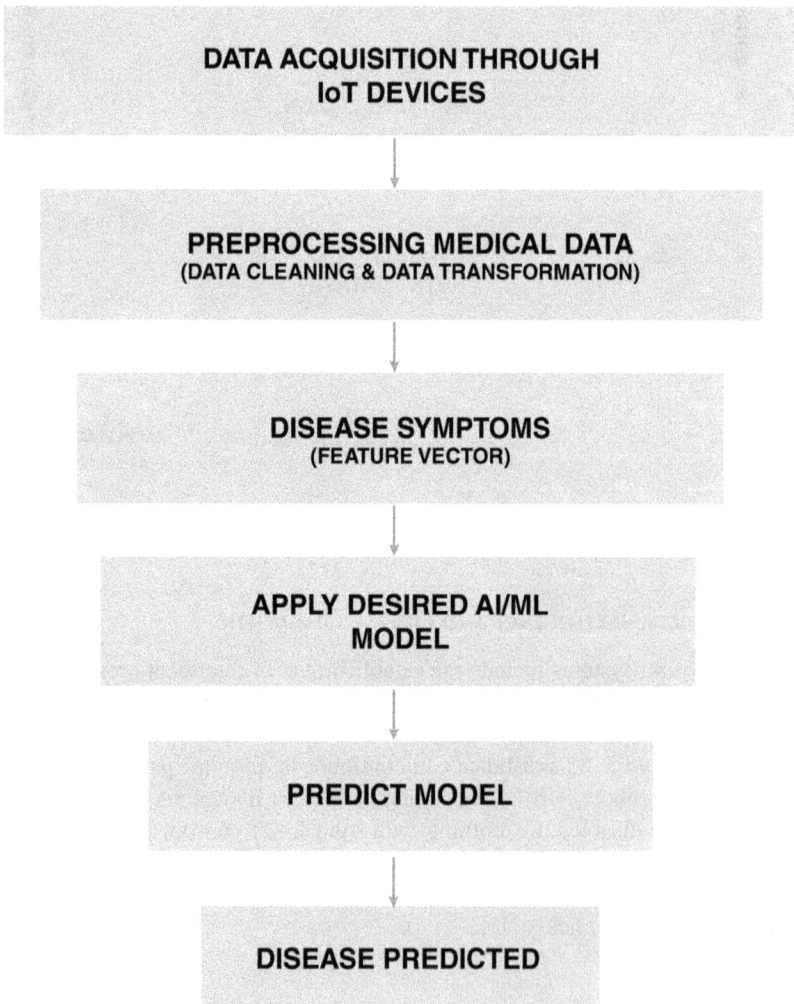

DATA ACQUISITION THROUGH IoT DEVICES

↓

PREPROCESSING MEDICAL DATA
(DATA CLEANING & DATA TRANSFORMATION)

↓

DISEASE SYMPTOMS
(FEATURE VECTOR)

↓

APPLY DESIRED AI/ML MODEL

↓

PREDICT MODEL

↓

DISEASE PREDICTED

FIGURE 6.4 IoMT-based AI/ML process to predict disease.

The IoMT makes it easier for patients to have always-connected medical devices thanks to affordable solutions and a comforting sense of 24-hour hospital support. Patients who are unable to travel to hospitals for regular checkups prefer to use IoMT devices. The IoMT devices' real-time statistics are also used by healthcare facilities to quickly diagnose issues and find remedies. Only when IoMT devices are utilized in conjunction with a strong convergence of technologies could this be accomplished.

6.5 IMPLEMENTATION OF AI/ML ALGORITHMS IN THE DIAGNOSIS AND PREDICTION OF DISEASE

The IoMT is a network of connected sensors, wearable technology, and clinical systems. The idea guarantees the smooth operation and inter-communication of many healthcare apps to improve the standard of medicinal care, deliver prompt medical responses, and lower the cost of healthcare. There are various instances where IoMT devices and online medicinal systems provide rapid healthcare advice to patients on the basis of symptoms.

In this section, we propose a design of a healthcare model that is able to take effective health decisions on the basis of a patient's health condition and their disease symptoms. This system examines the user's or patient's symptoms as input and outputs the most likely disease.

6.5.1 MATERIALS AND METHODS

This model includes Decision Tree, K-Nearest Neighbor (KNN), Naïve Bayes, and Random Forest ML algorithms for predictive data modeling tasks. We employ these methods in terms of their performance to determine which one is most accurate to predict the disease. The determined algorithm can be improved further for healthcare decision-making and diagnosis. In our testing, we use Python with different libraries like Numpy, Pandas, Scikit-Learn, Seaborn, Matplotlib, Tkinter, etc.

6.5.2 DECISION TREE

Decision trees are supervised learning-based techniques that can also be used to handle regression and classification issues. Decision trees' primary goal is to build a training model and forecast the class value of the target variable [12, 13]. This technique divides the data repeatedly into subsets according to the most important attribute at each node of the tree. The mathematical formula to find the entropy and information gain for Decision Tree can be represented as follows:

$$Entropy(Q|P) = -\sum_i p_i \sum_j \frac{p_{ij}}{p_i} \log_2\left(\frac{p_{ij}}{p_i}\right) = -\sum_{i,j} p_{ij} \log_2\left(\frac{p_{ij}}{p_i}\right)$$

$$Entropy(P|Q) = -\sum_j q_j \sum_i \frac{p_{ij}}{q_i} \log_2\left(\frac{p_{ij}}{q_j}\right) = -\sum_{i,j} p_{ij} \log_2\left(\frac{p_{ij}}{q_i}\right)$$

$$InformationGain_Y(X_i, D) = Entropy(P_Y(D)) - Entropy(P_Y(D)|P_{X_i}(D))$$

6.5.3   K-NEAREST NEIGHBOR (KNN)

KNN measures the distance between each instance and looks for the K examples with the smallest distance. The test instance receives the majority of the class that is present in k instances. Among all the ML algorithms, this is the simplest one [12]. KNN finds the distance between the two points as follows:

- Given two feature vectors with numeric values

$$A = (a_1, a_2, ..., a_n) \text{ and } B = (b_1, b_2, ..., b_n)$$

- Use the *distance measure*:

$$d = \sqrt{\sum_{i=1}^{n} \frac{(a_i - b_i)^2}{R_i^2}} = \sqrt{\frac{(a_1 - b_1)^2}{R_1^2} + \frac{(a_2 - b_2)^2}{R_2^2} + ... + \frac{(a_n - b_n)^2}{R_n^2}}$$

R_i is the *range* of the ith component

6.5.4   NAÏVE BAYES

The Naïve Bayes algorithm is a method of supervised learning for categorization issues that is pedestal on the Bayes theorem. It is generally used for text classification and has a sizable training set. One of the simplest and most effective classifiers now in use is the Naïve Bayes classifier [12]. Fast ML models that can anticipate outcomes accurately can be developed with its assistance. It bases its predictions on the probability that an object will happen because it is a probabilistic classifier. The Bayes theorem can be used to find the probability and implemented as follows:

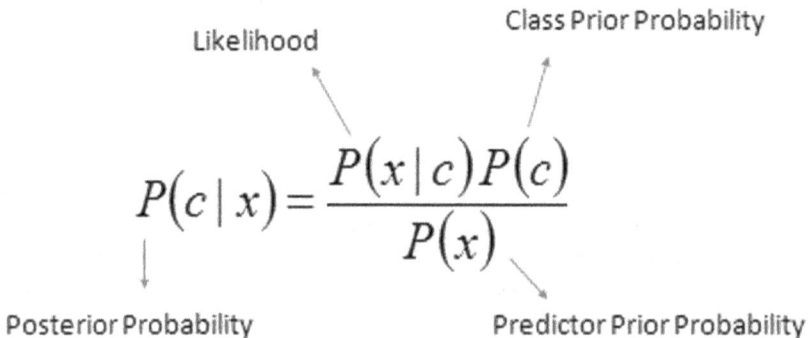

Likelihood   Class Prior Probability

$$P(c \mid x) = \frac{P(x \mid c) P(c)}{P(x)}$$

Posterior Probability   Predictor Prior Probability

$$P(c \mid X) = P(x_1 \mid c) \times P(x_2 \mid c) \times \cdots \times P(x_n \mid c) \times P(c)$$

6.5.5   RANDOM FOREST

Random Forest is a Decision Tree-based ensemble ML technique that combines various trees to make effective decisions. A Random Forest is proficient of performing classification and regression tasks mutually using multiple decision trees, which are called aggregation and bootstrapping [12, 13]. We simply need to determine the imperfection of our dataset and use that feature to determine which root node to use another term, which has the minimum Gini index, produces the minimum imperfection.

$$Gini\ Index\ =\ 1\ -\ \sum_{i=1}^{n}\left(P_i\right)^2$$
$$=\ 1\ -\ [(P_+)^2 + (P_-)^2]$$

The probability of a positive class represented as P+ and the probability of a negative class represented as P- can be used to represent the Gini index mathematically.

6.5.6 PYTHON

Guido van Rossum created the open-source general-purpose programming language Python in 1980. It supports a wide variety of applications that include ML, software development, web development, mathematics, statistics, and data analytics. Python provides an integrated development environment that works on different platforms like Windows, Linux, Mac OS, etc. [14, 15].

6.5.7 GRAPHICAL USER INTERFACE

Tkinter is used to build a graphical user interface in Python. It is the typical Python GUI library. A quick and simple way to construct a GUI application is to use Python and Tkinter. The Tk GUI toolkit's refined object-oriented framework is offered by Tkinter [15, 16.

6.5.8 DATASET

The dataset was taken from a study that was done at Colombia University [17]. This dataset is a knowledge base of disease–symptom relationships that was created automatically using data from textual discharge summaries of patients at New York Presbyterian Hospital. There are 150 diseases in it, and each one has an average of 8–10 symptoms. This dataset contains a variety of diseases like:

'Fungal infection', 'Allergy', 'GERD', 'Chronic cholestasis', 'Drug Reaction', 'Peptic ulcer disease', 'AIDS', 'Diabetes', 'Gastroenteritis', 'Bronchial Asthma', 'Hypertension', ' Migraine', 'Cervical spondylosis', 'Paralysis (brain hemorrhage)', 'Jaundice', 'Malaria', 'Chicken pox', 'Dengue', 'Typhoid', 'hepatitis A', 'Hepatitis B', 'Hepatitis C', 'Hepatitis D', 'Hepatitis E', 'Alcoholic hepatitis', 'Tuberculosis', 'Common Cold', 'Pneumonia', 'Dimorphic hemmorhoids(piles)', 'Heartattack', 'Varicoseveins', 'Hypothyroidism', 'Hyperthyroidism', 'Hypoglycemia', 'Osteoarthristis', 'Arthritis', '(vertigo) Paroymsal Positional Vertigo', 'Acne', 'Urinary tract infection', 'Psoriasis' and 'Impetigo'

Testing data will make up 30% of the dataset, with the remaining 70% being used for training. The dataset would be used for training and testing, and the desired output would be obtained. The disease's symptoms that were present were denoted as 1 and those that persisted as 0 (Figure 6.5).

6.5.9 Working Methodology

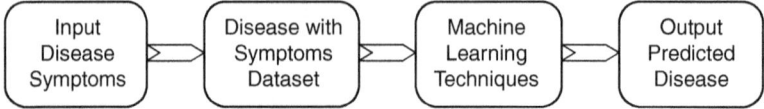

FIGURE 6.5 Workflow diagram.

6.5.10 Experimental Evaluation

In cases where the symptoms are provided as input, the system will forecast the disease. The four ML methods of Decision Tree, KNN, Naïve Bayes, and Random Forest will be used to forecast the disease. Diseases are labeled in the dataset, and symptoms are provided for each disease. The following symptoms were applied during experimental evaluation.

'back_pain', 'constipation', 'abdominal_pain', 'diarrhoea', 'mild_fever', 'yellow_urine', 'yellowing_of_eyes', 'acute_liver_failure', 'fluid_overload', 'swelling_of_stomach', 'swelled_lymph_nodes','malaise','blurred_and_distorted_vision','phlegm','throat_irritation', 'redness_of_eyes','sinus_pressure','runny_nose','congestion','chest_pain','weakness_in_lim bs', 'fast_heart_rate', 'pain_during_bowel_movements', 'pain_in_anal_region', 'bloody_stool', 'irritation_in_anus', 'neck_pain', 'dizziness', 'cramps', 'bruising', 'obesity', 'swollen_legs', 'swollen_blood_vessels', 'puffy_face_and_eyes', 'enlarged_thyroid', 'brittle_nails', 'swollen_extremeties', 'excessive_hunger', 'extra_marital_contacts', 'drying_and_tingling_lips', 'slurred_speech', 'knee_pain', 'hip_joint_pain', 'muscle_weakness', 'stiff_neck', 'swelling_joints', 'movement_stiffness', 'spinning_movements', 'loss_of_balance', 'unsteadiness', 'weakness_of_one_body_side', 'loss_of_smell', 'bladder_discomfort', 'foul_smell_of urine', 'continuous_feel_of_urine', 'passage_of_gases', 'internal_itching', 'toxic_look_(typhos)', 'depression', 'irritability', 'muscle_pain', 'altered_sensorium', 'red_spots_over_body', 'belly_pain', 'abnormal_menstruation', 'dischromic_patches', 'watering_from_eyes', 'increased_appetite', 'polyuria', 'family_history', 'mucoid_sputum', 'rusty_sputum', 'lack_of_concentration', 'visual_disturbances', 'receiving_blood_transfusion', 'receiving_unsterile_injections', 'coma', 'stomach_bleeding', 'distention_of_abdomen', 'history_of_alcohol_consumption', 'fluid_overload','blood_in_sputum', 'prominent_veins_on_calf', 'palpitations', 'painful_walking', 'pus_filled_pimples', 'blackheads', 'scurring', 'skin_peeling', 'silver_like_dusting', 'small_dents_in_nails', 'inflammatory_nails', 'blister', 'red_sore_around_nose' and 'yellow_crust_ooze's.

Before applying ML techniques on dataset, we make sure the data fit as follows: cleansing the data, changing object data's unknown values for null values in numeric data, making the substitutions for 0 and unknown, coding the information, getting rid of the pointless columns, and splitting the data into training and validation datasets after normalizing the data. The testing data will make up 30% of the dataset, with the remaining 70% being used for training.

6.5.11 RESULT ANALYSIS

The proposed system is set up so that it receives user symptoms as input and outputs a diagnosis of the condition. After passing a list of symptoms, there are five drop-down menu alternatives. The user can choose any five symptoms, and after pressing the "predict" button, the text box will show the probable disease (Figure 6.6).

The Decision Tree Classifier, KNN Classifier, Naïve Bayes Classifier, and Random Forest Classifier are employed to make predictions. A minimum of three and a maximum of five symptoms may be chosen by the user. If only one symptom is entered, less accuracy will be reached. The accuracy increases with the number of symptoms.

The following are the prediction buttons used for predicting the disease using different algorithms:

- Button 1 is used for predicting the disease through the Decision Tree algorithm.
- Button 2 is used for predicting the disease through the KNN algorithm.
- Button 3 is used for predicting the disease through the Naïve Bayes algorithm.
- Button 4 is used for predicting the disease through the Random Forest algorithm.

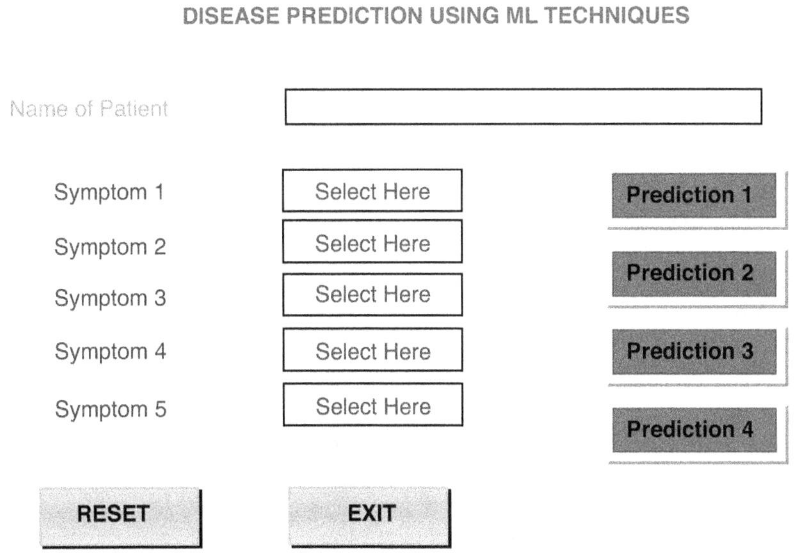

FIGURE 6.6 Disease prediction using ML techniques.

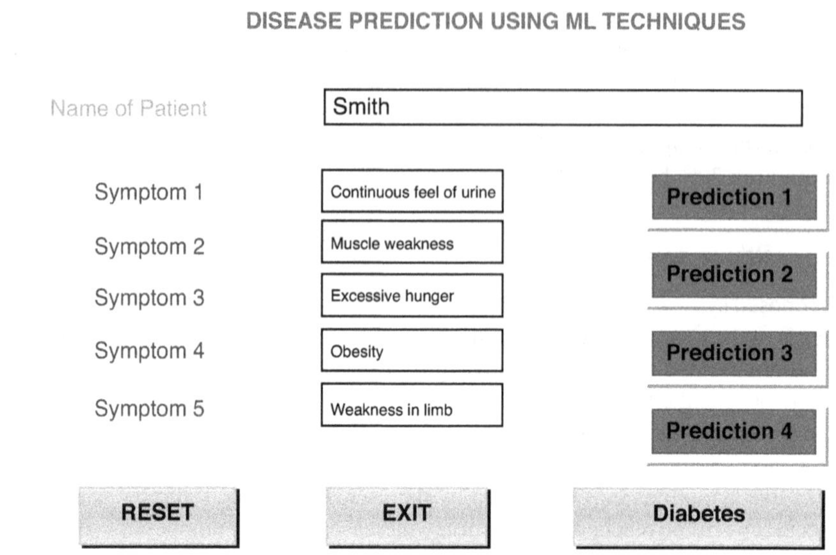

DISEASE PREDICTION USING ML TECHNIQUES

Name of Patient	Smith

Symptom 1	Continuous feel of urine	Prediction 1
Symptom 2	Muscle weakness	Prediction 2
Symptom 3	Excessive hunger	
Symptom 4	Obesity	Prediction 3
Symptom 5	Weakness in limb	Prediction 4

RESET	EXIT	Diabetes

FIGURE 6.7   Disease prediction based on input of symptoms.

- The reset button is used for resetting the inputs.
- The exit button is used for exiting from the system (Figure 6.7).

This method takes user input and forecasts the disease that is most likely to occur. The dataset and ML method are utilized to do this. The probabilistic algorithms include Decision Tree, KNN, Naïve Bayes, and Random Forest. The disease and its symptoms would be used for training and testing, and the desired output would be obtained.

6.6   APPLICATION OF AI/ML ALGORITHMS IN THE DIAGNOSIS AND PREDICTION OF DISEASE

The use of AI can make machine think and behave like a human. AI greatly increases the intelligence of the computer. According to a number of researchers, intelligence cannot develop in the absence of learning. AI greatly increases the intelligence of the computer. There are many kinds of ML and AI techniques that are available to use in disease diagnosis and prediction [10]. Here some eminent AI/ML algorithms are listed to illustrate the benefits and application areas (Table 6.1).

6.6.1   Disease Symptoms and Diagnostic Difficulties

The numerous signs of illness and related symptoms have been discussed in this section. On the other hand, medication errors associated with illness identification are quite common, are subject to severe fines, and have just recently started to significantly affect patient safety. Here, there are significant issues with the many diagnostic methods used to pinpoint the individual illnesses. The following diseases are included together with their signs, symptoms, and event markers [18]:

TABLE 6.1

Application Area of AI/ML Algorithms in Medical Science [10]

S. No.	AI/ML Algorithms	Medical Science Applications	Benefits
1.	Decision Tree (DT)	Glucose observation, surgery, medical diagnosis, performance management of systems, and healthcare monitoring	Very quick, effective, and easy to comprehend and explain. Able to handle a wide range of data formats. High classification, learning, and processing speeds.
2.	K-Nearest Neighbor (KNN)	Examine blood sugar for diabetes, resource managing for pandemics, health monitoring system, heart disease prognosis, computer-aided diagnosis, and disease prediction	Basic algorithm. There are no presumptions made about the dataset's properties or results. Effective at managing massive data and noisy data. Stable performance, quick learning, and effective overfitting control.
3.	Support Vector Machine (SVM)	Psychiatric and neurological illnesses, biomarker imaging, computer–human interaction, cancer detection, early Alzheimer's disease detection, cardiac surveillance, prognosis of surgical site infection, insulin testing, surgery, resource management for pandemics, and system for monitoring healthcare	Highly accurate, faster problem-solving convergence, ability to scale health for high-dimensional data, and small amount of training samples are required.
4.	Naïve Bayes (NB)	Medical diagnostics, system performance management, pandemic resource management, and disease prediction	High learning curve, rapid classification, and simple implementation. Capable of handling noisy data, overfitting, and missing values. Capable of predicting a test dataset's class. Helps with multi-class prediction issues.
5.	Neural Network (NN)	Cancer detection, recognizing Parkinson's illness, Alzheimer's disease, imaging-based heart monitoring, and surgery applications for sensors, diabetes forecast, computer–human interaction, resource management for pandemics, and visual computing	Flexible, effective, and quick algorithm. Produces calculations without the need of pre-programmed rules and continuously develops. Using multiple tasks at once has many uses. Complex and nonlinear databases are compatible with it.
6.	Logistic Regression	Cardiovascular image-based monitoring, diabetes glucose monitoring, pandemic resource management, and healthcare monitoring system	Easy in both execution and interpretation. A successful training program. The outputs are correctly calibrated and categorized. No need for empirical parameter tweaking. Decent accuracy for small sets of data.
7.	Random Forest (RF)	Healthcare monitoring system, heart disease prognosis, and illness forecasting	Excellent at handling noisy data. Rapid categorization. Good at managing vast, diverse databases. Automatic definition of a feature. Normalization of the input features is not necessary.

Heart attack signs: Shortness of breath, a cold sweat, heaving, and unsteadiness are all indications of a heart attack. Along with discomfort or anxiety in the core of the chest that lasts for longer than a few seconds, they also involve soreness or nervousness in various locations of the chest area.

Stroke signs: A stroke can cause facial listing, arm weakness, speech problems, abrupt contentment or equalization, unpredicted weakness or lack of impression, failure of visualization, mystification, or unbearable pain, to name a few symptoms.

Reproductive wellbeing: Symptoms like regular urination or urinary weight, blood loss or spotting between periods, tingling, copying, and disorder at the genital area, pain or discomfort during sexual activity, actual or painful feminine death, excruciating pelvic/stomach pain, strange vaginal releases, and a feeling of totality in the lower mid-region are all indications of reproductive health.

Breast issue: The negative repercussions of breast problems include areola release, abnormal areola tenderness or pain, changes in the skin around the areola or bosom, and a knot or thickening in the areola or under the arm.

Lung issue: Some of the negative impacts of lung problems include bloody hacking, shortness of breath, difficulty breathing, and recurring incidents of bronchitis or pneumonia, along with gasping.

Stomach-related issue: Some signs of stomach problems include rectal hemorrhage, blood in the stool or shady feces, alteration in the nerve structure, an inability to control the gut, obstruction, loose stools, heartburn or stomachache, or blood spattering (Kather et al. 2019).

Bladder issue: Constant urination, inability to control one's bladder, blood in the urine, getting up frequently in the middle of the night to urinate or wetting the bed in the middle of the night, and spilling urine are all indications of bladder issues.

Skin issue: Skin disorders can be identified by jaundice, thick, red skin with bright areas, persistent flushing and redness of the face and neck, new growths or moles on the skin, and sores that do not heal or go away.

Emotional issues: Anxiety, depression, weariness, tension, nightmares, skipping daily exercise, and experiencing self-destructive thoughts, mind flights, and fancy states are all symptoms of emotional issues.

Headache issues: The abrupt onset of migraines, "the most visibly horrible migraine of your life," and headaches accompanied by excessive energy, nausea, heaviness, and immobility are all signs of headache issues (apart from regular tension headaches).

It's possible for the illness to be nasty, serious, enduring, or benevolent. Mortal and commence refer to the probability that something will result in death, whereas importunate and rigorous refer to how long a situation has persisted. Additionally, certain symptoms that might not be relevant could serve as warning signs for a more serious condition or illness that requires more restorative measures.

6.6.2 Technological Perspective of IoMT in Disease Diagnosis and Prediction

Despite the fact that AI/ML-based techniques have become gradually more important in the detection of diseases, researchers still confront numerous impediments that must be conquered:

i. **Restricted Data Size**: The major problem most research ran into was a lack of data to train the model. A smaller training set is implied by a smaller sample size, which casts doubt on the efficacy of the offered strategies. The model can be trained more successfully with a larger sample set than a smaller individual.

ii. **High Dimensionality**: One more data-related concern that cancer research faces is high dimensionality. There are many more examples of high-dimensionality features than there are examples. However, this issue can be resolved using strategies for coping with several aspects [19].

iii. **Effective Feature Selection Procedure**: Several studies have produced excellent forecasting outcomes. Therefore, a computationally efficient feature selection approach is essential to do away with the data clean-up operations and generate great illness-forecasting precision.

iv. **Model Generalizability**: Study has to be refocused on perking up the generalizability of the model. Many researches have intended a prediction model that has been evaluated in one place. The models must be validated across multiple sites to improve generalizability.

v. **Clinical Implementation**: AI-based models have proven to be more effective than traditional models in the field of medicine, but they have not yet been applied in real-world clinical settings. These models should be assessed in a medical set in order to support the doctor in verifying the identification.

6.6.3 NEED TO OVERCOME IoMT CONSTRAINTS

IoMT constraints should be removed to implement the following solutions in the diagnosis and prediction of disease and offer improved healthcare solutions:

- Testing and tracing is essential for disease control, using IoMT devices.
- Medical devices with sensors can monitor physical and biological activities.
- Adoption of IoMT-enabled healthcare devices increased during COVID-19.
- Smart hospitals use IoMT to improve patient care.
- IoMT-enabled Smart Cyber Operating Theater (SCOT) with ORiN technology for connecting medical devices.
- Smart medications use IoMT sensors to monitor biomarkers, antibiotics, compliance, and dosage.
- IoMT-connected diagnostic devices provide real-time monitoring of biomarkers to distinguish between viral and bacterial infections.
- IoMT-aided robotic system uses cloud-robotic systems to process data.
- Tele-thermographic systems detect skin temperature accurately without physical contact.
- IoMT-based 3D scanners record intraoral impressions with accuracy.
- Image-guided surgery provides personalized simulation, planning, and practice drill.
- Mobile computing, sensors, communication technologies, and cloud computing provide efficient healthcare services.
- Electronic devices and software have revolutionized online teaching and learning.

- Tele-dentistry uses telecommunications to deliver oral healthcare and education services.
- Voice assistants provide remote healthcare during crisis and pandemic.
- Ambient assisted living provides real-time monitoring to provide assistance in medical emergencies.
- IoMT-based adverse drug reaction system uses unique identifier or barcode to verify compatibility with patient.

6.7 OPPORTUNITIES AND CHALLENGES

There have been reports on AI's benefits in a diversity of healthcare contexts, including the supervising of cardiac arrhythmia, the treatment of diabetes, and aided surgeries. Rapidly analyzing vast and composite sensor data enables additional investigation and enhances decision-making skills. The pulling out of investigative data from noisy data sources is another task that AI and ML aid in. The AI/ML method facilitates the IoMT devices to pull out the unknown information based on the correlation among the test parameters and observed signals through supervised machine learning (SML) techniques. The signal strength, sensitivity, specificity, and measurement time are all augmented using AI approaches.

Although AI and ML have the potential to transform healthcare practices and IoMT-integrated medical devices, a number of technical challenges still require resolution before these technologies can be commercialized and applied to clinics and society as a whole. Since AI and ML systems significantly rely on precise data for system design and preparation, the focal point should be on gathering a large amount of data on high-quality patient training and learning.

Heterogeneity in the data that has been attained is another main concern. There are contradiction in the training of AI due to bias and noise present in the health records assembled from assorted clinics. The datasets can be homogenized with the employment of sophisticated ML algorithms, which will enhance the correctness of the clinical diagnosis. The potential of surgery and the medical industry will thrive with AI-supported methods.

6.8 CONCLUSION AND FUTURE PROSPECTS

The conclusions of this chapter show that AI makes patients', doctors', and hospital administrators' lives better by performing tasks that would often be done by people, but at a small fraction of time and less cost. By analyzing a large amount of healthcare data, including electronic health records, symptom data, and physician reports, AI-based IoMT devices let doctors make recommendations that can improve patient health and possibly extend the patient's life. In addition, AI helps doctors diagnose illnesses more accurately by using intricate algorithms, hundreds of biomarkers, imaging data from thousands of patients, cumulative published medical studies, and thousands of doctor notes.

The collection, processing, and analysis of data have benefited from the growth of the IoT. The term IoMT refers to the combination of wireless body area networks and IoT technology in the healthcare sector. IoMT can be distinguished as a collection of medicinal tools and programs that can be networked to healthcare information technology systems. IoMT has many advantages, such as lowering medical expenses and improving people's quality of life by incessantly tracking their health. A variety of medical equipment and circuits, including sensors and actuators for physiological data like breathing rate, heart rate, and oxygen content, can be interacted with using the IoMT. Wi-Fi, 4G, and 5G are a few of the wireless communication technologies that the IoMT makes use of.

The IoMT systems will thrive in this era of deep AI and ML exploration in healthcare analysis, which will open up new opportunities for the industry. AI-based IoMT devices will develop multilevel functionality, high sensitivity, industrial-level production, downsizing, ultra-low power consumption, and affordability with the help of advances in nanotechnology and microelectronics. People will gradually have access to high-quality healthcare as AI/ML systems, IoMT devices, and integrated medicine are further integrated.

REFERENCES

1. Sun Y., Lo F. P. W., & Lo B. (2019). Security and privacy for the internet of medical things enabled healthcare systems: A survey, *IEEE Access*, *7*, 183339–183355.
2. Ahad A., Tahir M., and Yau K. -L. A. (2019). 5G-based smart healthcare network: Architecture, taxonomy, challenges and future research directions, *IEEE Access*, *7*, 100747–100762. https://doi.org/10.1109/ACCESS.2019.2930628.
3. Athmaja, S., & Hanumanthappa, M. (2017). A survey of machine learning algorithms for big data analytics. In *International Conference on Innovations in Information, Embedded and Communication Systems (ICIIECS)*. IEEE
4. Kaushik, A., Khan, R., Solanki, P., Gandhi, S., Gohel, H., & Mishra, Y. K. (2021). From nanosystems to a biosensing prototype for an efficient diagnostic: a special issue in Honor of Professor Bansi D. Malhotra, *Biosensors*, *11*, 359.
5. Suneetha, K. C., Shalini, R. S., Vadladi, V. K., & Mounica, M. (2020) Disease prediction and diagnosis system in cloud based IoT: a review on deep learning techniques, *Materials Today: Proceedings*. https://doi.org/10.1016/j.matpr.2020.09.519.
6. Uthayakumar, J., Metawa, N., Shankar, K., & Lakshmanaprabu, S. K. (2020). Intelligent hybrid model for financial crisis prediction using machine learning techniques, *Information Systems and e-Business Management*, *18*(4), 617–645.
7. Wagan S. A., Koo J., Siddiqui I. F., Attique M., Shin D. R., & Qureshi N. M. F. (2022). Internet of medical things and trending converged technologies: A comprehensive review on real-time applications, *Journal of King Saud University – Computer and Information Sciences*, *34*, 9228–9251.
8. Islam S. M. R., Kwak D., Kabir M. H., Hossain M., & Kwak K. -S. (2015). The Internet of Things for Health Care: A Comprehensive Survey, *IEEE Access*, *3*, 678–708, 2015. https://doi.org/10.1109/ACCESS.2015.2437951.
9. Shameer, K., et al. (2017). Translational bioinformatics in the era of real-time biomedical, health care and wellness data streams. *Brief Bioinform*, *18*(1), 105–124.
10. Manickam, P., Mariappan, S. A., Murugesan, S. M., Hansda, S., Kaushik, A., Shinde, R., & Thipperudraswamy, S. P. (2022). Artificial Intelligence (AI) and Internet of Medical Things (IoMT) assisted biomedical systems for intelligent healthcare. *Biosensors*, *12*, 562. https://doi.org/10.3390/bios12080562
11. Ullah Z., Al-Turjman F., Mostarda L., & Gagliardia R. (2020). Applications of Artificial Intelligence and Machine learning in smart cities, *Computer Communications*, *154*, 313–323

12. Wu, X., & Kumar, V. (2016). *The Top Ten Algorithms in Data Mining*, CRC Press Tylor and Franch Group.
13. Hartshorn, S. (2016). *Machine Learning with Random Forests and Decision Trees: A Visual Guide for Beginners*, Kindle Edition.
14. Muller, A., & Guido, S. (2016). *Introduction to Machine Learning with Python: A Guide for Data Scientists*, O'Reilly Media.
15. Python Tutorial (Accessed on 24 April 2023) https://docs.python.org/3/tutorial/
16. Python-GUI-Tkinter (Accessed on 24 April 2023) https://realpython.com/python-gui-tkinter/
17. Disease Symptoms Knowledge base (Accessed on 24 April 2023) https://people.dbmi.columbia.edu/~friedma/Projects/DiseaseSymptomKB/index.html
18. Kumar, Y., Koul, A., Singla, R., & Ijaz, M. F. (2022). Artificial intelligence in disease diagnosis: a systematic literature review, synthesizing framework and future research agenda, *Journal of Ambient Intelligence and Humanized Computing*, 14, 661–664. https://doi.org/10.1007/s12652-021-03612-z
19. Bibault J.-E., Xing L., Giraud P., El Ayachy R., Giraud N., Decazes P., Burgun A. (2020). Radiomics: A primer for the radiation oncologist, *Cancer/Radiothérapie*, 24(5), 403–410.

7 Predicting Cardiovascular Diseases Using Machine Learning
A Systematic Review of the Literature

Abhay Kumar Pathak, Bhupendra Kumar Dewangan, and Manjari Gupta
Banaras Hindu University

7.1 INTRODUCTION

Heart disease has been seen as one of the most complex and life-threatening human diseases in terms of high morbidity and mortality. It has the very serious effect on people's social and economic sides by going through medical treatment and diagnostics. According to the WHO static chart, cardiovascular diseases (CVD) are the global causes of death, an estimation of 17.9 million people died each year. There are several types of diseases like 'group of disordered heart and blood vessels', including coronary heart disease, cardiovascular disease, rheumatic heart disease and other conditions. Four out of five chances of death are due to heart attack and strokes [1].

Artificial intelligence (AI) has many applications in different domains such as medicine, pharmacy, agriculture, self-driving motor vehicles, e-commerce, education, life style, navigation, rotation, human resources, healthcare, agriculture, and gaming. But with respect to the healthcare domain dealing with patient data, it requires perfect or near-perfect results by models that we have applied. In this chapter, the idea of classification is thoroughly discussed in terms of heart diseases. By relying on the patient history, several studies have been done before. But the cumulative approach must be the best option.

There are numerous cardiovascular research labs located around the world, each working to advance the understanding of cardiovascular diseases (CVDs) and to develop new treatments and therapies. Some of the world-renowned cardiovascular research labs include the following [1]:

1. Harvard Stem Cell Institute, Cambridge, Massachusetts, USA
2. National Heart, Lung, and Blood Institute (NHLBI), Bethesda, Maryland, USA
3. Cardiovascular Research Institute, San Francisco, California, USA

DOI: 10.1201/9781003359951-9

4. German Center for Cardiovascular Research (DZHK), Berlin, Germany
5. Oxford Martin School, Oxford, United Kingdom
6. Pasteur Institute, Paris, France
7. Institute of Cardiovascular Sciences, University College London, United Kingdom
8. Center for Molecular Medicine, Stockholm, Sweden
9. Cardiovascular Research Center, University of Glasgow, Scotland
10. Instituto Nacional de Cardiología, Mexico City, Mexico

This chapter is designed to give better understanding and interpretation of clinical context. Our contribution in this chapter is to give a meta-analysis of different methods and techniques that have been used earlier for predicting CVDs. This chapter presents facts about the models and their efficiency to help in the diagnosis of heart diseases. Most information is tabulated for making it simple, abstract and easy to understand.

In this chapter, we give a brief description of data and its features. With the cumulative model approaches, a meta-analysis is included in which the results are compared with different machine learning (ML) models. There are several studies in which state-of-the-art techniques and preprocessing and dimensional reduction methods are used for creating an efficient classifier that can be used in the medical domain.

This review is done on research papers worked on four heart disease datasets. With reference to those datasets, the study that has been done in the area of ML are the actual articles which are focused on same problem. Most articles are from IEEE ACCESS, but for study and survey, few other papers are also considered from different publishing houses.

Section 7.2 discusses the methodology used in this chapter. Comparisons based on models and datasets used in the selected research articles are made in Section 7.3. These articles are compared based on evaluation metrics in Section 7.4. Finally, Section 7.5 concludes the chapter.

7.2 METHODOLOGY

Our approach for this review is to simply analyze and compare the selected research articles based on the datasets these used, ML, ensemble learning and other hybrid models these used as well as other methods based on the obtained performance metrics values of these models. Our research articles' selection strategy is discussed in detail in Section 7.2.1. Year-wise distribution of these articles is shown in Section 7.2.2.

7.2.1 STUDY SELECTION

Searching is performed only in IEEE ACCESS. Here we included only IEEE research articles. The searching of papers was performed using the keywords "Heart Diseases" and/or "Cardiovascular Diseases Prediction Using Machine Learning". By using these keywords, we found 698 papers. Among them, one paper was a duplicate, so the number of articles was reduced to 697. After excluding conference papers, only 60 articles were left, from which 48 articles were excluded due to time frame (we considered papers from 2019 to 2022 only), other diseases worked on or because of different fields of study. Thus, finally, there were 12 articles that were chosen for this review. This chapter's selection process is shown in Figure 7.1. Year-wise article distribution is shown in Figure 7.2.

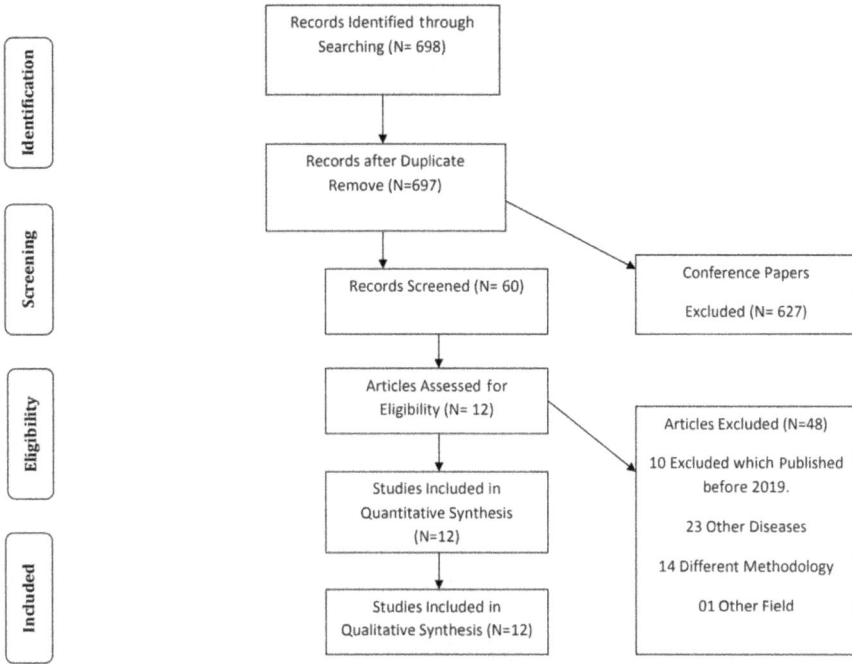

FIGURE 7.1   Study selection.

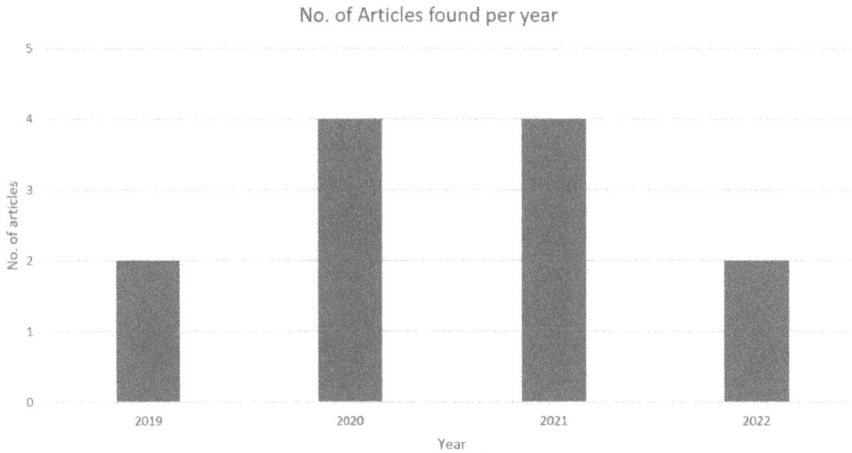

FIGURE 7.2   Total number of articles found per year.

7.2.2   YEAR-WISE DISTRIBUTION OF SELECTED ARTICLES

As discussed earlier, all research articles considered in this study were published from 2019 to 2022. Figure 7.2 shows the year-wise distribution of these research articles.

7.3   COMPARISON BASED ON MODELS AND DATASETS USED

In this section, we compare the selected research articles based on classification models used in these papers. Figure 7.3 shows the classification of these articles by the used state-of-the-art methods and hybrid methods.

7.3.1   Datasets Used

This study is done on papers worked on four datasets. A detailed description of these datasets is given below:

1. Statistics and attributes of the Dataset for Cleveland, Hungary, Switzerland, and Long Beach [2] are shown in Tables 7.1 and 7.2.

7.3.1.1   Attribute Information

2. Attributes considered in the Korea National Health and Nutrition Examination Survey (KNHANES) dataset to predict CHD risk [3] are shown in Table 7.3.
3. Heart-failure-clinical-records-dataset is derived from the UCI ML repository [4], and its statistics and attributes are shown in Tables 7.4 and 7.5. This dataset contains medical records of 299 patients who had heart failure, and these were collected during their follow-up period, where each patient profile has 13 clinical features.

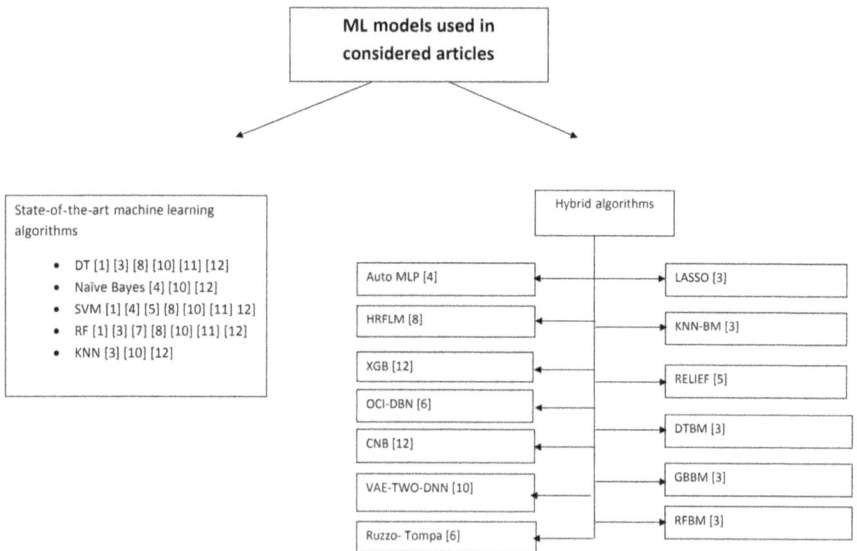

FIGURE 7.3   ML model used in considered articles.

TABLE 7.1

Dataset

Dataset Characteristics	Multivariate	Number of Instances	303	Area	Life
Attribute Characteristics	Categorical, Integer, Real	Number of Attributes	75	Date Donated	1988-07-01
Associated Tasks	Classification	Missing Values?	Yes	Number of Web Hits	2091675

TABLE 7.2

Dataset

Attributes	Range	Description
Sex	0,1	Sex of subject
Age	29–77	Age in years
CP	1–4	Chest pain
TrestBPS	94–200	Resting blood pressure
Chol	126–465	Serum cholesterol
Fbs	0,1	Fasting blood sugar
RestECG	0,2	Resting electrocardiographic result
Thalach	71–188	Maximum heart rate achieved
Exang	0,1	Exercise-induced angina
Oldpeack	0–6	ST depression induced by exercise
Slope	1–3	Slope of peak exercise
CA	5	Number of major vessels colored by fluoroscopy
Tha	4	Defect type
Number	0–4	Type of heart disease

TABLE 7.3

Dataset

Attributes	Range
Age (year)	–
BMI (kg/m²)	22.27–24.30
Total cholesterol (TC) (mg/dl)	86.32–192.78
High-density lipoprotein cholesterol (HDL) (mg)	54.22–47.68
Systolic blood pressure (SBP) (mmHg)	111.35–122.80
Waist circumference (WC) (cm)	75.04–83.80
Natural fat (NF) (mg/dl)	86.32–154.84

(Continued)

TABLE 7.3 (*Continued*)
Dataset

Attributes	Range
Obesity status	–
1. Underweight	
2. Normal	
3. Obesity	
Knee-joint pain status	–
1. Yes	
2. No	
3. Non-applicable (below 50 years of age)	
Weight change in 1 year	–
1. Weight gain	
2. Weight loss	
3. No change	
4. No responses	
Frequency of eating out	–
1. More than twice a day	
2. Once a day	
3. 5–6 times a week	
4. 3–4 times a week	
5. 1–3 times a month	
6. Less than once a month	
7. No response	
Marital status	–
1. Married living together	
2. Married living separately	
3. Bereavement	
4. Divorced	
5. Response refused	
6. No response	
7. Non-applicable	

TABLE 7.4
Dataset

Dataset Characteristics	Multivariate	Number of Instances	299	Area	Life
Attribute Characteristics	Integer, Real	Number of Attributes	13	Date Donated	2020–02–05
Associated Tasks	Classification, Regression, Clustering	Missing Values?	N/A	Number of Web Hits	163354

TABLE 7.5
Dataset

Attributes	Range	Description
Age	–	Age in years
Anemia	Boolean	Decrease of red blood cells or hemoglobin
BP	Boolean	High blood pressure
CPK	–	Creatinine phosphokinase enzyme in blood
Diabetes	Boolean	If the patient has diabetes
Ejection fraction	%	Of blood leaving the heart at each contraction
Platelets		Platelets in the blood (kiloplatelets/mL)
Sex	0,1	Woman or man
Serum creatinine	–	Level of serum creatinine in the blood
Serum sodium	–	Level of serum sodium in the blood
Smoking	Boolean	If the patient smokes or not
Time	–	Follow-up period (days)
[Target] death event	Boolean	–

TABLE 7.6
Dataset

Dataset Characteristics	N/A	Number of Instances	303	Area	Life
Attribute Characteristics	Integer, Real	Number of Attributes	56	Date Donated	2017-11-17
Associated Tasks	Classification	Missing Values?	N/A	Number of Web Hits	35878

7.3.1.2   Attribute Information
Thirteen (13) clinical features of this dataset listed below

4. UCI Machine Learning Repository, The Z-Alizadeh Sani dataset contains the records of 303 patients, each of which have 54 features. The features are arranged in four groups: demographic, symptom and examination, ECG, and laboratory and echo features [5]. Its characteristics are shown in Table 7.6.

7.4   COMPARISON BASED ON PERFORMANCE METRICS

For reviewing the articles, several state-of-the-art ML techniques have been used along with the hybrid approaches. In the article titled "Categorical Study of Heart Disease with Machine Learning Techniques", linear discriminant analysis got an accuracy of 84.00%, Random Forest got 99.00%, Decision Tree Classifier got

98.00%, and Support Vector Machine (SVM) got 95.00% of accurate results [6], and these results are very promising in the context of predicting heart disease. Synthetic minority oversampling techniques (SMOTE) with hyper parameter optimization (HPO) along with ensemble learning methods gave an accuracy of 95.73% of results so far with Cleveland dataset [7]. In another paper, least absolute shrinkage and selection operator (LASSO) technique that is applied only to 11 features gave improved results over the state-of-the-art algorithms, with LASSO: DT – 88.66%, RF – 89.07%, KNN – 93.00%, AB – 90.75%, GB – 92.85%, DTBM – 8865%, RFBM – 97.65%, and GBBM – 97.85% [8].

There is another technique called data acquisition and processing with SVM. Naïve Bayes (NB) got an accuracy 83.49% and DT+NB+Auto MLP gave an accuracy of 94.85% [9]. When used with FCMIM feature selection techniques, artificial neural network, decision tree, and LASSO gave an accuracy of 93.27%, with minimum redundancy and maximum relevance (MRMR) along with SVM – provided 87.00% of accuracy [10].

A new method called the Ruzzo–Tompa algorithm (RTA) with optimally configured and improved deep belief network (OCI- DBN) collectively gave an accuracy of 94.61%, whereas DNN (Deep Neural Network) and ANN (Artificial Neural Network) gave an accuracy of 90.57% and 86.20%, respectively [11]. Recursion-enhanced random forest with an improved linear model (RFRF-ILM) got an accuracy of 95.20% and the database is used 50 times [12]. Genetic Algorithm-85.10%, computer-aided decision support system, hybrid random forest with linear model (HRF-LM) is the novel model presented in the article provided accuracy of 87.40% [13]. Hybrid grid search algorithm and extra tree are new concepts proposed that have an accuracy of 92.22% and 88.00% consecutively [14]. In addition with PCA (principle component analysis) using variation auto encoders along with deep neural network (VAE-TWO-DNN) given the result of 89.20% accuracy [15].

Data mining techniques have been used in the context of finding proper class. SMOTE gave results with an accuracy of 92.62%, and Gaussian Naïve Bayes classifier gave results with an accuracy of 86.67% [16]. Multilayer perceptron gave results with an accuracy of 71.40%, adaboost 71.90%, gradient boost Naïve Bayes 63.30%, and XGBoost 72.60% [16]. An important note is to track the right dataset that has been used in a different model. When we do modeling of different datasets, the results may vary. There are a lot of things such as feature selection, feature extraction, dimensionality reduction, PCA approaches, and data cleaning and preprocessing techniques that make methods more dependent on the data that have been dealt and preprocessed before for a particular algorithm.

All the comparative results of research papers are considered along with author names. The dataset and techniques used, and their respected accuracy are mentioned in Table 7.7 (papers worked on Cleveland, Hungary, Switzerland, and Long Beach dataset), Table 7.8 (The Korea National Health and Nutrition Examination Survey (KNHANES) dataset to predict CHD risk), Table 7.9 (papers worked on heart-failure-clinical-records-dataset, derived from the UCI ML repository), and Table 7.10 (papers worked on the Z-Alizadeh Sani dataset).

TABLE 7.7

Comparison of Research Articles Worked on Dataset for Cleveland, Hungary, Switzerland, and Long Beach [12]

Sr. No.	Year	Author(s)	Focus of the Paper	Technique(s) Used	Parameters Analyzed	Accuracy
1	2022	Ghulab Nabi Ahmad	Categorical study of heart disease with machine learning techniques	Linear Discriminants Analysis (LDA), Random Forest Classifier (RFC), Decision Tree Classifier (DTC), Gradient Boosting Classifier (GBC), Support Vector Classifier (SVC)	Cp, Sex, Age, Chol, Fbps, Restbps, Restecga, Oldpeack, Exang, Slope, Ca, Thal, Thalach	LDA: 84% RFC: 100% DTC: 100% GBC: 98% SVC: 95%
2	2022	Abdallah Abdellatif	Heart disease detection and severity level classification model using hyperparameter optimization methods and machine learning	Synthetic Minority Oversampling Technique (SMOTE), Hyperparameter Optimization (HPO), Hyperband (HB), Ensemble Learning Technics (ELT), Machine Learning Classifiers	Cp, Sex, Age, Chol, Fbps, Restbps, Restecga, Oldpeack, Exang, Slope, Ca, Thal, Thalach	Proposed Method: HB+SMOTE+ET= 95.73%

(*Continued*)

TABLE 7.7 (Continued)

Comparison of Research Articles Worked on Dataset for Cleveland, Hungary, Switzerland, and Long Beach [12]

Sr. No.	Year	Author(s)	Focus of the Paper	Technique(s) Used	Parameters Analyzed	Accuracy
3	2021	Pronab Ghosh	Efficient prediction of cardiovascular disease using machine learning algorithms with relief and LASSO feature selection techniques	Least Absolute Shrinkage and Selection Operator (LASSO) Techniques, Relief Feature Selection, Adaboost (AB), Decision Tree (DT), Gradient Boosting (GB), K-Nearest Neighbors (KNN), Random Forest (RF), Random Forest Bagging Method (RFBM), Gradient Boosting Method (GBBM), K-Nearest Neighbors Bagging Method (KNNBM), Decision Tree Bagging Method (DTBM), Adaboost Boost Method (ABBM)	Cp, Sex, Age, Chol, Fbps, Restbps, Restecga, Oldpeack, Exang, Slope, Ca, Thal, Thalach	**Accuracy of all 13 features:** DT: 86.95% RF: 88.65% KNN: 88.61% AB: 89.07% GB: 86.97% DTBM: 87.97% RFBM: 92.65% KNNBM: 89.63% ABBM: 89.07% GBBM: 90.97% LASSO model with 11 features: DT: 88.6% RF: 89.07% KNN: 93% AB: 90.75% GB: 92.85% DTBM: 88.65% RFBM: 97.65% KNNBM: 96.6% ABBM: 90.75% GBBM: 97.85% **Relief model with 10 features:** DT: 89.12% RF: 97.89%

(Continued)

TABLE 7.7 (Continued)
Comparison of Research Articles Worked on Dataset for Cleveland, Hungary, Switzerland, and Long Beach [12]

Sr. No.	Year	Author(s)	Focus of the Paper	Technique(s) Used	Parameters Analyzed	Accuracy
4	2021	Saba Bashir	Cardiovascular disease prediction and an intelligent ensemble voting scheme	Data Acquisition and Preprocessing, Support Vector Machine (SVM), Naïve Bayes (NB), Decision Tree (DT), Neural Network (NN), Perceptron, Auto MLP	Cp, Sex, Age, Chol, Fbps, Restbps, Restecga, Oldpeack, Exang, Slope, Ca, Thal, Thalach	KNN: 94.11% AB: 92.85% GB: 96.22% DTBM: 90.22% RFBM: 99.05% KNNBM: 98.05% ABBM: 95.38% GBBM: 98.32% **Ensemble Scheme:** SVM+NB+Auto MLP: 83.00% SVM+NB+NN: 83.49% SVM+Perceptron+NN: 94.85% DT+NB+Auto MLP: 75.00%
5	2020	Jian Ping Li	Machine learning classification for heart disease identification	Artificial Neural Network (ANN), Decision Tree (DT), K-Nearest Neighbor (KNN), Relief FS Algorithm, Least Absolute Shrinkage and Selection Operator Algorithm (LASSO), FCMIM Feature Selection Algorithms, Minimal Redundancy Maximal Relevance (MRMR)	Cp, Sex, Age, Chol, Fbps, Restbps, Restecga, Oldpeack, Exang, Slope, Ca, Thal, Thalach	**Accuracy produced by used feature selection model** 1. Relief (SVM): 86% 2. MRMR (SVM): 87% 3. LASSO (LR): 87% 4. LLBFS (LR): 88% 5. FCMIM (SVM): 93.27%

(Continued)

TABLE 7.7 (Continued)

Comparison of Research Articles Worked on Dataset for Cleveland, Hungary, Switzerland, and Long Beach [12]

Sr. No.	Year	Author(s)	Focus of the Paper	Technique(s) Used	Parameters Analyzed	Accuracy
6	2020	Syed Arslan Ali	Heart Disease Prediction	Ruzzo–Tompa Algorithm (RTA), Optimally Configured and Improved Deep Belief Network (OCI-DBN), Stacked Genetic Algorithm (SGA), Deep Neural Network (DNN), Artificial Neural Network (ANN)	Cp, Sex, Age, Chol, Fbps, Restbps, Restecga, Oldpeack, Exang, Slope, Ca, Thal, Thalach	By Using Ruzzo–Tompa Algorithm (RTA) OCI-DBN: 94.61% DNN: 90.57% ANN: 86.20%
7	2020	Chunyan Guo	Heart disease detection	Recursion-Enhanced Random Forest with an Improved Linear Model (RFRF-ILM), Decision Tree (DT), Random Forest	Cp, Sex, Age, Chol, Fbps, Restbps, Restecga, Oldpeack, Exang, Slope, Ca, Thal, Thalach	No. of dataset used 10 RFRF-ILM: 69.5% No. of dataset used 20 RFRF-ILM: 74.9% No. of dataset used 30 RFRF-ILM: 89.4% No. of dataset used 40 RFRF-ILM: 87.8% No. of dataset used 50 RFRF-ILM: 95.2%

(Continued)

TABLE 7.7 (Continued)
Comparison of Research Articles Worked on Dataset for Cleveland, Hungary, Switzerland, and Long Beach [12]

Sr. No.	Year	Author(s)	Focus of the Paper	Technique(s) Used	Parameters Analyzed	Accuracy
8	2019	Senthilkum ar Mohan	Heart disease prediction	Machine Learning (ML), Genetic Algorithm (GA), K-Nearest Neighbor Algorithm (KNN), Computer-Aided Decision Support System (CADSS), Hybrid Random Forest with a Linear Model (HRFLM), Naïve Bayes (NB), Generalized Linear Model (GLM), Decision Tree (DT), Deep Learning (DL), Linear Regression (LR), Support Vector Machine (SVM), Gradient-Boosted Tree (GBT)	Cp, Sex, Age, Chol, Fbps, Restbps, Restecga, Oldpeack, Exang, Slope, Ca, Thal, Thalach	Proposed Model (HRFLM): 87.4% NB: 75.8% GLM: 85.1% DL: 87.4% LR: 82.9% DT: 85% GBT: 78.3% RF: 86.1% SVM: 86.1%
9	2019	Liaqat Ali	Effective prediction of heart failure	Hybrid Grid Search Algorithm (HGSA), Support Vector Machine (SVM), Fuzzy Logic, Artificial Neural Network (ANN), Adaboost, Random Forest (RF), Extra Tree (ET)	Cp, Sex, Age, Chol, Fbps, Restbps, Restecga, Oldpeack, Exang, Slope, Ca, Thal, Thalach	Accuracy of adaboost: 88.00% RF:88.00% ET: 88.00% Proposed Model L1 Linear SVM+L2 Linear & RBF SVM: 92.22%

TABLE 7.8

Comparison of Research Articles Worked on the Korea National Health and Nutrition Examination Survey (KNHANES) Dataset to Predict CHD Risk [13]

Sr. No.	Year	Author(s)	Focus of the Paper	Technique(s) Used	Parameters Analyzed	Accuracy
1	2021	Tsatsral	Coronary Heart Disease	Variational Autoencoders (VA), Principal Component Analysis (PCA), Adaboost, K-Nearest Neighbors (KNN), Random Forest (RF), Support Vector Machine (SVM), Decision Tree (DT), Naïve Bayes (NB), Deep Neural Network (DNN)	Age, BMI, TC, HDL, SBP, WC, NF, Obesity status, Knee-joint pain status, Weight change in 1 year status, Frequency of eating out of year status, Marital status	Proposed Model (VAE-TWO -
		Amarbayasgalan	Prediction			
						DNN): 89.2%
						KNN: 76.5%
						DT: 74.5%
						SVM: 75.1%
						Adaboost: 80.9%
						RF: 80.1%
						NB: 73.2%

TABLE 7.9

Comparison of Research Articles Worked on Heart-Failure-Clinical-Records-Dataset Is Derived from the UCI Machine Learning Repository [14]

Sr. No.	Year	Author(s)	Focus of the Paper	Technique(s) Used	Parameters Analyzed	Accuracy
1	2021	Abid Ishaq	Heart Failure Prediction	Data Mining Techniques (DMT), Gaussian Naïve Bayes Classifier (GNB), Stochastic Gradient Classifier (SGD), Synthetic Minority Oversampling Technique (SMOTE), Adaptive Boosting Classifier (Adaboost), Extra Tree Classifier (ETC), Gradient Boosting Machine (GBM)	Time, Event, Gender, Smoking, Diabetic, BP, Anemia, Age, Ejection Fraction, Sodium, Creatinine, Platelets, CPK	SMOTE Accuracy: 0.9262 DT: 78.89% Adaboost: 82.23% LR: 85.56% ETC: 83.34% GBM: 84.44% SGD: 66.67% RF: 88.89% SVM: 86.67% G-NB: 86.67%

TABLE 7.10

Comparison of Research Articles Worked on The Z-Alizadeh Sani Dataset [15]

Sr. No.	Year	Author(s)	Focus of the Paper	Technique(s) Used	Parameters Analyzed	Accuracy
1	2020	Jikuo Wang	Detection of Coronary Heart Disease	Logistic Regression (LR), Gaussian Process Classification (GPC), Multilayer Perceptron (MLP), Adaboost (ADB), Random Forest (RF), Linear Regression (LR), Gradient Boosting Naïve Byes (GNB), Support Vector Classification (SVC), Decision Tree (DT), K- Nearest Neighbor (KNN), Adaboost (ADB), Gradient Boosting (GB), Extra Tree (ET), Machine Learning Programming (MLP), XGboost (XGB)	Age, Weight, Length, Sex, BMI, Diabetes Mellitus, HTN, Current Smoker, Ex-Smoker, FH Obesity, CRF CVA, Airway Disease, Thyroid Disease, CHF, DLP, BP, PR, Edema, Weak Peripheral Pulse, Lung Rales, Systolic Murmur, Diastolic Murmur, Typical Chest Pain, Rhythm, Q Wave, ST Elevation, ST, Depression, T-Inversion, LVH, Poor R Progression, FBS, Cr, TG, LDL, HDL, BUN, ESR, HB, K, Na, WBC, Lymph, Neut, PLT, EF, RWMA, VHD	Obtains An Accuracy Of Proposed Method: 96.3% LR: 68.00% RF: 69.7% GNB: 63.3% SVC: 69.8% DT: 68.2% KNN: 69.3% ADB: 71.9% GB: 73.1% ET: 68.9% MLP: 71.4% XGB: 72.6%

7.5   CONCLUSION

In this chapter, we reviewed research articles in the area of CVDs where ML, deep learning, and other AI techniques are used. Most of these articles were published by IEEE ACCESS. We considered articles published from 2019 to 2022. Year-wise distribution of these articles was shown. Twelve research articles passed our selection criteria. All the datasets, solution approaches, and their performance in terms of evaluation metrics were discussed further in detail. In future, this review can be improved in many ways: articles published before 2019 and after 2022 should be included, and articles from other publishers should also be included.

REFERENCES

[1] World Health Organization, Accessed: Jan 9, 2023 [online]. https://www.who.int/health-topics/cardiovascular-diseases

[2] Janosi, A., Steinbrunn, W., Pfisterer, M., & Detrano, R. (1988). UCI Machine Learning Repository, Databases: Cleveland, Hungary, Switzerland, and the VA Long Beach. Accessed: Jan 9, 2023 [online]. Available: https://archive.ics.uci.edu/ml/datasets/heart+disease.

[3] Statistics Korea. Causes of Death Statistics in 2019. Accessed: Jan 9, 2023 [Online]. Available: https://kostat.go.kr/portal/eng/pressReleases/1/index.board?bmode=read&bSeq=&aSeq=385629&pageNo=1&rowNum=10&navCount=10&currPg=&searchInfo=srch&sTarget=title&sTxt=death

[4] Janosi, A., Steinbrunn, W., Pfisterer, M., & Detrano, R. (1988). UCI Machine Learning Repository, Heart-Failure-Clinical-Records-Dataset is Derived from the UCI Machine Learning Repository. Accessed: Jan 9, 2023 [online]. Available: https://archive.ics.uci.edu/ml/datasets/heart+disease

[5] Sani, Z. A., Alizadehsani, R., & Roshanzamir, M. (2017). UCI Machine Learning Repositor, Z-Alizadeh Sani Data Set. Accessed: Jan 9, 2023 [online]. Available: https://archive.ics.uci.edu/ml/datasets/Z-Alizadeh+Sani

[6] Ahmad, G. N., Shafiullah, N., Algethami, A. A., Fatima, H., & Akhter, S. H. (2022). Comparative study of optimum medical diagnosis of human heart disease using machine learning technique with and without sequential feature selection. *IEEE Access*, *10*, 23808 23828. https://doi.org/10.1109/access.2022.3153047

[7] Abdellatif, A., Abdellatef, H., Kanesan, J., Chow, C. O., Chuah, J. H., & Gheni, H. M. (2022). An effective heart disease detection and severity level classification model using machine learning and hyperparameter optimization methods. *IEEE Access*, *10*, 79974–79985. https://doi.org/10.1109/access.2022.3191669

[8] Ghosh, P., Azam, S., Jonkman, M., Karim, A., Shamrat, F. M. J. M., Ignatious, E., Shultana, S., Beeravolu, A. R., & De Boer, F. (2021). Efficient prediction of cardiovascular disease using machine learning algorithms with relief and LASSO feature selection techniques. *IEEE Access*, *9*, 19304–19326. https://doi.org/10.1109/access.2021.3053759

[9] Bashir, S., Almazroi, A. A., Ashfaq, S., Almazroi, A. A., & Khan, F. H. (2021). A knowledge-based clinical decision support system utilizing an intelligent ensemble voting scheme for improved cardiovascular disease prediction. *IEEE Access*, *9*, 130805–130822. https://doi.org/10.1109/access.2021.3110604

[10] Li, J. P., Haq, A. U., Din, S. U., Khan, J., Khan, A., & Saboor, A. (2020). Heart disease identification method using machine learning classification in E-healthcare. *IEEE Access*, *8*, 107562–107582. https://doi.org/10.1109/access.2020.3001149

[11] Ali, S. A., Raza, B., Malik, A. K., Shahid, A. R., Faheem, M., Alquhayz, H., & Kumar, Y. J. (2020). An optimally configured and improved deep belief network (OCI-DBN) approach for heart disease prediction based on Ruzzo-Tompa and stacked genetic algorithm. *IEEE Access, 8*, 65947–65958. https://doi.org/10.1109/access.2020.2985646

[12] Guo, C., Zhang, J., Liu, Y., Xie, Y., Han, Z., & Yu, J. (2020). Recursion enhanced random forest with an improved linear model (RERF-ILM) for heart disease detection on the internet of medical things platform. *IEEE Access, 8*, 59247–59256. https://doi.org/10.1109/access.2020.2981159

[13] Mohan, S., Thirumalai, C., & Srivastava, G. (2019). Effective heart disease prediction using hybrid machine learning techniques. *IEEE Access, 7*, 81542–81554. https://doi.org/10.1109/access.2019.2923707

[14] Ali, L., Niamat, A., Khan, J. A., Golilarz, N. A., Xingzhong, X., Noor, A., Nour, R., & Bukhari, S. A. C. (2019). An optimized stacked support vector machines based expert system for the effective prediction of heart failure. *IEEE Access, 7*, 54007–54014. https://doi.org/10.1109/access.2019.2909969

[15] Amarbayasgalan, T., Pham, V. H., Theera-Umpon, N., Piao, Y., & Ryu, K. H. (2021). An efficient prediction method for coronary heart disease risk based on two deep neural networks trained on well-ordered training datasets. *IEEE Access, 9*, 135210–135223. https://doi.org/10.1109/access.2021.3116974

[16] Ishaq, A., Sadiq, S., Umer, M., Ullah, S., Mirjalili, S., Rupapara, V., & Nappi, M. (2021). Improving the prediction of heart failure patients' survival using SMOTE and effective data mining techniques. *IEEE Access, 9*, 39707–39716. https://doi.org/10.1109/access.2021.3064084

8 Identification of Unipolar Depression Using Boosting Algorithms

Parul Verma
Amity University

Roopam Srivastava and Shikha Srivastava
Mahatama Gandhi Post Graduate College

8.1 INTRODUCTION

Exposure to environmental pollutants, unstable employment, poor sleep, poor hygiene, poor diet, substance abuse, and other factors are key contributors to depression [1,2]. Students in schools and colleges are more likely to experience severe depression, which often results in early mortality. Depression comes in different forms, which medical professionals identify based on symptoms and root causes. Various forms of depression include the following:

Depression of major proportions (MDD): This type of depression persists for more than 1–2 weeks. The severe symptoms of such depression make life difficult for a person.

Bipolar disorder: Manic episodes are intermingled with depressive episodes in the course of this illness. At this time, they may suffer depression symptoms including melancholy, hopelessness, or exhaustion.

Perinatal and postpartum depression: This kind of depression occurs during pregnancy period and even after delivery as well. It is also termed as "baby blues" that result in stress and anxiety.

Persistent depressive disorder (PDD): Another common name of PDD is dysthymia. It is quite common, although its symptoms are not as severe as actual major depression.

Premenstrual dysphoric disorder (PMDD): This type of disorder occurs due to premenstrual disorder. Many women face this disorder.

Psychotic depression: It is a kind of delusions where person feels some hallucinations.

Seasonal affective disorder (SAD): It is a kind of stress where seasonal changes are the key drivers.

DOI: 10.1201/9781003359951-10

Your emotions, intellect, and body can all be impacted by depression. Indications of depression are as follows:

- Losing enjoyment for activities that once made you happy.
- Having a short fuse or being easily annoyed.
- Consuming excessive or inadequate food.
- A change in the duration of sleep.
- Finding it difficult to focus or remember things.
- Experiencing bodily discomforts, such as headache, stomachache, or erectile dysfunction.
- Consciousness of self-harm or self-death.

Various factors which cause depression are as follows:

- **Chemistry of the brain**: Anxiety may be brought on by abnormalities in brain chemical levels.
- **Aspects of genetics**: You may have a higher chance of developing melancholy if a family member does.
- **Life events**: There are many events in our lives that lead us to anxiety or stress like loss of loved one, loneliness, and lack of support.
- **Conditions that affect health**: Chronic sickness and physical discomfort can contribute to depression. Diabetes, cancer, Parkinson's disease, and other illnesses frequently co-exist with insomnia in people.
- **Personality**: People who are not able to cope with problems and get overwhelmed quite easily are prone to depression.

There are various types of therapies available. The most used ones are cognitive, behavioral, interpersonal, and psychodynamic therapies. Mostly a blend of all these therapies is used. The smart healthcare in today's scenario is supporting well the complete system of detection, monitoring, and collection of data and its analysis with respect to various diseases. The whole system is made simple by making use of IoT and AI techniques. It helps in remote monitoring of patients and immediately send alert to the healthcare system in adverse situations [3]. Figure 8.1 illustrates how clinical and non-clinical data are used in smart healthcare. Any patient participating in a clinical study for an illness must go to the hospital and be accessible for routine physical exams. In smart healthcare system, patient health can be monitored from a remote location by accessing vital symptoms through wearable devices or implants. The non-clinical data related to the daily activity of the person can be tracked by bio-signals in smart healthcare systems. Higher patient participation, better patient outcomes, prompt diagnosis, and treatment are a few advantages of integrating IoT in healthcare.

Wearable technology (Figure 8.2) assists in collecting and analyzing real-time personal data that teach us everything from our fitness to our health. A wearable is exactly what? Electronic devices that are worn on a person, typically next to the skin, to accurately transmit critical biological, medical, and exercise data to a database are called wearable. The wearable business alone is predicted to grow to $111 billion by

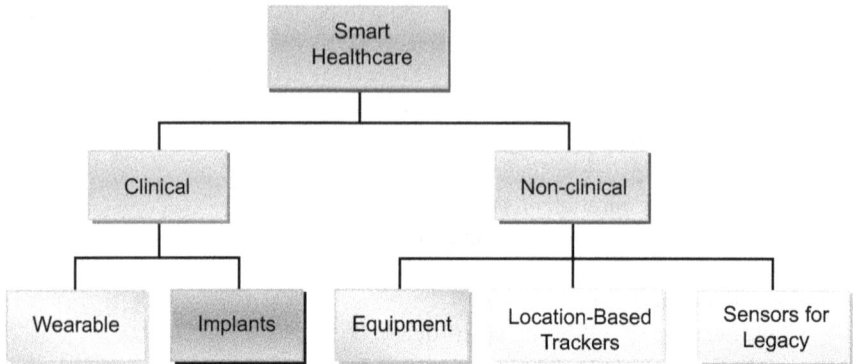

FIGURE 8.1   IoT-based sensors in healthcare [4].

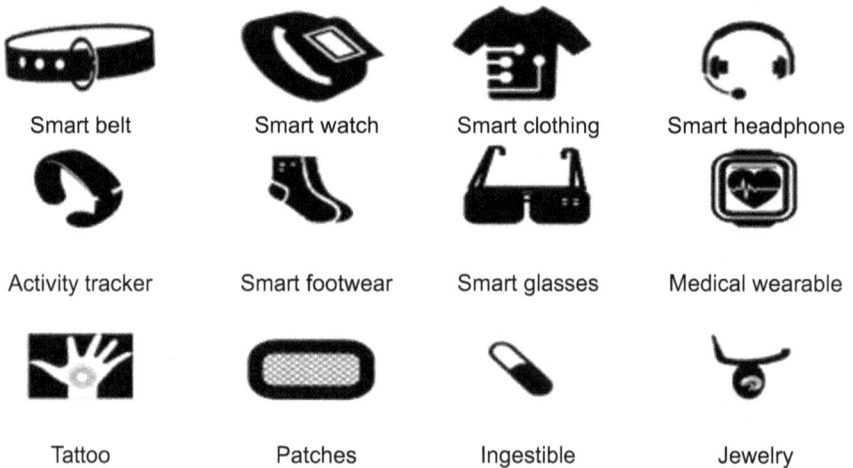

FIGURE 8.2   Wearable devices [5].

2027, contributing to the $594 billion behemoth that is the IoT market today. Although Fit bits and Apple Watches are iconic examples of wearable technology, other gadgets are also being created now. Smartwatches, Virtual Reality (VR), Augmented Reality (AR), smart jackets, and several other devices are taking us into the future. Probably the sector benefiting most from wearable technology is the healthcare sector. Patients can assess everything from their body temperature to their blood pressure while wearing these smart gadgets, and these data are subsequently transmitted in real time to their medical staff. Medical professionals can diagnose and treat patients more quickly if something seems awry. Due to the data gathered by a wearable device, doctors may now diagnose a sickness or disease without having to do a battery of tests, which has sped up the entire treatment process. To identify the source of the medical error immediately, they can consult the data gathered by a wearable.

Our ability to "Affect" the feelings we encounter daily is one of the three fundamental and interrelated human capacities that are required for us to live healthy lives. "Behavior" refers to how we behave and express ourselves in social situations. Lastly, "Cognition" refers to how we think and relate that thinking to the outside environment [6]. In social psychology, these three make up the Tripartite Attitude Model. Our emotions and moods are examples of how we express our feelings, and these changes are dependent on external circumstances. The affective state of a person is a neurophysiologic state that allows them to deliberately access their basic, primal emotions. The nature of this condition could also be non-reflective. A person generally has two affective states: a functional emotional state and an aesthetic psychological state. A person's affective state adapts to changes in emotion, mood, or affect. An individual's affective psychological state reveals their mental health, as only those who are in a healthy affective mental state can carry out their daily tasks without interruption and concentrate on their work. A person's affective, psychological, or mental state may change because of stress in reaction to daily activities. The tone of voice, a frown, pursed lips, tears, and eye gazing are just a few ways that the affective state might alter, along with internal body changes such as a shift in heart rate. It is challenging to determine a person's true emotional and psychological states because mental health is invisible to others. Contrarily, when our physical health declines, we can see the signs and symptoms, and we go to a doctor or other healthcare provider right away to get a diagnosis and start therapy. Additionally, because certain mental health diseases, such as depression, acute stress syndrome, anxiety, and insomnia, share symptoms, identifying the correct type can be difficult. A dry mouth, abrupt sweating, and faster breathing, for instance, are indicators of both stress and anxiety. Because an anxiety attack only lasts a few minutes and is frequently brought on by outside stimuli, it requires an extensive evaluation of the person to categorize them as stress or anxiety. The effects on the body and mind can be cyclically interconnected because emotional discomfort and poor mental health can set off or exacerbate physical health issues, which in turn can lead to additional suffering. The chance of having mental health issues might also increase with poor physical health. Physical infirmity may also enhance the risk of mental illness. A little bit of stress might be advantageous because people do virtually as well at work when under mild stress. Constructive stress is typically linked to a positive challenge, unlike agony, which has negative impacts. However, sustained and chronic stress can negatively influence a person's health, harm the entire body, and raise their risk of contracting certain diseases. It may manifest with a variety of psychological symptoms that might make day-to-day activities more difficult. A standard procedure is used to assess a person's mental and emotional functioning during a mental status examination (MSE) performed by a professional. It entails a detailed set of observations as well as certain targeted inquiries. Seven major evaluation criteria, including appearance, psychomotor behavior, attitude, speech features, affect, and mood, are used to analyze the entire diagnostic picture. It's important to assess how someone behaves in various situations, such as how they act while driving, watching a movie, or working at a desk. Only then can a psychologist determine the subject's true mental state. Each psychiatric disorder has distinct symptoms that can be assessed using biomarkers, as well as some basic warning indicators that indicate a need for professional assistance [7]. Nowadays, wearable sensors that monitor many aspects of human activity are a widespread technology. Data gathered with these devices are also of interest in the study. Nevertheless, datasets including sensor data are uncommon in the world of medicine. Frequently, data are private, and only the outcomes are made public. Other researchers may find it challenging to collaborate or even to replicate and compare the results as a result. Without human intervention or an emotional approach, vast amounts of data can now be collected and analyzed thanks to machine learning and data science. Based on the data that were scored and classified, the researchers use a machine-learning system to identify depression [8, 9]. This chapter aims to detect depression based on the actigraph watch measures by using two different machine-learning algorithms, namely, Adaboost and XGboost. The rest of this chapter is organized as follows: Section 8.2 describes the related work. Section 8.3 presents the proposed framework system. Finally, results are summarized and concluded in Sections 8.4 and 8.5, respectively.

8.2   RELATED WORK

Dautov R et al. in their work proposed a distributed hierarchical data fusion approach. Their approach exploits the potential processing power of various nodes in the smart healthcare architecture. The three-level processing paradigms are conveniently derived from and correspond to the contemporary data fusion taxonomies, which frequently use a hierarchical structure to allow data fusion at various levels. Data fusion was implemented with the help of CEP technology, which naturally facilitates the hierarchical processing of streaming data. A smart healthcare case study, in which the entire IoT was utilized, served as a means of demonstrating the feasibility and effectiveness of the suggested strategy [3].

Kumar A. in his research developed a distributed hierarchical model to deal with the increasing number of heterogeneous data sources and handle the time limitations. Answering sites' work here proposes remedies to the major issue of dishonesty in the medical profession (Medical CQAs). It suggests using a hybrid deep learning model to solve the semantic question-matching problem for the detection of duplicate question pairs. To assess the similarity between two queries, the model combines Bi-LSTM neural network with multi-layer perceptron classifier. The Euclidean distance function is then used to calculate how similar the questions are to one another. A total of 100 question-and-answer pairs from the three major categories of "Irritable Bowel Syndrome," "Anxiety Disorder," and "Menopause" are used to test the suggested model [6]. Lee S. et al. in their research survey focused on the creation and application of wearable technology and sensors in patients with depression. They collected work of 18 researchers that also worked on the use of wearable technology to assess, monitor, or predict depressive symptoms. They examined the sensors of various wearable gadgets (such as wristbands, fitness trackers, smartwatches, and actigraphy units) and sensor-measured metrics in depressed individuals in their proposed report. They observed that various parameters collected through wearable devices can be used positively for identification of depression [10]. To identify and diagnose depression, the review work of Shumaila A. et al. includes various machine-learning (ML) methods. They described a generalized framework for depression diagnostics. The steps included data extraction, preprocessing, training of ML classifier, and performance evaluation. Additionally, it provides a summary of the goals and restrictions of the various research papers that have been presented in the field of depression detection. A group categorization method is developed using precise passive sensing of data to assist depressed individuals in quickly determining the severity of their depression [9].

The purpose of the study by Gutierrez L. J. et al. was to identify user depression using social media user data. The Naïve Bayes classifier and the hybrid model NBTree are then fed with the Twitter data in two separate ways. To choose the most effective algorithm to identify depression, the results were compared based on the highest accuracy value. Results showed that both algorithms performed equally by demonstrating the same level of accuracy [11]. Kulkarni M. et al in their research work included designing of a wearable to record the physiologic markers that a clinically depressed individual experiences when under stress. The data collection, analysis, and processing all benefited from IoT. The automated mobile application is built which displays the specifics of a person's health band information. The mobile application notifies the person's caregiver about crises [12]. Thati R. P. et al. done an in-depth research on the junction of IoT, and mental health diseases are surveyed in this publication. It assesses various computational frameworks, techniques, and tools, as well as research findings and potential unresolved problems for the efficient application of IoT systems in mental health [13].

Kumbhar P. Y. et al. in their investigation focused on the fundamental analysis of the strategies employed to forecast human depressive episodes. To do so, a system with several methods of diagnosing depression disorder has been constructed. The system is divided into four sections: sentiment analysis, EEG signal processing, diagnosis, and question and answer. ML algorithms like Naïve Bayes and Neural Networks are used here for the classification of data. The system can be accessed remotely without going to psychiatrists and thus saves the cost of hospital visits and EEG checkups [14]. Pandey R. et al. in their work introduced a healthcare framework for expecting mothers and their fetus, which employs the Internet of Medical Things (IoMT). The data gathered from IoMT gadgets will be sent to a server and examined through an AI/ML module. If any unusual activity is detected, the central system will alert the healthcare providers responsible for the mother and fetus to take appropriate action [15].

Priya A. et al. in their work used ML algorithms for prediction of depression and anxiety. They tested various ML algorithms on the dataset collected with the help of questionnaire prepared by them. They classified the severity of anxiety, depression, and stress at five different levels. In their research work, they applied five popular ML algorithms. Out of all, Random Forest Classifier performed best [16]. Uddin M. Z. et al. created a better mental healthcare technology, such as intelligent chatbots, a multimodal human depression predictor. Their work proposed an effective method using Long Short-Term Memory (LSTM)-based Recurrent Neural Network (RNN) to identify texts that may have some symptoms of depression [17].

Helwani Hasni K. et al. in their research developed a forecast model to predict future depression cases by the help of classification and regression tree (CART), one of the data mining approaches. They used a dataset of students' mental health to test their prediction model [18]. Dabhane S. et al. emphasized on depression in their work with numerous ways of spotting it. By evaluating the user's data and actions using various ML approaches, their system can determine whether a person using social media is experiencing depression. They used the ensemble learning method, treating the earlier individual algorithms as base learners, and showcased improved accuracy on their dataset. Their system made impact in early detection of depression [19].

Sharma A. et al. in their research aimed to explore the detection of depression cases among a sample of 11,081 Dutch citizen datasets. To balance the dataset, they created many samples using the oversampling, under-sampling, and ROSE sampling strategies. Next, they applied the ML algorithm "Extreme Gradient Boosting" on each sample to distinguish between cases of mental illness and healthy cases. The goal of the study was to see whether biomarkers can be used to diagnose and differentiate between cases of depression and healthy cases using the XGBoost algorithm [20]. Wanqing X. et al. in their work suggested screening of depression and anxiety disorders using the Self-Reported Anxiety Scale (SAS) and Self-Reported Depression Scale (SDS). They extracted features from the facial expression and movements that were created from the videos that were simultaneously recorded as the subjects filled out the scale in this study. After that, data on facial expressions, gestures, and scales are gathered to create a multimodal framework for enhancing the reliability and accuracy of the diagnosis of depression and anxiety [21]. Ogunseye E. O. et al. in their work studied use of ML implementations to address the rise in mental disease cases around the world and the need for effective mental healthcare (MHC). The datasets used in this study are from a Kaggle repository called "Mental Health Tech Survey." Classifiers like Linear Regression, K-Nearest Neighbor, Tree Class, Neural Network, Random Forest, and AdaBoost were employed. AdaBoost performed well in comparison to all other models tested to forecast the results of mental health treatment [22].

8.3   METHODOLOGY

There is a strong link between depression and suicide. The number of people suffering from depression worldwide exceeds 264 million. The disease is the second leading cause of death and disability among 15–29-year olds. A variety of psychological and pharmacological treatments are available to treat depression. Obtaining clinical datasets with quality data samples, size, and modern artificial intelligence training models is one of the most difficult tasks. It is not always simple to acquire health information because it must be safeguarded and protected by privacy legislation. Preprocessing the data involves a lot of work due to the numerous difficulties with data quality and format.

8.3.1   DATASET

The dataset is downloaded from the site https://doi.org/10.5281/zenodo.1219550 or can be accessed directly from http://datasets.simula.no/depresjon/. The complete dataset is divided into two files: control file and condition file. The data collected over time are kept in CSV file. The dataset consists of recordings of the moto activity of 23 unipolar and bipolar depressed patients as well as 32 healthy controls [23]. Actigraphy data from 23 patients with unipolar and bipolar depression make up the collection (condition group). The MADRS scores are also included in the scores. csv file. The columns in this table are number (patient identifier), gender (1 or 2 for female or male), age (age in age groups), melanch (1: melancholia, 2: no melancholia), afftype (1: bipolar II, 2: unipolar depressive, 3: bipolar I), inpatient (1: inpatient, 2: outpatient), edu (education grouped in years), days (number of days of measurements), marriage (1: married or cohabiting, 2: single), work (1: working or studying, 2: unemployed/sick leave/pension), madrs1 (MADRS score when measurement started), madrs2 (MADRS when measurement stopped) as shown in (Figure 8.3).

8.3.2 Clean/Explore Data

In the score dataset, Control rows have empty columns except for number (id), days, gender, and age. Therefore, let's split between condition and control for observations. We explore the condition dataset column with its shape in Figure 8.4.

Data before cleaning are shown in Figure 8.3. Data cleansing starts with a comparison to replicate, non-numeric, and null values. After the acquisition, we eliminate duplicates and non-numeric values (Figure 8.5).

Some text in the column shows missing values, and we handle it with a standard value. Furthermore, we use type conversion over these columns and convert them into categorical data forms. The missing text is replaced by some text as shown in Figure 8.6.

	number	days	gender	age	afftype	melanch	inpatient	edu	marriage	work	madrs1	madrs2
0	condition_1	11	2	35-39	2.0	2.0	2.0	6-10	1.0	2.0	19.0	19.0
1	condition_2	18	2	40-44	1.0	2.0	2.0	6-10	2.0	2.0	24.0	11.0
2	condition_3	13	1	45-49	2.0	2.0	2.0	6-10	2.0	2.0	24.0	25.0
3	condition_4	13	2	25-29	2.0	2.0	2.0	11-15	1.0	1.0	20.0	16.0
4	condition_5	13	2	50-54	2.0	2.0	2.0	11-15	2.0	2.0	26.0	26.0
5	condition_6	7	1	35-39	2.0	2.0	2.0	6-10	1.0	2.0	18.0	15.0
6	condition_7	11	1	20-24	1.0	NaN	2.0	11-15	2.0	1.0	24.0	25.0
7	condition_8	5	2	25-29	2.0	NaN	2.0	11-15	1.0	2.0	20.0	16.0
8	condition_9	13	2	45-49	1.0	NaN	2.0	6-10	1.0	2.0	26.0	26.0

FIGURE 8.3 Dataset exploration.

```
df_condition = df[df.number.str.contains('condition')].copy()
```

	number	days	gender	age	afftype	melanch	inpatient	edu	marriage	work	madrs1	madrs2
0	condition_1	11	2	35-39	2.0	2.0	2.0	6-10	1.0	2.0	19.0	19.0
1	condition_2	18	2	40-44	1.0	2.0	2.0	6-10	2.0	2.0	24.0	11.0
2	condition_3	13	1	45-49	2.0	2.0	2.0	6-10	2.0	2.0	24.0	25.0
3	condition_4	13	2	25-29	2.0	2.0	2.0	11-15	1.0	1.0	20.0	16.0

```
df_condition.shape
```

```
(23, 12)
```

FIGURE 8.4 Details of patient condition features.

```
df_condition.info()

<class 'pandas.core.frame.DataFrame'>
RangeIndex: 55 entries, 0 to 54
Data columns (total 12 columns):
 #    Column      Non-Null Count    Dtype
---   ------      --------------    -----
 0    number      55 non-null       object
 1    days        55 non-null       int64
 2    gender      55 non-null       int64
 3    age         55 non-null       object
 4    afftype     23 non-null       float64
 5    melanch     20 non-null       float64
 6    inpatient   23 non-null       float64
 7    edu         53 non-null       object
 8    marriage    23 non-null       float64
 9    work        23 non-null       float64
 10   madrs1      23 non-null       float64
 11   madrs2      23 non-null       float64

dtypes: float64(7), int64(2), object(3)
memory usage: 5.3+ KB
```

FIGURE 8.5   Data before cleaning.

```
df_condition.info()

<class 'pandas.core.frame.DataFrame'>
Int64Index: 23 entries, 0 to 22
Data columns (total 12 columns):
 #    Column      Non-Null Count    Dtype
---   ------      --------------    -----
 0    number      23 non-null       int32
 1    days        23 non-null       int64
 2    gender      23 non-null       category
 3    age         23 non-null       category
 4    afftype     23 non-null       category
 5    melanch     23 non-null       category
 6    inpatient   23 non-null       category
 7    edu         23 non-null       category
 8    marriage    23 non-null       category
 9    work        23 non-null       category
 10   madrs1      23 non-null       int64
 11   madrs2      23 non-null       int64

dtypes: category(8), int32(1), int64(3)
memory usage: 2.4 KB
```

FIGURE 8.6   Data after cleaning.

We receive a cleaned dataset after preprocessing the data as illustrated in Figure 8.7.

8.3.3 Feature Categorization and Data Visualization

We have analyzed 23 cases of unipolar and bipolar depressive patients using actigraphy, where Figure 8.8 presents the descriptive statistics along with information on the condition dataset's parameters and the range of values for each parameter. The goal of feature extraction is to transform the gathered data into features that may be used to enhance the data while enhancing the accuracy and performance of the forecasting algorithm. Hence, numerical and categorical features were extracted from the dataset. Now visualize the numerical and categorical features.

features_num=['days','madrs1',madrs2']

features_cat=['age','gender','afftype', 'melanch','inpatient','edu','marriage','work']

Development of MADRS score (before activity recording/after activity recording): A rating scale for antipsychotic treatment effects must meet several key criteria, including being quick and simple to use in a clinical environment, relevant for depressed disease, and offering a sensitive and accurate measure of change. The Montgomery-Asberg Depression Rating Scale (MADRS) is used to gauge how severe a persistent depression is. Here, the start of the measurement is denoted by MADRS1, while the end is denoted by MADRS2, as shown in Figure 8.9a, and the correlation between these scores are shown in Figure 8.9b.

	number	days	gender	age	afftype	melanch	inpatient	edu	marriage	work	madrs1	madrs2
0	0	3	1	3	1	1	1	2	0	1	4	6
1	11	8	1	4	0	1	1	2	1	1	6	0
2	16	5	0	5	1	1	1	2	1	1	6	10
3	17	5	1	1	1	1	1	0	0	0	5	3
4	18	5	1	6	1	1	1	0	1	1	8	11

FIGURE 8.7 Data after cleaning and conversion.

	days	madrs1	madrs2
count	23.000000	23.000000	23.000000
mean	12.652174	22.739130	20.000000
std	2.773391	4.797892	4.729021
min	5.000000	13.000000	11.000000
25%	12.500000	18.500000	16.000000
50%	13.000000	24.000000	21.000000
75%	14.000000	26.000000	24.500000
max	18.000000	29.000000	28.000000

FIGURE 8.8 Statistical description of numerical feature.

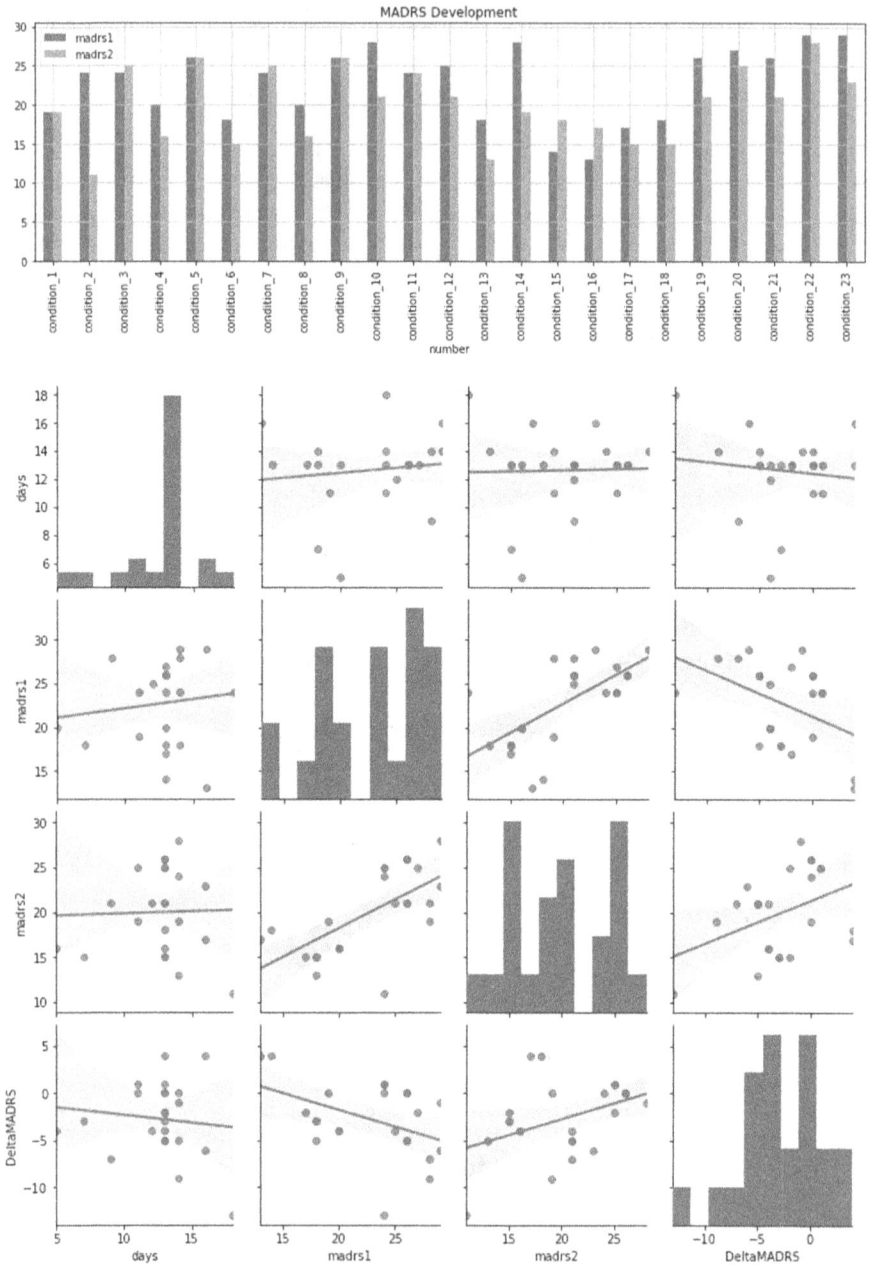

FIGURE 8.9   (a) MADRS scale for condition dataset. (b) Correlations between MADRS scores.

Impact of categorical features on scores: To illustrate our categorical feature, we use some visuals shown in Figure 8.10. These graphs display the effects of a feature on the score Madrs1 at the start of an activity measurement and the score Madrs2 at the end of an activity measurement.

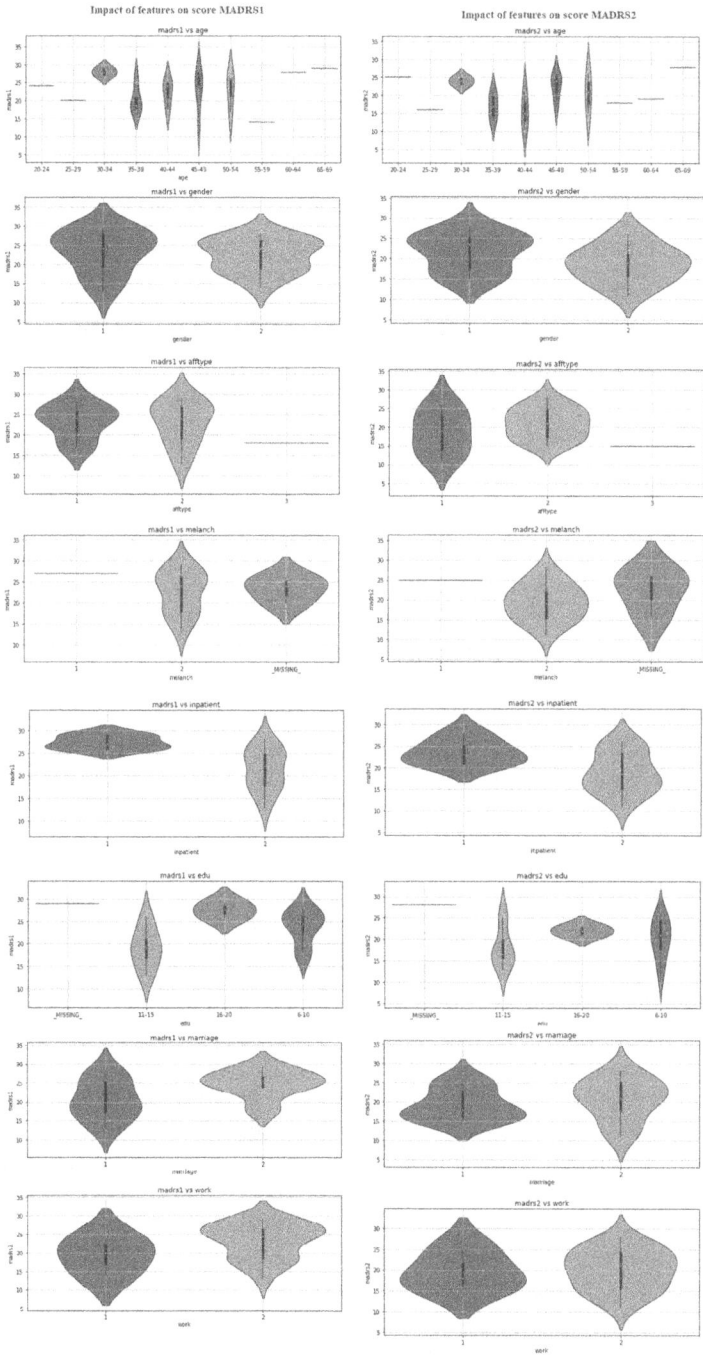

FIGURE 8.10 Influence of a feature on the MADRS score.

8.4 TRAIN/TEST SPLIT METHOD

With this technique, the dataset is split into a training set and a testing set. While the testing set is used to test the input data and assess its accuracy, the training set is used to train the dataset. The dataset utilized for this work was divided into train and test data, with 25% of the data used as test data. 'afftype' is the target dataset.

8.5 CALCULATING ACCURACY USING CLASSIFIERS

Utilizing classification techniques comes next, after feature selection and preprocessing. The goal behind the ML strategy known as "boosting" is to combine a number of very weak and faulty prediction rules to produce a highly accurate prediction rule. In this work, AdaBoost and XGBoost classifiers are used for accuracy calculation.

AdaBoost: This algorithm by Freund and Schapire, which has applications in many other industries, was the first real-world boosting algorithm and continues to be one of the most popular and extensively studied algorithms [24]. Its emphasis is on improving performance in those areas where the fundamental learner struggles. The model's initial iteration is called a base learner. Being a weak learner, it combines the forecasts from decision stumps, which are small trees with only one level. The term 'stagewise, additive modeling' was used, where "additive" did not refer to a model fit that included covariates but rather a linear combination of estimators.

Extreme gradient boosting algorithm (XGBoost): To handle a range of cutting-edge ML challenges, data scientists typically use the scalable end-to-end tree boosting system XGBoost [25]. The algorithm made it to CERN (European Organization for Nuclear Research), where statistical physicists thought it was the best method for categorizing signals from the Large Hadron Collider. With this algorithm, decision trees with nodes that have weights created with less evidence are severely condensed. The method's performance and memory use both benefit from this ingenious approach of removing insignificant nodes. Additionally, it has a parameter for randomization that reduces the association between the trees. The performance of the ensemble classifiers improves with lower correlation between classifier trees.

To create the model for the accuracy computation, we employed the above classifiers. Using these functions as a baseline, objects were created to fit our model on the train set using fit () and to predict on the test set using prediction, and their training and testing scores are compared. Table 8.1 showcases the results. As seen from Table 8.1, the XGBoost model outperforms AdaBoost classifier with 83% accuracy.

TABLE 1
Results

Classifiers	Score
AdaBoost	0.80
XGBoost	0.83

8.6   CONCLUSION

To predict unipolar depression in the proposed study, we applied ML boosting techniques. It resembles major depressive disorder somewhat. Both physical and mental health is affected by this condition. The visualization illustrates how numerous attributes relate to the MADRS score. The database has been taken into consideration for the input data for the ML algorithms during training and testing. We demonstrated the effectiveness of XGBoost classifiers on the provided dataset. To secure IoT-based mental health solutions and safeguard the wider ecosystem as well as sensitive patient and stakeholder data, an effective identity management system is the key. With transparency, portability, and simple management, these techniques could provide identity bearers complete ownership and control over their identities. Future research will examine verbal semantic cues, visual head attitude and eye-gaze cues, and physiological cues including skin conductance and heartbeat, and also for the similar case.

REFERENCES

1. Depressive disorder (dipression). (2020). World Health Organization. https://www.who.int/news-room/fact-sheets/detail/depression.
2. Murthy, R. S. (2017). National Mental Health Survey (NHMS), depression in India and healthcare gap. *Indian Journal of Psychiatry*, *59*(1), 21–26. https://doi.org/10.4103/psychiatry.IndianJPsychiatry_102_17
3. Dautov, R., Distefano, S., & Buyya, R. (2019). Hierarchical data fusion for smart healthcare. *Journal of Big Data*, *6*, 19. https://doi.org/10.1186/s40537-019-0183-6
4. Kumar, A., Sharma, K., & Sharma, A. (2021). Hierarchical deep neural network for mental stress state detection using IoT based biomarker. *Pattern Recognition Letters*, *145*, 81–87. https://doi.org/10.1016/j.patrec.2021.01.030
5. John Dian, F., Vahidnia, R., & Rahmati, A. (2020). Wearables and the Internet of Things (IoT), applications, opportunities, and challenges: a survey. *IEEE Access*, *8*, 69200–69211. https://doi.org/10.1109/ACCESS.2020.2986329
6. Kumar, A. (2020). Using cognition to resolve duplicacy issues in socially connected healthcare for smart cities. *Computer Communications*, *52*, 272–281. https://doi.org/10.1016/j.com.2020.01.041
7. Govindasamy, K., & Palanichamy, N. (2021). Depression detection using machine learning techniques on twitter data. In *Proceedings of the Fifth International Conference on Intelligent Computing and Control Systems*. ICICCS. IEEE Xplore Part Number: CFP21K74-ART. ISBN: 978-0-7381-1327-2.
8. Ranaware, M. V., & Joshi, K. (2020). Depression monitoring using wearable sensors. *(JETIR) Journal of Emerging Technologies and Innovative Research*, *7*, 599–604. https://www.jetir.org/papers/JETIR2003391
9. Aleem, S., Huda, N. U., Amin, R., Khalid, S., Alshamrani, S. S., Alshehri, A. (2022). Machine learning algorithms for depression: diagnosis, insights, and research directions. *Electronics*, *11*(7), 1111. https://doi.org/10.3390/electronics11071111
10. Lee, S., Hyewon, K., Park, M. J., & Jeon, H. J. (2021). Current advances in wearable devices and their sensors in patients with depression. *Frontiers in Psychiatry*, *12*, 672347. https://doi.org/10.3389/fpsyt.2021.672347
11. Gutierrez, L. J., Rabbani, K., Oluwashina, J. A., Gebresilassie, S. K., Rafferty, J., Castro, L. A., & Banos, O. (2021). Internet of things for mental health: open issues in data acquisition, self-organization, service level agreement, and identity management.

International Journal of Environmental Research and Public Health, 18(3), 1327. https://doi.org/10.3390/ijerph18031327

12. Kulkarni, M., & Wadhekar, A. R. (2019). Depression prediction system using different methods. (IRJET) International Research Journal of Engineering and Technology, 6(11), https://www.irjet.net/archives/V6/i11/IRJET-V6I11237.pdf

13. Thati, R. P., Dhadwal, A. S, Kumar, P., & Sainaba, P. (2022). A novel multi-modal depression detection approach based on mobile crowd sensing and task-based mechanisms. In Multimodal Tools and Applications. Springer. https://doi.org/10.1007/s11042-022-12315-21215

14. Kumbhar, P. Y., Dube, R., Barbade, S., Kulkarni, G., Konda, N., & Konkati, M. (2021). Depression detection using machine learning. In Proceedings of the International Conference on Smart Data Intelligence. ICSMDI. https://dx.doi.org/10.2139/ssrn.3851975

15. Pandey, R., Pandey, A., Maurya, P., & Singh, G. D. (2023). Prenatal healthcare framework using IoMT data analytics. In The Internet of Medical Things (IoMT) and Telemedicine Frameworks and Applications (pp. 76–104). IGI Global. https://www.igi-global.com/chapter/prenatal-healthcare-framework-using-iomt-data-analytics/313070

16. Priya, A., Garg, S., & Tigga, N. P. (2020). Predicting anxiety, depression and stress in modern life using machine learning algorithms. Procedia Computer Science, 167, 1258–1267. https://doi.org/10.1016/j.procs.2020.03.442

17. Uddin, M. Z., Dysthe, K. K., Følstad, A., & Brandtzaeg, P. B. (2021). Deep learning for prediction of depressive symptoms in a large textual dataset. Neural Computing and Applications, 34(3), 1–24. https://doi.org/10.1007/s00521-021-06426-4

18. Hasni, K. H., & Mohd Yasin, S. (2022). Depression prediction using the classification and regression tree (CART). Journal of Soft Computing and Data Mining, 3(1), 28–33. Retrieved from https://publisher.uthm.edu.my/ojs/index.php/jscdm/article/view/11659

19. Dabhane, S., & Chawan, P. M. (2021). Depression detection on social media using machine learning techniques. International Journal for Scientific Research & Development, 9(4). https://www.irjet.net/archives/V7/i11/IRJET-V7I1116.pdf

20. Sharma, A., & Willem, J. M. (2020). Improving diagnosis of depression with XGBOOST machine learning model and a large biomarkers dutch dataset (n = 11,081). In Sec. Medicine and Public Health Front. Frontiers in Big Data, 3(2), 11–26. https://doi.org/10.3389/fdata.2020.00015

21. Wanqing, X., Zhixiong, X., Manzhu, X., Lizhon, L., Liu, X., Wang, Y., Luo, H. & Cheng, M. (2022). Multimodal fusion diagnosis of depression and anxiety based on CNN-LSTM model. Computerized Medical Imaging and Graphics, 102, 102128. https://doi.org/10.1016/j.compmedimag.2022.102128

22. Ogunseye, E. O., Adenusi, C. A., Nwanakwaugwu, A. C., Ajagbe, S. A., & Akinola, S. O. (2022). Predictive analysis of mental health conditions using ADABOOST algorithm. Paradigmplus, 3(2), 11–26. https://doi.org/10.55969/paradigmplus.v3n2a2

23. Garcia-Ceja E., Riegler M., Jakobsen P., Tørresen J., Nordgreen T., Ketil, J. O., & BerntFasmer, O. (2018). Depresjon: a motor activity database of depression episodes in unipolar and bipolar patients. In MMSys '18: Proceedings of the 9th ACM Multimedia Systems Conference (pp. 472–477). Association for Computing Machinery. https://doi.org/10.1145/3204949.3208125

24. Freund, Y., & Schapire, R. E. (1997). A decision-theoretic generalization of on-line learning and an application to boosting. Journal of Computer and System Sciences, 55(1), 119–139.

25. Chen, T., & Guestrin, C. (2016). XGBoost: A Scalable Tree Boosting System. arXiv: 1603.02754v3

9 Development of EEG-Based Identification of Learning Disability Using Machine Learning Algorithms

Nitin Ahire and R. N. Awale
VJTI

Abhay Wagh
DTE

9.1 INTRODUCTION

A neurological disorder contributes to a learning deficit [1]. Reading, writing, spelling, forming things, and other academic tasks might be challenging for a youngster with a learning disability (LD). It has nothing to do with a child's IQ, according to Ref. [1]. But with the right identification and support, the kids can be helped in realizing their potential and choosing their future careers. Today, a wide range of industries employ machine learning to predict upcoming events. Predicting LDs in children and identifying the root causes are two of the most helpful uses of machine learning. Numerous variables can contribute to learning difficulties [1, 2]. One's ability to understand information is impacted by a LD. Understanding the nature of these disabilities is difficult. However, significant progress has been made in mapping particular brain structures and regions as well as some of the difficulties that some types of learning impairments experience.

A visual, auditory, or motor impairment is not regarded as a LD. Learning impairments do not include conditions like mental retardation, unstable emotions, or cultural challenges [3].

9.1.1 ELECTROENCEPHALOGRAPHY

Electroencephalography (EEG) is one of the newer approaches being investigated for identifying specific brain activation patterns in LDs [1–5]. EEG detects electrical activity present in our brain with the help of metal caps (electrodes) that are placed on our scalp top view by using a 10–20 system as shown in Figure 9.1. Even when

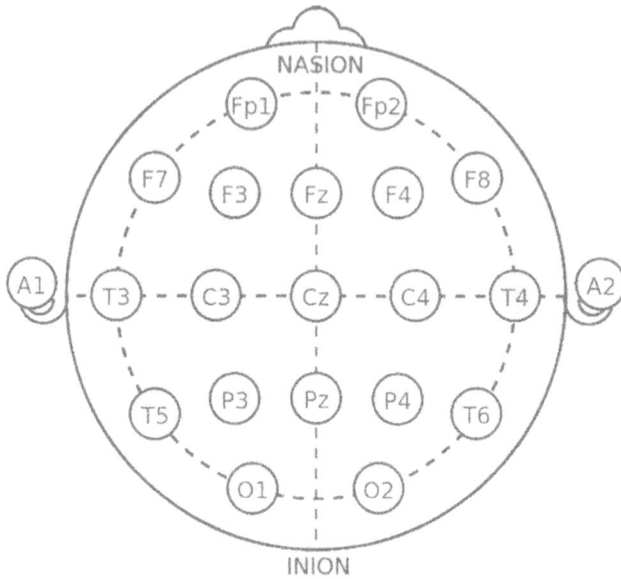

FIGURE 9.1 The 10–20 system top view for 19 electrodes in numbers.

you are sleeping, electrical impulses interact between your brain cells. This activity appears as wavy lines on an EEG recording.

Machine learning is being used to predict future events in a variety of sectors. The most useful areas of machine learning applications are predicting learning problems in children, diagnosing the actual disability, and determining how it may be recognized.

EEG signal features can detect dyslexia, attention-deficit hyper disorder (ADHD), dementia, sleeping problems, depression, and other brain illnesses. EEG headsets monitor brain activity by placing electrodes in an array along the user's or research subject's scalp.

This research comprises creating a machine learning model that will assess a dataset obtained from the EEG signals of different learning difficulties like dyslexia and ADHD and offer accurate results in minutes.

9.2 METHOD

9.2.1 Literature Review

The past studies have shown that those with dyslexia and ADHD have changed patterns of brain activity associated to control individuals. These changes are mostly in the theta and alpha bands, which are related to attention and working memory processes. EEG-based studies have also shown that these variations can be perceived even during rest, suggesting that they are stable features of these disabilities. However, most of the earlier studies used small example sizes and lacked generalizability.

9.2.1.1　Advanced Machine Learning Techniques to Assist Dyslexic Children for Easy Readability [1]

This research focuses on a review of software and hardware choices that can assist dyslexic children. Machine learning techniques used include K Means, K-Nearest Neighbor (K-NN), Adaptive Clustering, LS Algorithm, Support Vector Machine (SVM), and Human Markov Model. To extract speech features and measure accuracy and performance improvement in children, machine learning methods are often used. As a result, the research will concentrate on children aged 5–7 who have trouble reading Hindi words. Machine learning techniques will be employed in the design process to assist dyslexic children. The letters two and three are utilized. With Hindi words as input, the system is trained with a dynamic time wrapping method. Once you have programmed the words into the system, you're ready to go. Words will be taught to a dyslexic child. A child will be given another chance to read if he or she pronounces a word wrong the first time. If the same event occurs three times, the system will repeat the word out, along with an image, to ensure that the child learns the term. For the child, the same lesson will be repeated with new words. To recognize speech in this scenario, machine learning techniques will be used. If the system is functioning properly, the child said the word. For another 20 minutes, the same program will be performed with different words for the child. In this situation, machine learning techniques will be used to recognize speech. The dynamic time wrapping technique yields 90%–100% accuracy if the system is evaluated with the same user, and 30% accuracy if the system is tested with a new user.

9.2.1.2　Machine Learning–Based Learning Disability Detection Using Learning Management System (LMS) [2]

This research focuses on an E-learning system built with Moodle, an open-source learning management system (LMS) that enables tutors and students to work together more efficiently. This method recognizes a couple of student profiles: students with and without LD by using keen courses constructed based on the topic of numerous features of an LD student (non-LD). Several elements of our informal testing approach for collecting learning aspects for dyslexic students are also included in this chapter. The first stage, data collection, is divided into two parts: the first group is the group having age between 8 and 10 years with a smaller number of constraints, and the other one is in the age group of 11–13 years (Class of 6–8) with a greater number of parameters. On the computer, a speech-to-text (STT) conversion was done. Natural Language Processing was used to record the consumers' audio responses (NLP). The Python programming language was used to analyze the responses. To detect whether or not a user has LD, machine learning (ML) is used (dyslexia in this case). The dataset's two classifications are LD and non-LD, and the binary classification is done using ML techniques like SVM and LR (Logistic Regression). The results are shown for both approaches, and a comparison of the datasets generated in another method for collecting parameters using NLP shows that the dataset generated in the final technique is far better and accurate in the case of accuracy level of comparisons. The LR method for ML outperforms SVM when it comes to detection based on the created dataset.

9.2.1.3 A Survey Paper on Learning Disability Prediction Using Machine Learning [3]

This is a survey essay designed to learn about prior work on the subject and to find gaps that are operational for various ML systems. A variety of ML algorithms are used to predict whether a child has an LD. It is easier to deal with such problems if a prognosis is given sooner rather than later. A comparison of existing ML algorithms for a particular dataset can be used to see which one offers the best accurate prediction results. It is impossible to overestimate the importance of the pre-processing stage in preparing data for prediction. Several new strategies can be formalized to help forecast better outcomes.

9.2.1.4 Machine Learning and Dyslexia: Diagnostic and Classification System (DCS) for Kids with Learning Disabilities [4]

In this study, they proposed an automated diagnosis and categorization method. The system was trained using pre-classified data from 857 school children's spelling and reading scores. The twenty-fifth percentile was applied to the scores in order to categorize them. Dyslexia was diagnosed in children who scored below the twenty-fifth percentile, while non-dyslexic children scored above the twenty-fifth percentile. The diagnostic module is a prescreening tool that experts, trained users, and parents can use to detect dyslexia symptoms. The students are divided into two groups in the second module, classification: non-dyslexics and dyslexics with suspicion of dyslexia in spelling and reading. A research study implementation is the third unit. In the final results, it implies that 23% of children were at risk for dyslexia (LD) in the training data and 20.7% in the testing data, with a 98% accuracy.

9.2.1.5 Diagnosis of Dyslexia Using Computation Analysis [5]

This study examines the use of a computer system to diagnose dyslexia, considering people's difficulties in either writing, reading, or speaking. As a result, a computer-based classifier can be built using dyslexia system of measurement approaches. As a result, the Gibson brain skills test will be used, which will consider the effects of working memory, auditory and filmic memory and thought, filmic and hearing sensitivities, inscription and motorized aids, math and time organization, performance, well-being, growth, and personality, and mental skill in people with LDs, particularly reading problems. Computational study with classifiers will be used to analyze the recommended dataset, which contains around 80 archives of children. This computing model was created and used to aid in uncovering underlying issues that may interfere with learning to read and then write, as well as concerns that may cause difficulty with memorized understanding. This approach is utilized to help therapists and paternities know the problem and guide children down the right road to academic achievement.

9.2.1.6 ADHD and Learning Disabilities: Research Findings and Clinical Implications [6]

This study shows what is ADHD and LDs. The authors mentioned that although not every kid that has ADHD feels difficulties in academics, the association or connection between attention deficits and LDs is stronger. The authors explain the same with a sample and a work from the past, and at least 17 distinct trainings were showed between 1978 and 1993 to determine the ratio of children with ADHD who also had LDs in one or more than one subject ranges. They found that occurrence rates in LDs have wide-ranging across educations, fluctuating from a minimum value of 7% to a maximum value of 92%. According to these surveys, about one of every three children with ADHD was delectated to have an LD (mean, 31.1%; median, 27%).

9.2.1.7 Children with Attention-Deficit/Hyperactivity Disorder and Reading Disability: A Review of the Efficacy of Medication Treatments [7]

ADHD is characterized by inattentiveness, operational and instinctual control impairments, and poor decision-making. Among the regular studies practiced by the children having Attention-Deficit/Hyperactivity Disorder are met with the term interpretation or analysis understanding. Similarly, children with Specific Learning Disorders with Impairment in Reading (SLD-R) often meet problems in the development of suitable understanding skills. SLD-R includes dysfunctions in basic visual and auditory processes that result in difficulties with decoding and spelling words. There has been inadequate practical exploration of the success of participation in recovering the understanding or reading capacity of kids with both Attention-Deficit/Hyperactivity Disorder and specific learning disorders with impairment in reading.

9.2.2 LIMITATIONS OF EXISTING SYSTEMS

The papers referred to above have the following limitations: Behavioral elements of members during homogenous tests, such as writing, reading, or working memory, are examined by psychologists in traditional dyslexia diagnosis procedures. Although ADHD-suffering people have problems with visual attention, they are identified by their low scores on these exams. However, because symptoms differ between people, these procedures are generally time-consuming and useless for a wide group of people.

9.2.3 PROBLEM STATEMENT

Identify and evaluate the most commonly practiced techniques used to classify the EEG data associated with LDs, and to evaluate the accuracy for various ML algorithms, and identify which model is well suited for the specified dataset.

9.2.3.1 Dyslexia

The primary step is to identify which procedure to conduct for a user survey and get information from them. In traditional dyslexia diagnosis techniques, psychologists assess participants' behavior throughout uniform exams, such as writing skills along with phonological awareness and working memory [1, 5]. Low results on these tests are used to identify dyslexics. However, because people's symptoms vary, these methods are frequently time intense and unsuccessful for a large group of people. As a result, academics are increasingly turning to ML techniques, which are less time intense and low-cost [8–12].

9.2.3.2   Attention-Deficit Hyper Disorder

Patients suffering from ADHD may also have trouble focused on a solo task or sitting ideal for long time. Anyone of any age can be affected by ADHD. One of the widely studied matters in small and teen-age children's cerebral well-being is ADHD. The cause of the disease is still a mystery even after many years of study. There are indications that ADHD is inherited. It is an organic disorder that is based in the brain. Children with ADHD have reduced levels of dopamine, a neuro-transmitter that is one of the brain's biochemicals [6, 7, 13–15, 17].

9.2.4   PROPOSED SYSTEM

9.2.4.1   Data Collection

It was not anticipated that we would create our dataset utilizing a sample of EEG signals from patients who were suffering from LDs because of the pandemic condition. Instead, we look for a source of online benchmark data. The first dataset contains data from kids who are dyslexic and those who are not, and it is accessible through Skit Learn. The dataset we have chosen is an impartial dataset of 28 dyslexic and 21 non-dyslexic individuals for dyslexia identification (25 male and 24 female). Memory, vocabulary, speed, visual discrimination, and audio discrimination features are chosen as test objectives for differentiating between dyslexic and non-dyslexic people.

The second dataset for ADHD and non-ADHD students was submitted by Ali Motie Nasrabadi [16]. It consists of unbiased data of 61 children with ADHD and 60 healthy controls (boys and girls, aged 7–12) on IEEE data port. For ADHD identification, children with and without ADHD are classified based on EEG recordings. The data are divided into four sections: ADHD part 1, ADHD part 2, Control part 1 and Control part 2. Each portion is made up of several mat files, each of which corresponds to a person's EEG data.

In the performance of mental math counting ability, the number of cartoon characters was exposed to the kids and they were requested to count the number of cartoon characters present in the shown pictures. The number of characters in each copy was arbitrarily selected from 5 to 16 in numbers, and the magnitude of the pictures was big enough to be just noticeable and numerable by these kids. The sampling of EEG recording through this mental pictorial job was reliant on the child's counting performance (i.e., visual attention and response speed).

9.2.4.2   Pre-Processing, Feature Extraction and Feature Selection

The dataset must be preprocessed and screened previously applying ML algorithms. This entails converting the data into a number or qualitative/textual format. To locate useful attributes and eliminate nulls, pre-processing is utilized. Following the initial pre-processing step, the feature elimination/extraction technique for relevant features is recognized, and a variety of standards are assigned. Here each frequency channel band is a feature, and the threshold value of each wave is its respective frequency as stated.

We accomplish three types of component analysis on this dataset, and the outcomes are specified later in this paper:

Data gathering	→	Pre-processing and feature extraction	→	System training and classification	→	Performance evaluation

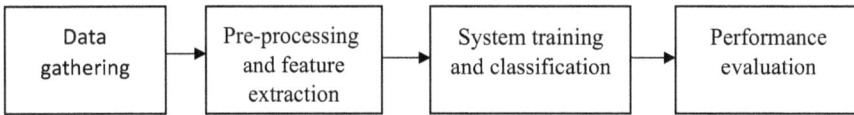

FIGURE 9.2   Proposed system.

- Principal component analysis
- Independent component analysis
- Linear discriminant analysis

9.2.4.3   System Training and Classification

ML algorithms such as Random Forest (RF) Classifier, Decision Tree Classifier, Linear Regression, XGBoot (XGB) Classifier, SVM and K-NN are employed in the component analysis (Figure 9.2).

9.2.4.4   Performance Evaluation

In the performance evaluation, Python-based tools are employed. In this scenario, accuracy score is utilized to assess the performance of ML based on detection of LD systems.

Different ML algorithms are applied to the database to identify the accuracy of their results while classifying problems associated with LDs or non-LDs subjects.

9.3   RESULTS

The results showed that the RF, DT, and XGB models performed best with an accuracy of 98% in identifying individuals with dyslexia and 83% in identifying individuals with ADHD using SVM. The EEG features that contributed most to the classification were those related to theta and alpha band power, which are associated with cognitive processes such as attention and working memory.

    A. **Dyslexia**: The following results give the accuracy score for different ML-based algorithms (Figure 9.3)
    B. **ADHD**: The following results give the accuracy score for different ML-based algorithms (Figure 9.4).
    C. **Dyslexia:** Figure 9.5 shows the topographic map of a child diagnosed with dyslexia and a normal child.
    D. **ADHD**: Figure 9.6 shows the topographic map of a child diagnosed with ADHD and a normal child.

9.4   DISCUSSION AND CONCLUSIONS

9.4.1   Discussion

The basic idea of our project is to increase the accuracy of the LD assessment and reduce the time used for LD assessment. ML is used in a wide variety of fields and applications where certain outcomes have to be predicted.

MODEL ACCURACY SCORE FOR DYSLEXIA

■ Accuracy   ▨ Number of EEG Channels

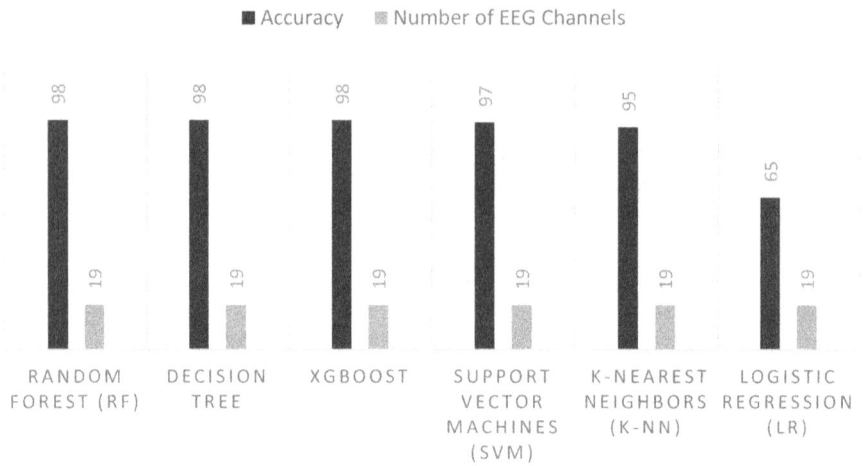

FIGURE 9.3   Model accuracy score for dyslexia.

MODEL ACCURACY SCORE FOR ADHD

■ Accuracy   ▨ Number of EEG Channels

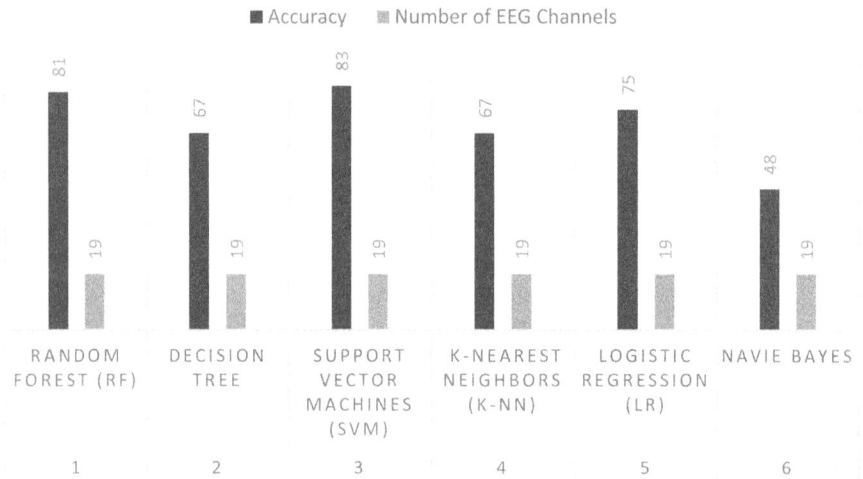

FIGURE 9.4   Model accuracy score for ADHD.

9.4.2   INNOVATION

ML outcomes allow software applications to learn and become more accurate in predicting outcomes. A comparison of ML algorithms is used for a particular dataset to discover which one produces accurate prediction results. We tried to find which

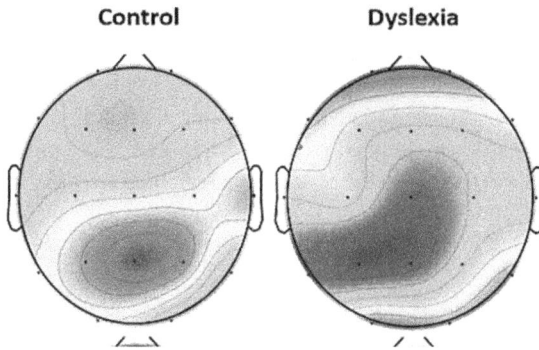

FIGURE 9.5   Topographic map of dyslexia children.

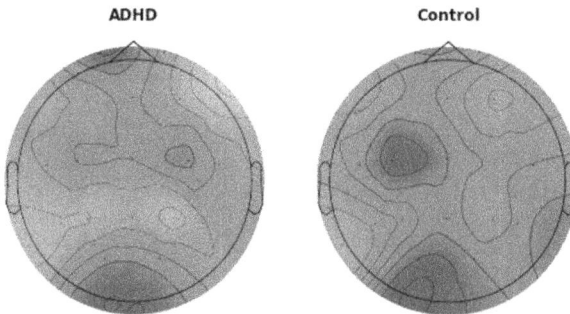

FIGURE 9.6   Topographic map of ADHD children.

ML algorithm works best for identifying a specific disability. After the outcomes obtained, it is understood in what way inventively the ML models classify dyslexia and ADHD datasets.

By reducing features using various feature elimination techniques such as variance threshold, correlation matrix and recursive elimination, we observed no significant increase in the test accuracy. Feature generation of degree 2 was performed on the 19 columns leading to a total of 210 columns, out of which best 30 were selected. Even this proved useless as the final accuracy only fallen. After selecting the best 8 features out of the 19, only 61% of accuracy was obtained, which proves that all of the features are substantially correlated to the output and they cannot be ignored.

9.4.2.1   Inferences Made for Tomographic Graphs

From Figure 9.5, topographic mapping of dyslexic brain's electrical activity revealed four discrete regions of difference between the two groups involving both hemispheres, left more than right. Aberrant dyslexic physiology was not restricted to a single locus but was found in much of the cortical region ordinarily involved in reading and speech.

From Figure 9.6, we can understand that visual cortex regions of both ADHD and normal children were activated. In a normal (control) child, we can see the left side of the head (central lateral sensory motor area) (F3–Fpl) lights up. This indicates that the child is responding to the visual stimulus. However, for a child with ADHD, this region doesn't light up. This is because visual attention is one of the deficits in ADHD children; therefore, they are slower to respond to visual stimulus in comparison to normal children.

9.4.3 CONCLUSION

The result shows that the accuracies of RF, Decision Tree, and XGB algorithms are 98% in the identification of dyslexia diagnosis, whereas SVM gives an accuracy of 83% in the identification of ADHD diagnosis.

After performing feature reduction on the given dataset to reduce its size, it is understood that the accuracy levels go on decreasing, as all the features are highly corrected to each other; hence, feature reduction technique is not useful here.

The tomographic graphs give a clear graphical representation between control and un-controlled subjects.

9.4.4 FUTURE SCOPE

Here we have considered accuracy as the only parameter to identify LDs. In future, we should consider precision, recall and $F1$-score along with the accuracy score for better results:

$$\text{Accuracy} = \frac{\text{TP} + \text{FP}}{\text{TP} + \text{TN} + \text{FP} + \text{FN}} \tag{1}$$

$$\text{Precision} = \frac{\text{TP}}{\text{TP} + \text{FP}} \tag{2}$$

$$\text{Recall} = \frac{\text{TP}}{\text{TP} + \text{FN}} \tag{3}$$

$$F1\text{-Score} = 2 * \frac{\text{Recall} * \text{Precision}}{\text{Recall} + \text{Precision}} \tag{4}$$

Although the technique achieves reasonable accuracy and success rates, there is room for improvement. In this case, data from many bases (e.g., pictures, text, games, and scans) may be integrated for improvement and the prediction models' performance. Text as comprehension, arithmetic, including calculation and problem-solving, and written expression, including writing, predicting, and structuring games, which provide a gesture-based interface, computer-based video-game-type tests like word identification and word attack that measure children's reasoning attributes while they are using their motorized skills to relate to the game.

9.4.5 Declarations

Ethical approval: We will conduct ourselves with integrity, fidelity and honesty. We will openly take responsibility for my actions, and only make agreements, which we intend to keep. We will not intentionally engage or participate in any form of malicious activities that can cause harm to another person or animal.

Competing interests: The authors did not have any competing interests.

Authors' contributions: The first draft of the manuscript was written by Nitin Ahire, and all authors commented on previous versions of the manuscript. All authors read and approved the final manuscript.

Availability of data and materials: Both the datasets are freely available on following links:

Dyslexia dataset Link: Predicting Risk of Dyslexia – PLOS ONE|Kaggle

ADHD dataset is on IEEE data port link: https://ieee-dataport.org/open-access/eeg-data-adhd-control-children

REFERENCES

[1] Geeta, A, & Priyadarshini, J. (2020). Advanced machine learning techniques to assist dyslexic children for easy readability, *International Journal of Scientific &Technology Research*, *9*(03).

[2] Modak, M., Warade, O., Saiprasad, G., & Shekhar, S. (2020). Machine learning based learning disability detection using LMS. In *2020 IEEE 5th International Conference on Computing Communication and Automation (ICCCA)*. IEEE. https://doi.org/10.1109/iccca49541.2020.9250761.

[3] Chakraborty, M. V. (2019). A survey paper on learning disability prediction using machine learning. *International Journal of Information and Computing Science*, 6(5).

[4] Khan, R. U., Chang, J. L. A., & Bee, O. Y. (2018) Machine learning and Dyslexia: diagnostic and classification system (DCS) for kids with learning disabilities, *International Journal of Engineering and Technology*, *3*(18) 97–100.

[5] Al-Barhamtoshy, H. M., & Motaweh, D. M. (2017). Diagnosis of Dyslexia using computation analysis. In *2017 International Conference on Informatics, Health & Technology (ICIHT)*. IEEE. https://doi.org/10.1109/iciht.2017.7899141

[6] DuPaul, G. J., & Volpe, R. J. (2009). ADHD and learning disabilities: Research findings and clinical implications, *Current Attention Disorders Reports*, *1*(4), 152–155. https://doi.org/10.1007/s12618-009-0021-4

[7] Gray, C., & Climie, E. A. (2016). Children with attention deficit/hyperactivity disorder and reading disability: a review of the efficacy of medication treatments, *Frontiers in Psychology*, *7*, 988. https://doi.org/10.3389/fpsyg.2016.00988

[8] Usman, O. L., Muniyandi, R. C., Omar, K., & Mohamad, M. (2021b). Advance machine learning methods for Dyslexia biomarker detection: a review of implementation details and challenges, *IEEE Access*, *9*, 36879–36897. https://doi.org/10.1109/access.2021.3062709

[9] Perera, H., Shiratuddin, M. F., Wong, K. W., & Fullarton, K. (2017). EEG signal analysis of real-word reading and nonsense-word reading between adults with Dyslexia and without Dyslexia. In *2017 IEEE 30th International Symposium on Computer-Based Medical Systems (CBMS)*. IEEE. https://doi.org/10.1109/cbms.2017.108

[10] Hayes, A., Dombrowski, E., Shefcyk, A., & Bulat, J. (2018). Learning Disabilities Screening and Evaluation Guide for Low- and Middle-Income Countries. RTI Press. https://doi.org/10.3768/rtipress.2018.op.0052.1804

[11] David, J. M., & Balakrishnan, K. (2010). machine learning approach for prediction of learning disabilities in school-age children, *International Journal of Computer Applications*, *9*(11), 7–14. https://doi.org/10.5120/1432-1931

[12] Selvi, H., & Saravanan, M. (2019). Diagnosis of Dyslexia students us- ing classification mining techniques, *International Journal of Computer Sciences and Engineering*, *7*(5), 28–33.

[13] Mishra, P. M., & Kulkarni, S. (2013). Classification of data using semi-supervised learning (a learning disability case study), In *International Journal of Computer Engineering and Technology (IJCET)* (Vol 4, issue 4).

[14] Anuradha, J., Tisha, R. V., Arulalan, K. V., & Tripathy, B. K. (2010). *Diagnosis of ADHD using SVM algorithm*. https://doi.org/10.1145/1754288.1754317

[15] Medicalnewstoday.com. (2022). Dyslexia: Symptoms, treatment, and types. [online] Available at: https://www.medicalnewstoday.com/articles/186787what-is-dyslexia

[16] Nasrabadi, A. M., Allahverdy, A. Samavati, M. & Mohammadi, M. R. (2020) EEG data for ADHD/Control children, *IEEE Data Port,* https://dx. doi.org/10.21227/rzfh-zn36

[17] Rojas, G. M., Alvarez, C., Montoya, C. A., De La Iglesia-Vayá, M., Cisternas, J., & Gálvez, M. G. (2018b). Study of resting-state functional connectivity networks using EEG electrodes position as seed. *Frontiers in Neuroscience*, *12*. https://doi.org/10.3389/fnins.2018.00235

10 Deep Learning Approaches for IoMT

Farjana Farvin Sahapudeen and T. Vigneswari
Anjalai Ammal Mahalingam Engineering College

S. Krishna Mohan
E.G.S. Pillay Engineering College

10.1 INTRODUCTION

The Internet of Medical Things (IoMT) enables doctors to treat more patients and save many lives through remote monitoring of patients. IoMT refers to the combination of Internet of Things (IoT) with medical equipment (IoMT). The IoMT enables medical equipment to communicate sensitive data with one another. These developments enable the healthcare sector to interact with and care for its patients more effectively [1]. IoMT demands more advanced techniques to process large volumes of data to enhance the quality of patient care. A robust classification mechanism is required due to the growing number of data as well as the increased dimensions and dynamics of medical data.

Deep learning (DL) is a technique that may give extremely accurate predictions and handle more complex issues that even machine learning (ML) cannot handle. Multiple processing layers in DL allow the computing process to learn the input at several abstraction levels. There are two major key areas where the efficiency of DL could be utilized in IoMT. The major concern is data processing for detecting health anomalies accurately. This is followed by protection of the sensitive medical data from unauthorized users.

DL-driven methodologies are utilized to check and keep track of the vast collection of medical data using the IoT based on wearable devices. It can handle huge amounts of data, including patient information, insurance details, and medical reports, to provide accurate predictions which assist doctors in deciding on the best course of therapy. Figure 10.1 depicts the various applications of DL in IoMT.

Data sharing in IoMT leads to numerous security issues and difficulties with privacy like virus assaults, password cracking, privileged insider attacks, man-in-the-middle attacks, etc. There is a possibility that the attacker will obtain authorized data if any of these attacks on sensitive data are successful. In the IoMT environment, a deep learning neural network (DNN) is assigned to provide an effective and structured approach to categorizing and predicting unanticipated cyberattacks.

DOI: 10.1201/9781003359951-12

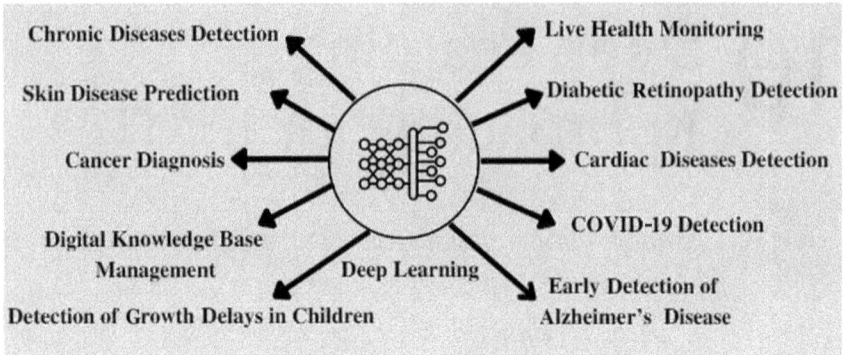

FIGURE 10.1 Deep learning in IoMT.

Auto-encoders, Boltzmann machines, recurrent neural networks, and Convolutional Neural Network (CNN) variants were used to develop different DL methodologies for machine health monitoring. IoMT intelligent medical devices collect data that are utilized in DL systems to identify heart diseases. ECG signals are categorized using a CNN classifier in order to track and find arrhythmia. DL and cloud-based analytics methods are designed to screen patients with chronic diseases and lifestyle diseases. It is considered to use cloud computing and DL models to detect mosquito-borne diseases (MBDs) in their early stages.

Deep neural networks and naive Bayes classifiers are used to determine the stage order of ulcerative colitis. To manage contactless sleep tracking systems and analyze body movement and posture, DL algorithms have been developed. Using abdominal signals, a DL approach may be used in an IoMT context to find the fetal QRS (Q wave, R wave and S wave) complex. The cardiac image–enabled elderly patient monitoring used by DL-IoMT gives a comprehensive, accurate, and thorough view of the emergency situation. Using DNNs, cancers in the brain, liver, lungs, and other organs may be recognized and diagnosed [2].

The key benefits of DL approaches for the IoMT domain are as follows:

- DL IoMT might provide early illness identification, allowing for faster disease treatment.
- DNN methodologies provide fundamental knowledge to non-experts and localize Region of Interest (ROI) in medical images for further investigation.

This chapter aims to analyze three key emphasis areas where DL is used along with IoMT:

- Healthcare monitoring using DL models in IoMT
- Healthcare diagnosis using DL models in IoMT
- Security management in IoMT using DL approaches

10.2   DL FRAMEWORKS FOR IoMT

Huge development in DL makes it possible to provide individualized patient care by analyzing test findings, medical histories, and symptoms of patients' diseases. Drug development is a difficult process, but DL can make it simpler, more efficient, and quicker. DL algorithms are able to forecast complicated tasks like drug discovery, drug characteristics, and the prediction of drug–target interactions. DL expedites, lowers the cost, and simplifies the administration of healthcare. Clinical trials, which are difficult and expensive, are made simpler by it. So, in this research era, investigation and research into DL are very crucial.

DL is a branch of ML techniques which has a data-driven methodology. DL is composed of several layers of complicated nonlinear processing structures. It mimics how the human brain works. Each layer is responsible for segmenting and analyzing a discrete portion of the source image. The DNN comprises many layers of fully connected nodes, each of which improves and classifies the dataset samples depending on the output of the layer before it. The final layer includes the prediction results, whereas the input (first) layer perceives data from the original medical image. Hidden layers are those available in between the aforementioned layers [2]. Different layers of DL models may be used to extract various details from the input images. Each pixel in a medical image contains a separate set of characteristics and attributes.

Medical image analysis and evaluation can be characterized by various neural networks for DL such as deep autoencoder (DAN), CNNs, recurrent neural networks (RNN), U-Net, and generative adversarial networks (GAN). The following sections describe the importance and functionalities of some DL techniques used in IoMT systems. Table 10.1 depicts some commonly used DL algorithms in the IoMT.

TABLE 10.1
Commonly Used Deep Learning Algorithms Applied in IoMT

Deep Learning Algorithm	Working Principle	Potential Applications in IoMT
CNN	The ResNet18 with 17 fully connected layers with learning parameters.	Detection of fetal QRS complexity [3]
RNN	Attention-based Recurrent Neural Network	Multiple diagnosis prediction based on patient records [4]
DAN	Deep autoencoder using spectral data augmentation	Brain tumor diagnosis and classification [5]
U-Net	Modified U-Net with learning hyperparameters	Liver cancer segmentation from abdominal CT scan images [6]
GAN	GAN model with gradient penalty (CCWGAN-GP) added to classifier	Synthetic fetal heart rate monitoring [7]

10.2.1 Convolutional Neural Network (CNN)

One of the most popular and established DL methods in image processing is CNN. The CNN model training requires a large number of labeled data in order to prevent overfitting. Deep CNN focuses on the effectiveness of object recognition and image classification [8]. CNN frequently uses the pre-trained models ResNet, GoogleLeNet, AlexNet, and VGGNet. It requires a thorough grasp of infrastructure and optimization strategies to deploy CNN to numerous applications. Each of the aforementioned CNNs had different convolution layers, pooling layers, dropout approaches, data augmentation, and activation functions. Convolutional layers were used to extract the characteristics of the medical image [9]. The feature maps are then reduced by filters using a pooling layer. Nonlinear parameters are handled using activation layers to activate the neural network. CNN's fully linked layers have the ability to reduce dimensionality and improve the classification. CNN often has several convolutional layers. The first layer of CNN derives some fundamental features such as lines and edges of ROI, and later layers extract some advanced features from previous layer results. For example, if we wish to identify liver cancer, the boundaries of the liver may be recognized by CNN's first layer, whereas the liver's internal cancer patterns will be found by CNN's second layer.

10.2.2 Recurrent Neural Network (RNN)

RNN, a particular type of artificial neural network, forms a loop with a single direction as a result of connections made between the layers. RNNs, as compared to traditional neural networks like feed-forward networks, can employ network memory and produce significant performance increases in image processing [2]. RNN receives the patient's previous visit data as input and learns about the patient's medical history, and then multiple diagnoses will be predicted using recurrent neural networks [4]. In order to progressively infer the medical frames, a bidirectional RNN can be created, with the backward RNN refining the forward RNN's findings in order to enhance the clinical video reconstruction [10].

10.2.3 Deep Autoencoder (DAN)

The training technique for autoencoder DNN is significantly simpler. When perceptual loss is included, it becomes more effective than a GAN [11]. By extracting patterns from the medical images, an auto encoder DNN can restore the images. The broad spatial formats of the output images are compared to the input images to determine how different their content is. Perceptual loss is included in Deep Perceptual Autoencoder to enhance learning.

Individual imaging domains would actually prefer auto-encoded DNN over more generic algorithms since it is a sensitive anomaly detection method, capable of differentiating against similarity in images, and appropriate for a specific imaging modality. Even for experienced doctors, it can be challenging to spot abnormalities on medical imaging since abnormalities can only take up a small portion of the human body or are nearly invisible. Even in those areas also DAN plays a vital role in detection and classification [11]. Even to segregate character-istics and classify images of the tumor-affected brain, Dense layer added DAN approach is used[5].

10.2.4 U-Net

U-Net is a DNN that comprises a decoder (expanding) pathway that permits accurate image localization and an encoder (contracting) pathway to capture the necessary characteristics. U-Net plays a vital role in medical segmentation. Architecture of U-Net follows the "U" structure with an encoder, bottleneck, and decoder. Both encoder and decoder are combinations of convolutional, pooling layers with their respective activation functions [12]. The variants of U-Net such as Attention U-Net, U-Net ++, Residual U-Net, and AWEUNet explored for various cancer image segmentations [13]. A Modified U-Net model is employed for the segmentation of liver regions affected by tumors in abdominal CT scan images, distinguishing them from normal liver tissue [6]. The spatial and channel attention–based blocks may also be added to this U-Net mechanism to improve it and use the encoder path's learning for the decoder path's reconstruction.

10.2.5 Generative Adversarial Network (GAN)

GAN is an unsupervised neural network which is used to develop high-quality, pro-ductive models of stable, healthy, or diseased anatomical shapes. Based on general-ization and discrimination, a deep convolutional adversarial pair learns a hierarchy of semantic representations from objects to the scenario [14]. It makes it possible to process the collected image information as vector arithmetic [15]. During diago-nalization, keeping track of how the disease is progressing or how well a therapy is working must be found and quantified in medical images. The main issue of develop-ing a model of appearance that is successfully representational can be resolved by GANs. This particular sort of learning is an effective algorithm for mapping from the input feature space to the hidden feature space [15]. GANs are applied in the health monitoring of fetal heart rates in clinical observations [7].

10.3 DEEP LEARNING HEALTHCARE MONITORING

Healthcare services by wearable devices and smart phones have turned traditional healthcare services into digital healthcare services. Monitoring vital parameters of humans such as temperature motion, heart pulses, and respiration rate, etc. plays an important role in human health monitoring. There are two categories of wearable sensors:

(i) Physical sensors
(ii) Electrochemical sensors [16].

IoMT health monitoring systems make it possible for real-time behavioral analysis, health monitoring, activities of daily living, elderly care, rehabilitation, and home surveillance, among other things. IoMT systems get signals through wearables, analyze them through DL algorithms, and produce alert messages to respective healthcare professionals and family members in an emergency. Remote health monitoring can limit frequent visits to doctors during recovery periods and provide an emergency note at the needed time. Wearables can be used to continuously monitor health metrics at a lower cost than hospitalization or frequent hospital visits. Many studies have been done to forecast the probability of senior citizens falling and dying as a result of an accident [17]. Screened sensory data may be divided into three categories:

- Falling measurement of elder people
- Screening for falling risk of elders
- Planning for preventive care.

DL in medical imaging mostly assists in the detection and recognition of sickness states or anomalies. Using multiple DL approaches, many physiological parameters, including voice, posture, skin, movements, and environmental circumstances, may be tracked. In pandemic scenarios, drones, robotics, intelligent medical equipment, and other IoMT smart devices are useful for health monitoring. On the battlefield, soldiers who become lost or the health status of their injuries can be tracked using DL techniques.

A platform for quick and inexpensive diagnosis of IoMT-based medical devices available for fetal health monitoring. A portable ultrasound device, a small wearable worn by the mother, and the Internet of Things (IoMT) are used for remote mother and fetus observation throughout pregnancy. The devices will provide monitoring for both the mother and the fetus by capturing data on the mother's health and wellness. Upon finding any irregularity, the system will notify the attending clinicians to take the appropriate action [18]. Objectively assessing the various DL techniques and analyzing the applicability of the CNN algorithm generate the possibility for medical image analysis and disease identification using DL and ML techniques [19]. The real-time DL-enabled IoMT emergency response system affords end users direct access to various medical help providers, including wearable sensors in human health monitoring. Incurable chronic diseases require ongoing observation. Monitoring chronic patients with a DL system can help to avert emergency scenarios.

As shown in Figure 10.2, the Health Monitoring System using Deep Learning (HMSDL) [20] processes data from two sources. The first one is sensory data collected using wireless networks such as Bluetooth or WiFi. The other one is electronic medical records (EMRs). Medical sensors gather critical information from people including electrocardiogram (ECG), heart rate, blood pressure, sweating rate, cholesterol level, body posture, daily activities, electroencephalogram (EEG), sweating level, electromyogram (EMG), and blood sugar for regular health monitoring. The second source, EMRs, contains data regarding medical history, diet history, diabetes details, family health history, and clinical observations (lab reports). After sensing patient data, the proposed system forwards the details for further processing.

Medical databases properly store and manage EMRs and sensory data. The primary goal of the DL algorithm is to forecast the patients' risk factors based on the collected data. Unstructured data can be acquired as signals, images, videos, or text

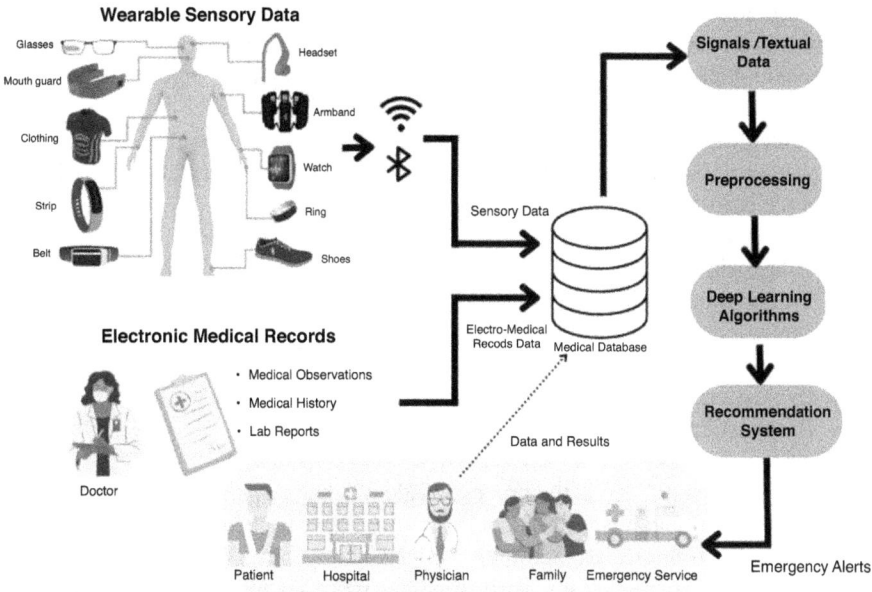

FIGURE 10.2   Structure of health monitoring system using deep learning approach.

records. In order to extract the salient features in the proposed systems, the data must either be noise-free or have undergone some form of preparation. Preprocessing techniques that are appropriate are chosen based on the needs. The filtered data are then sent to a DNN for additional processing after preprocessing. Deep neural networks' projected outcomes are sent to the recommendation system. Depending upon the conditional DNN outcome, the recommendation system should provide the appropriate decision message.

An alert message can be forwarded to the patient, emergency services, hospital, patient's family, and physician for further diagnosis. The physician can refer to further data from the medical server for an immediate understanding of the patient's health history. Based on this kind of approach, even serious health issues like heart disease can be predicted from medical sources. The development of wearable technology makes it possible to provide these sensory inputs for real-time data processing. Smartphones may be utilized in health monitoring systems in addition to Wi-Fi and Bluetooth for precise prediction and remote medical diagnosis.

10.4   DEEP LEARNING HEALTHCARE DIAGNOSIS SYSTEM (DL-HDS)

The IoMT offers wireless-based device operation and communication with medical experts and diagnostic centers by sharing the reports among them for further expert opinion. Despite the fact that the IoMT system for diagnosis has made great development, there are still numerous obstacles to overcome. In the medical field, radiologists or doctors have to review the health images and recommend treatments; hence, medical images account for the majority of healthcare data. The data generated from such IoMT frameworks are noisy, diverse, varied in size, and irregularly gathered. Integrating those data from multiple sources in automated prediction is a major issue in IoMT. Because of this consequence, one of the most significant sources of evidence for clinical examination and medical therapy is medical imaging (MI). The cardinal task in the field of medical image analysis involves image registration. Medical image registration is applied in clinical scenarios like

- Diagnosis of disease
- Proper planning for treatment
- Image-guided therapy
- Surgical interventions
- Evaluation of treatment given to patients
- Prognostication of patients.

Image registration [21–23] is the process of identifying a spatial transformation that maps dual or multiple images to a common coordinate frame such that corresponding anatomical structures are optimally aligned.

The feature extraction framework to gather more features for preprocessing in order to merge medical information from diverse sources is discussed in Ref. [24]. IoMT inputs come in a variety of formats, necessitating a thorough preprocessing phase to remove the noise. Depending on the type of inputs such as signals, audio, text, or images, preprocessing techniques such as median filter, Gaussian filter, skimming, or tokenization have been applied. This step will increase the accuracy of the DL algorithm.

Patients can search for diagnoses of the disease based on their queries using the deep learning–based medical diagnosis system (DLMDS). Our proposed DLMDS contains three important components:

(i) Disease diagnosis prediction module
(ii) Explanation modules that provide an explanation to patients
(iii) Query processing module that processes patient query

Disease diagnosis prediction module will predict the presence and severity of disease based on the DL algorithm. Figure 10.3 displays the structure of healthcare diagnosis using DL approach. Whenever the patient posts a query to our Deep Learning Healthcare Diagnosis System (DLHDS), the query processing module converts the posted query with the keyword extraction process. The output of the query processing module is taken as input to the explanation module, and the corresponding explanation will be replied to from the explanation module. The explanation module is designed over the decision-making rules which are collected from doctor experts. Medical knowledge of experts can be transformed into a knowledge base that contains a set of symptom-disease prediction rules. The explanation module gets support from the diagnosis prediction module and knowledge base for an explanation. This in turn helps the user to get further clarification from the prediction made by the DL algorithm.

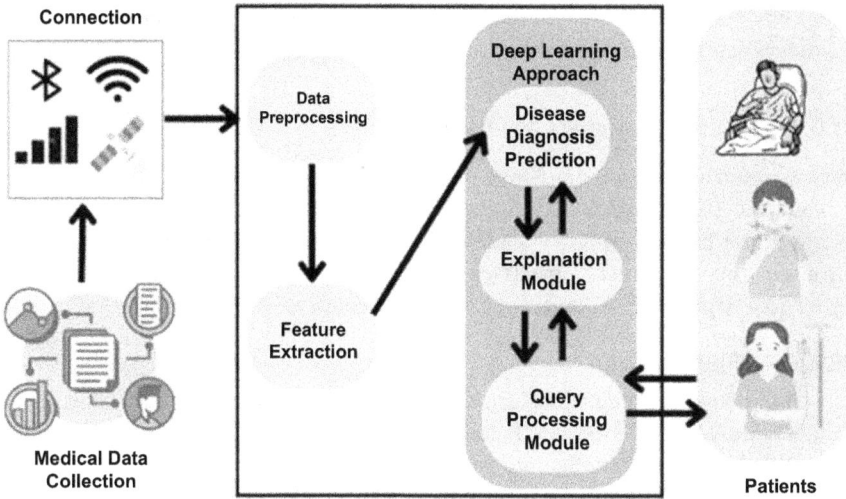

FIGURE 10.3 Structure of healthcare diagnosis using deep learning approach.

10.5 DEEP LEARNING IN SECURITY

The IoMT is an emerging paradigm which aims to provide improved quality of treatment by facilitating customized e-health services to the patient without considering the patient's locality. IoMT devices, which are made up of several core components of the IoMT network, are vulnerable to numerous security risks, and as a result, there may be a serious risk to the safety and privacy of patients.

Whenever sensitive information is attacked, there is a chance that the attacker will modify or gain access to unauthorized data. To avoid these kinds of unwanted attacks, in the IoMT domain, DNN algorithms are used in the development of reliable intrusion detection systems (IDS) to identify and predict unexpected attacks. IoMT's distributed ledger technology (DLT) makes it possible to share health data in a secure and scalable manner. An IDS is responsible for monitoring data traffic for malicious actions and notifies the user upon identifying the same. Particle swarm optimization deep neural network-based (PSO-DNN) models can reliably identify attacks by monitoring data traffic and biometrics data of users.

IoMT systems contain several linked devices, each of which may have built-in vulnerabilities. So it is complex and insufficient to use the current security protection techniques, such as authentication, encryption-decryption, network, and application security mechanisms for maintaining security in IoMT. IoMT components are handled only in unsupervised environments, which make them vulnerable for the attackers to target them. IoMT devices are typically connected by wireless means, where a breach might allow an attacker to obtain sensitive data from the communication network [25].

The main advantages of DL in handling IoMT security issues are as follows:

- Large-scale malware detection with less computational cost
- High accuracy in malware detection
- Predict unexpected attacks in IoMT

If any malicious activity is detected in the IoMT framework, DL-enabled security techniques send a warning message to prompt users to take appropriate action.

10.5.1 SECURITY ESSENTIALS FOR IoMT

Security is a key consideration for healthcare services when using wireless or remote connections. If any information is leaked from the IoMT components, the entire system might be dangerous for patients and healthcare professionals. This section introduces the six primary objectives to ensure security in the IoMT environment. Figure 10.4 depicts the six objectives of IoMT security.

10.5.1.1 Confidentiality

This ensures that private medical information won't be shared or made accessible to unauthorized parties. In the IoMT domain, confidentiality refers to the safeguarding of a patient's medical data that are given to a doctor, therapist, and other healthcare professionals. Unauthorized users may misuse the shared data which may cause harm to the patient.

10.5.1.2 Integrity

This assures the patient's data have not been tampered with or modified without permission. In the IoMT environment, integrity ensures that patient details remain unaltered during transmission. Healthcare departments are paying more attention to the need for integrity while transferring patients' data to their intended destination [26].

FIGURE 10.4   Security objectives of IoMT environment.

10.5.1.3 Authentication

Every time a user logs in to an IoMT system, their identity must be verified by the appropriate authorities. Mutual authentication is the process of determining the source and recipient's originality before transferring data. Due to lack of processing power (Central Processing Unit) and memory storage, lightweight authentication protocols have emerged to handle cryptographic operations.

10.5.1.4 Authorization

Access to network resources and services, and the ability to update the software on IoMT domain components, are restricted to authorized users. Typically, access control is one of the best security techniques that ensure authorization. The authorization ensures that only trusted entities can access and update the patient data stored in IoMT ledgers.

10.5.1.5 Availability

In order to ensure availability in the IoMT domain, continuous monitoring and updating of patient health must be required. In case of emergency, services and data should not be denied to authorized users. IoMT servers and equipment should be available for providing uninterrupted services to patients and healthcare professionals when needed. Table 10.2 displays the research contributions of various security-based mechanisms for IoMT

10.5.2 SECURITY THREATS IN IoMT DOMAIN

Attacks over IoMT components are categorized into attacks on IoMT wearables, IoMT network services, IoMT cloud services, and web-mobile application interfaces.

Due to security threats in installed applications, IoMT wearables are most vulnerable to sensor-based attacks. Attackers are able to access sensitive patient information through wearable sensors (such as an accelerometer, sphygmomanometer, body temperature sensor, and pulse and oxygen sensor) and send it to unauthorized parties [27].

Figure 10.5 describes the potential attacks on IoMT components. Android versions below 4.4 have cross-site scripting vulnerabilities in the web view. The lack of options for the URL bar in mobile web apps makes users unaware of the website they are browsing. Users are cautioned that they cannot enter sensitive credentials as a result [28]. In a Denial of Service (DoS) attack, the attacker uses servers as a means of sending a huge number of network packets to exhaust the available resources. As a result, authorized users cannot access the network services.

Attacks on data exploitation are the process of stealing sensitive information from users and trying to misuse them. The insider attacks can occur when authorized users try to abuse their privileges. Attackers steal critical data and can divulge information to rivals [29]. Phishing attacks have the chance of resulting in identity theft and

TABLE 10.2
Research Contributions of Security-Based Mechanisms for IoMT

S. No.	Authors	Description	Confidentiality	Integrity	Authentication	Authorization	Availability
1	M. Abomhara et al., 2015 [25]	Define threat categories, as well as examine and describe attackers and intrusions against IoT services and devices [25].	✓	✓	✓	✓	✓
2	Hireche et al., 2022 [26]	Current status of security and privacy policies in IoMT domain [26].	✓	✓	✓	✓	✓
3	Sikde et al., 2021 [27]	In-depth investigation of current threats to IoT devices' sensors and solutions designed specifically to safeguard them [27].	✓	✓			✓
4	Allouzi et al., 2021 [30]	Computes the Markov transition probability matrix–based probability distribution of IoMT threats [30].	✓	✓	✓	✓	✓
5	Rajasekhar Chaganti et al., 2022 [32]	Particle swarm optimization deep neural network (PSO-DNN) for implementing an effective and accurate intrusion detection system in IoMT [32].	✓	✓			
6	S. Aneja et al., 2018 [34]	Deep learning approach for device fingerprinting [34].	✓		✓	✓	✓

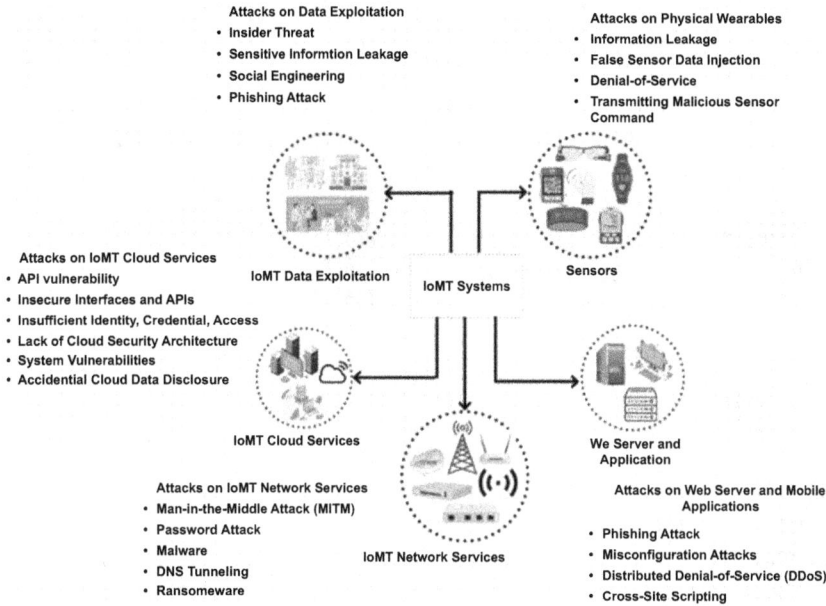

FIGURE 10.5 Potential attacks on IoMT components.

the loss of patient health information. Phishing assaults may have significant consequences for the victim beyond only compromising their health information and also causing harm to their reputation or national security. Social engineering is a threat of manipulating patients and their records to obtain private information such as Aadhar card, PAN card, and bank details [30].

Networking components in IoMT may be under threat of password attacks, malware, DNS tunneling, etc. Malware attacks are designing a virus or harmful code to violate a system's security. It successfully infects a system that includes worms, Trojan horses, malicious mobile codes, virus programs, and other malicious entities. Threats from ransomware include threats to release private information that a person or business doesn't want to share. Threats of this sort demand payment in order to protect their personal information.

Cloud services in IoMT might lead to attacks such as API vulnerability, insecure interfaces, and system vulnerabilities. Attacks on API Authentication Services can take place by taking advantage of the defects in the cloud API that offers strong authentication for cloud infrastructure [31]. Another DoS attack known as an API Exhaustion Attack will send massive amounts of data to overwhelm cloud services. Maintaining data in the cloud might create an opportunity for intruders to use them illegally. A vulnerable cloud system and lacking security features in its architecture might invite the attackers to collapse the system.

10.5.3 Deep Learning in Intrusion Detection

DL approaches are most acceptable and generally used in instruction detection in IoMT networks [32]. Figure 10.6 depicts the process for applying DL models to predict intrusion in IoMT networks using preprocessed medical data. In order to enhance the security along with privacy features of the IoMT system, DL algorithms are used as follows:

- Using network traffic information and patient health records, deep neural networks can be applied to detect IoMT attacks.
- To enhance the IoMT IDS assault detection, conduct a thorough examination of the detection performance of several DL models.
- Investigate and contrast various DL approaches to eliminate inconsistencies between them and to produce hybrid models that combine the best features of each.

Approaches to intrusion classification/prediction may include deep supervised neural network (NN), unsupervised NN, hybrid NN, and reinforcement NN learning. Preprocessed medical data are fed in the process of training the DNN. Following that, supervised DL techniques provide predictions based on the mapping between an example and a specific label or category. IoMT includes a wide range of components, network connections, protocols, many end users, and a huge amount of information. It might be difficult to locate or anticipate vulnerabilities inside an IoMT.

Therefore, rather than securing each individual layer or system, it is obvious to investigate the entire IoMT system and its data flow. The following three major issues should be the focus of intrusion detection in the IoMT system:

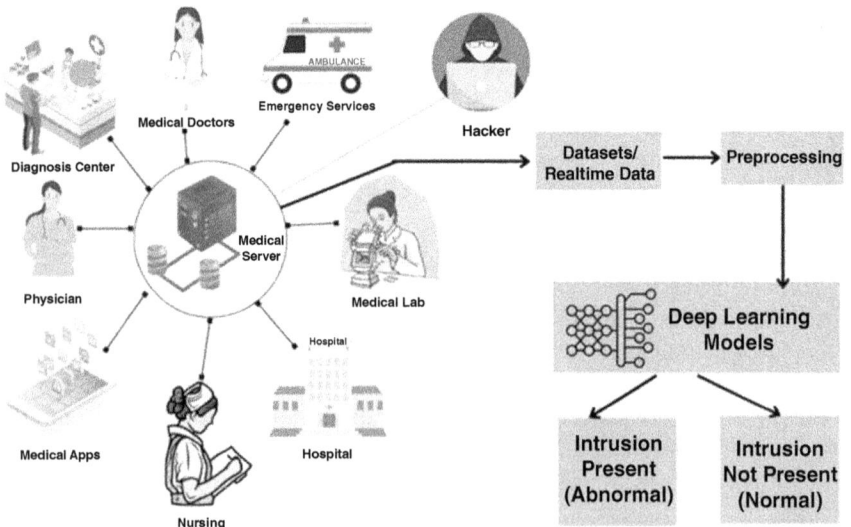

FIGURE 10.6 General architecture of IoMT–deep learning (DL) intrusion detection system.

- Identifying each IoMT device's uniqueness through classification
- IoMT device fingerprint extraction
- Examining IoMT network behaviors.

10.5.3.1 Deep Learning Device Identification

In the past, network devices might have used MAC addresses, static serial numbers, or other fixed identifiers to identify devices. These might be misleading or being abused by an attacker. When training DL models, a large feature set makes it simple to identify the type of attached component in an IoMT network. In order to identify a device, DL approaches collect information from the signals and traffic generated by the connected devices. Deep CNNs can extract relevant attributes to identify the capture device [33], and a Support Vector Machine (SVM) can forecast the camera model.

10.5.3.2 Deep Learning for Service Fingerprint Extraction

Due to constant changes in IoMT networks, it is challenging to preserve static service fingerprints of the connected devices. As a result, it is increasingly crucial to building dynamic service fingerprints of heterogeneous devices. Devices may now be identified based on their services using DL that indirectly creates dynamic service fingerprints for them [34]. District packet inter-arrival times for IoMT devices are used by CNN to extract features that aid in device identification.

10.5.3.3 Deep Learning for Device Integrity

The number of devices connected in IoMT increases the possibility of direct or indirect attacks. When untrusted third parties access IoMT components, Trojan attacks pose a serious security risk. To identify anomalies in systems being attacked by Trojans, a basic neural network can be tuned. Artificial neural networks are deployed to locate the variance in chip power usage where the Trojan is present. Principal Component Analysis (PCA) is used for feature extraction from a self-organizing feature map (SOFM). In order to disseminate information with a sort of security issues, experts have also developed an IoT architecture connected with a distributed digital ledger [35]. To avoid potential interoperability problems in a sensor network, even an analysis of the sensors' API through a semantic graph representation is employed. This provides a tool to enable the coordination of the sensors [36].

10.5.3.4 Deep Learning in Protecting Healthcare Records (HCRs)

Data may be safeguarded using context-aware access control techniques. Context-aware access control policies in IoMT improve the security of medical records [37]. With hybrid logical security architecture, HCR may be safeguarded in terms of registration, authentication, and data security [38]. The construction of a system that can prevent unwanted access is required to protect the IoMT system from unauthorized access. To restrict unauthorized access to medical records, an IoMT gateway with a firewall system can be deployed [39]. With the rapid development of Internet technologies and the widespread use of smart devices like smartphones, laptops, and tablets, security has become a difficult problem for researchers. To safeguard HCRs and increase their security, several DL techniques have been found and implemented [40].

10.6   PERFORMANCE METRICS

The efficiency of DL algorithms in various applications of IoMT is determined with the performance parameters. The performance metrics used in DL algorithms are precision, accuracy, specificity, $F1$ score, and recall. The True Positive and False Negative are defined as follows:

- **True Positive**: Number of samples that are unhealthy and are classified as unhealthy
- **False Negative**: Number of samples that are unhealthy and are classified as healthy

The algorithm predicts the illness by the presence or absence of features by true positive (TP) and true negative (TN). The false identification of features is measured by False Positive (FP) and False Negative (FN). The precision measures the proportion of actual positive samples with total samples positively classified. Recall calculates the proportion of total positive samples with the total number of positive samples. $F1$ score combines both recall and precision metrics together for effective evaluation. Specificity is a measure of the ratio between healthy patients and the ones who are not affected by any illness.

10.7   CONCLUSION

The IoMT enables medical equipment to communicate sensitive data with one another. These developments enable the healthcare sector to interact with stakeholders more effectively by sharing the medical data among them to provide precise diagnosis and treatment. IoMT demands more advanced techniques to process large volumes of data to enhance the quality of patient care. A robust mechanism is required for analyzing the data due to the growing number of data as well as the increased dimensions and dynamics of medical data. DL is a technique that may give extremely accurate predictions and handle more complex issues. This chapter provided a brief overview of the IoMT framework which involves DL techniques for the most commonly used application fields in IoMT such as healthcare monitoring, healthcare diagnosis, and IDSs.

REFERENCES

[1] Secinaro, S., Calandra, D., Secinaro, A., Muthurangu, V., & Biancone, P. (2021). The role of artificial intelligence in healthcare: a structured literature review. *BMC Medical Informatics and Decision Making, 21*, 1–23.
[2] Lee, J. G., Jun, S., Cho, Y. W., Lee, H., Kim, G. B., Seo, J. B., & Kim, N. (2017). Deep learning in medical imaging: general overview. *Korean Journal of Radiology, 18*(4), 570–584.
[3] Krupa, A. J. D., Dhanalakshmi, S., Lai, K. W., Tan, Y., & Wu, X. (2022). An IoMT enabled deep learning framework for automatic detection of fetal QRS: a solution to remote prenatal care. *Journal of King Saud University-Computer and Information Sciences, 34*(9), 7200–7211.
[4] Suo, Q., Ma, F., Canino, G., Gao, J., Zhang, A., Veltri, P., & Agostino, G. (2017). A multi-task framework for monitoring health conditions via attention-based recurrent neural networks. In *AMIA Annual Symposium Proceedings* (Vol. 2017, p. 1665). American Medical Informatics Association.

[5] Nayak, D. R., Padhy, N., Mallick, P. K., & Singh, A. (2022). A deep autoencoder approach for detection of brain tumor images. *Computers and Electrical Engineering, 102*, 108238.

[6] Ayalew, Y. A., Fante, K. A., & Mohammed, M. A. (2021). Modified U-Net for liver cancer segmentation from computed tomography images with a new class balancing method. *BMC Biomedical Engineering, 3*, 1–13.

[7] Zheng, M., Li, T., Zhu, R., Tang, Y., Tang, M., Lin, L., & Ma, Z. (2020). Conditional Wasserstein generative adversarial network-gradient penalty-based approach to alleviating imbalanced data classification. *Information Sciences, 512*, 1009–1023.

[8] Wang, J., Zhu, H., Wang, S. H., & Zhang, Y. D. (2021). A review of deep learning on medical image analysis. *Mobile Networks and Applications, 26*, 351–380.

[9] Zhu, N., Najafi, M., Han, B., Hancock, S., & Hristov, D. (2019). Feasibility of image registration for ultrasound-guided prostate radiotherapy based on similarity measurement by a convolutional neural network. *Technology in Cancer Research & Treatment, 18*, 53303381882196

[10] Cheng, Z., Lu, R., Wang, Z., Zhang, H., Chen, B., Meng, Z., & Yuan, X. (2020, November). BIRNAT: bidirectional recurrent neural networks with adversarial training for video snapshot compressive imaging. *In Computer Vision-ECCV 2020: 16th European Conference, Glasgow, UK, August 23–28, 2020, Proceedings, Part XXIV* (pp. 258–275). Springer International Publishing.

[11] Shvetsova, N., Bakker, B., Fedulova, I., Schulz, H., & Dylov, D. V. (2021). Anomaly detection in medical imaging with deep perceptual autoencoders. *IEEE Access, 9*, 118571–118583.

[12] Ronneberger, O., Fischer, P., & Brox, T. (2015). U-net: convolutional networks for biomedical image segmentation. In *Medical Image Computing and Computer-Assisted Intervention-MICCAI 2015: 18th International Conference, Munich, Germany, October 5–9, 2015, Proceedings, Part III 18* (pp. 234–241). Springer International Publishing.

[13] Oktay, O., Schlemper, J., Folgoc, L. L., Lee, M., Heinrich, M., Misawa, K., & Rueckert, D. (2018). Attention u-net: Learning where to look for the pancreas. *arXiv preprint arXiv:1804.03999.*

[14] Schlegl, T., Seeböck, P., Waldstein, S. M., Schmidt-Erfurth, U., & Langs, G. (2017, May). Unsupervised anomaly detection with generative adversarial networks to guide marker discovery. In *Information Processing in Medical Imaging: 25th International Conference, IPMI 2017, Boone, NC, USA, June 25–30, 2017, Proceedings* (pp. 146–157). Springer International Publishing.

[15] Radford, A., Metz, L., & Chintala, S. (2015). Unsupervised representation learning with deep convolutional generative adversarial networks. *arXiv preprint arXiv:1511.06434.*

[16] Nasiri, S., & Khosravani, M. R. (2020). Progress and challenges in fabrication of wearable sensors for health monitoring. *Sensors and Actuators A: Physical, 312*, 112105.

[17] Sun, R., & Sosnoff, J. J. (2018). Novel sensing technology in fall risk assessment in older adults: a systematic review. *BMC Geriatrics, 18*, 1–10.

[18] Pandey, R., Pandey, A., Maurya, P., & Singh, G. D. (2023). Prenatal healthcare framework using IoMT data analytics. In *The Internet of Medical Things (IoMT) and Telemedicine Frameworks and Applications* (pp. 76–104). IGI Global.

[19] Pandey, R., Sahai, A., & Kashyap, H. (2022). Implementing convolutional neural network model for prediction in medical imaging. In *Artificial Intelligence and Machine Learning for EDGE Computing* (pp. 189–206). Academic Press.

[20] Ali, F., El-Sappagh, S., Islam, S. R., Kwak, D., Ali, A., Imran, M., & Kwak, K. S. (2020). A smart healthcare monitoring system for heart disease prediction based on ensemble deep learning and feature fusion. *Information Fusion, 63*, 208–222.

[21] Fu, Y., Lei, Y., Wang, T., Curran, W. J., Liu, T., & Yang, X. (2020). Deep learning in medical image registration: a review. *Physics in Medicine & Biology, 65*(20), 20TR01.

[22] Chen, X., Diaz-Pinto, A., Ravikumar, N., & Frangi, A. F. (2021). Deep learning in medical image registration. *Progress in Biomedical Engineering, 3*(1), 012003.

[23] Haskins, G., Kruecker, J., Kruger, U., Xu, S., Pinto, P. A., Wood, B. J., & Yan, P. (2019). Learning deep similarity metric for 3D MR-TRUS image registration. *International Journal of Computer Assisted Radiology and Surgery*, *14*, 417–425.

[24] Shivanna, D. B., Stephan, T., Al-Turjman, F., Kolhar, M., & Alturjman, S. (2022). IoMT-based automated diagnosis of autoimmune diseases using multistage classification scheme for sustainable smart cities. *Sustainability*, *14*(21), 13891.

[25] Abomhara, M., & Køien, G. M. (2015). Cyber security and the internet of things: vulnerabilities, threats, intruders and attacks. Journal of Cyber Security and Mobility, *4*, 65–88.

[26] Hireche, R., Mansouri, H., & Pathan, A. S. K. (2022). Security and privacy management in Internet of Medical Things (IoMT): a synthesis. *Journal of Cybersecurity and Privacy, 2*(3), 640–661.

[27] Sikder, A. K., Petracca, G., Aksu, H., Jaeger, T., & Uluagac, A. S. (2021). A survey on sensor-based threats and attacks to smart devices and applications. *IEEE Communications Surveys & Tutorials, 23*(2), 1125–1159.

[28] Mutchler, P., Doupé, A., Mitchell, J., Kruegel, C., & Vigna, G. (2015, May). A large-scale study of mobile web app security. In *Proceedings of the Mobile Security Technologies Workshop (MoST)* (Vol. 50). Association for Computing Machinery.

[29] Shibli, A., Habiba, U., & Masood, R. (2013). Intrusion detection system in cloud computing: challenges and opportunities. In *Conference Paper-December*. IEEE.

[30] Allouzi, M. A., & Khan, J. I. (2021). Identifying and modeling security threats for IoMT edge network using Markov chain and common vulnerability scoring system (CVSS). *arXiv preprint arXiv:2104.11580*.

[31] Ariffin, M. A. M., Ibrahim, M. F., & Kasiran, Z. (2020). API vulnerabilities in cloud computing platform: attack and detection. *International Journal of Engineering Trends and Technology, 1*, 8–14.

[32] Chaganti, R., Mourade, A., Ravi, V., Vemprala, N., Dua, A., & Bhushan, B. (2022). A particle swarm optimization and deep learning approach for intrusion detection system in internet of medical things. *Sustainability*, *14*(19), 12828.

[33] Bondi, L., Baroffio, L., Güera, D., Bestagini, P., Delp, E. J., & Tubaro, S. (2016). First steps toward camera model identification with convolutional neural networks. *IEEE Signal Processing Letters, 24*(3), 259–263.

[34] Aneja, S., Aneja, N., & Islam, M. S. (2018, November). IoT device fingerprint using deep learning. In *2018 IEEE International Conference on Internet of Things and Intelligence System (IOTAIS)* (pp. 174–179). IEEE.

[35] Gupta, M., Sharma, S., & Sharma, C. (2022). Security and privacy issues in blockchained IoT: principles, challenges and counteracting actions. In *Blockchain Technology* (pp. 27–56). CRC Press.

[36] Pandey, R., Paprzycki, M., Srivastava, N., Bhalla, S., & Wasielewska-Michniewska, K. (2021). *Semantic IoT: Theory and Applications*. Springer Nature, Switzerland.

[37] Kaur, A., Rai, G., & Malik, A. (2018, January). Authentication and context awareness access control in Internet of Things: a review. In *2018 8th International Conference on Cloud Computing, Data Science & Engineering (Confluence)* (pp. 630–635). IEEE.

[38] Batra, I., Verma, S., Malik, A., Ghosh, U., Rodrigues, J. J., Nguyen, G. N., & Mariappan, V. (2020). Hybrid logical security framework for privacy preservation in the green internet of things. *Sustainability*, *12*(14), 5542.

[39] Sharma, S., Manuja, M., & Kishore, K. (2019). Vulnerabilities, attacks and their mitigation: an implementation on internet of things (IoT). *International Journal of* Innovative Technology and Exploring Engineering, *8*(10), 146–150.

[40] Sharma, C., & Sharma, S. (2022). Latent DIRICHLET allocation (LDA) based information modelling on BLOCKCHAIN technology: a review of trends and research patterns used in integration. *Multimedia Tools and Applications*, *81*(25), 36805–36831.

11 Machine Learning and Deep Learning Techniques to Classify Depressed Patients from Healthy, by Using Brain Signals from Electroencephalogram (EEG)

Gosala Bethany
Banaras Hindu University

Gosala Emmanuel Raj
Osmania University

Manjari Gupta
Banaras Hindu University

11.1 INTRODUCTION

A psychiatric or mental disorder is a disturbance in an individual's cognition, emotion regulation, or behavior. These disorders are usually associated with distress or impairment in important areas of functioning like anxiety, eating disorders, split and multiple personality disorder, mood swings, post-traumatic stress disorder (PTSD), bipolar disorder, schizophrenia, and depression. According to the WHO, one in every eight people in the world lives with a mental disorder, and a study done in 2019 revealed that around 280 million people suffer from depression, among which 23 million are children and adolescents [1]. Based on a study from the WHO, by 2030, depression will be the primary cause of ill health in the world. Studying the signals from the brain can help in diagnosing mental illness. The biometric information from the human brain can be measured by signals which are generated from the brain, and these signals reveal about the active or passive mental state. The information from the human brain is processed by millions of brain neurons. To measure these information from the brain, many invasive and non-invasive techniques are available. Electroencephalograph (EEG), magneto-encephalography (MEG), magnetic resonance imaging (MRI), and computed tomography (CT) are a few of the non-invasive techniques. Here, in this study, we will mainly focus on the signals collected from electroencephalography (EEG). These signals have high temporal resolutions and also have low signal-to-noise ratio which is an advantage over other brain signal techniques [2, 3].

Machine learning techniques have benefited diverse domains in our modern society. Deep learning, a subcategory of AI technology, is drastically improving classification accuracy. ML and DL techniques have been applied in almost all fields of research like pattern recognition image classification, voice recognition, computer vision, automatic driving vehicles, medicine, etc., and are also showing excellent performance [4–8]. In one of our previous works, we used supervised machine learning techniques and EEG data for classifying the diseased signals of schizophrenia and healthy control [9].

A feature represents an exclusive property, a recognizable metric, and a functional component obtained from a section of a pattern. Extracting valid features from data is crucial in optimizing the loss of important information embedded in the signal. If optimal features are extracted, they can reduce the number of resources that are required to describe a huge set of data, which in turn reduces the space complexity and computational complexity and may also contribute to increasing accuracy [10]. In our work, we will discuss different ML and DL algorithms in application to MDD. We used logistic regression, support vector machine (SVM), and one-dimensional convolutional neural network (1DCNN) algorithms for the classification, and in this work, we extracted eleven statistical features from the signals.

Our main contributions through this chapter are as follows:

1. We have developed both ML and DL models for the classification of MDD EEG signals from HC by building LR, SVM, and 1DCNN classifiers.
2. We have done the time domain feature extraction by extracting statistical features from the EEG signal to get the relevant features and also to remove the redundant features and reduce the dimension.
3. We investigated the signal collected from three different types of EC, EO, and TASK, and conducted experiments separately and found that the signal which is collected when a subject is doing some tasks gives better results than the EC and EO signals.

The rest of the chapter is organized in the following manner: in Section 11.2, we give the background and related work of the study we have done; in Section 11.3, we give a detailed description of different models that were built and the methodology which we followed during our experiments; in Section 11.4, we discuss the results that we got; and finally in Section 11.5, conclusion of the work and future exploration are provided.

11.2   RELATED WORK

The authors in Ref. [11] have proposed a machine learning-based approach for the diagnosis of major depressive disorder (MDD), and they have used resting-state EEG data of 33 MDD patients and 30 healthy controls for this purpose. The input features such as power and EEG alpha were investigated in the model. They proposed also to reduce the irrelevant features. Z-score standardization is used concerning mean and variance. They developed three classifiers, namely, logistic regression (LR), support vector machine (SVM), and naïve Bayesian (NB) for the classification of MDD signals from a healthy control. In the last step, a 10-fold cross-validation is used for validation of the developed model, and accuracy, sensitivity, and specificity were calculated. Among all classifiers, SVM provided the best results given an accuracy of 98.4%, a sensitivity of 96.66%, and a specificity of 100%.

The authors in Ref. [12] have used multi-model data from EEG, eye tracking, and galvanic skin response (GSR) for classifying MDD patients from healthy control, and they also used multiple machine learning algorithms including Random Forests, Logistic Regression, and support vector machine (SVM) for training and to build a dichotomous classification model. The dataset used in the study is a combination of 201 HC and 144 MDD patients. The best results were acquired by LR with an accuracy of 79.63%, precision of 76.67%, recall value of 85.19% and $f1$ score of 80.70% [13]. Maie Bachmann et al. have built a machine learning classification model by using logistic regression analysis and leave-one-out cross-validation for classifying MDD patients from healthy controls. The EEG signals were collected from 13 medication-free depressive patients and 13 healthy controls whose age and gender were matched with MDD patients. The EEG data are analyzed, and both linear and nonlinear features were extracted. Linear methods like spectral asymmetry index, alpha power variability, and relative gamma power and nonlinear methods such as Higuchi's fractal dimension, detrended fluctuation analysis, and Lempel-Ziv complexity were extracted. After the feature extraction, logistic regression is used for the classification of EEG signals, and the ML model they developed gave the best accuracy of 92% which is achieved by mixing both linear and nonlinear together.

Akar, S. et al. [14] have done a statistical study on MDD patients when they are at rest and emotional state, and calculated various statistical features and did statistical analysis on the signal. They have extracted features like Shannon entropy (ShEn), Higuchi's fractal dimension (HDF), Lempel-Ziv complexity (LZC), Kolmogorov complexity (KC), and Katz's fractal dimension (KFD). They found that three features, namely, KFD, HFD, and LZC values, were more sensitive in detecting EEG complexities than the ShEn and KC values, by using ANOVA.

A computer-aided depression diagnosis system has been proposed for the detection of depression by Sharma, M. et al. [15] using bandwidth-duration localized (BDL), three-channel orthogonal wavelet filter bank (TCOWFB), and EEG signal. They used six-length TCOWFB to decompose the signal into seven wavelet sub-bands (WSBs). The logarithm of the L2 norm (LL2N) of six detailed WSBs and one approximate WSB is used as discriminating features. They developed a hybrid least square support vector machine (LS-SVM) model for classifying normal and depressed EEG signals and the developed model achieved an accuracy of 99.58%.

Mumtaz, W. et al. [16] built a machine learning framework that involves EEG-derived synchronization likelihood (SL) features as input data for the automatic diagnosis of MDD. They built three classification models, namely, support vector machine (SVM), logistic regression (LR), and naïve Bayesian (NB), for classification of MDD patients from healthy controls. They have achieved the best results by SVM classification with 98%, 99.9%, 95%, and 0.97 of accuracy, sensitivity, specificity, and f-measure, respectively.

11.3 MATERIAL AND METHODS

11.3.1 Dataset

The dataset we used in this study is an open-source dataset provided in Ref. [11], which is available at https://figshare.com/articles/EEG_Data_New/4244171. The dataset comprises signals recorded from 34 patients with MDD, ages ranging from 27 to 53 (mean $= 40.3 \pm 12.9$), and 30 healthy subjects with ages ranging from 22 to 53 (mean $= 38.3 \pm 15.6$). EEG data recording involved 20-channel EEG with 10–20 electrode positioning system. The signal data obtained three times from each subject: (i) when their eyes are closed (EO), (ii) when their eyes are open (EO), and (iii) while they are doing some task (TAST). The EEG sampling frequency was set to 256 Hz. A notch filter was applied to reject 50 Hz power line noise. An amplifier was simultaneously used to magnify weak signals. Collected EEG signals were applied to a band-pass filter with cutoff frequencies at 0.1 and 70 Hz. Table 11.1 gives the labeling of signal, and we round off duration of the signal to 5 minutes because of the variation in the duration of the collected signal (Figure 11.1).

11.3.2 Preprocessing

The dataset consists of signals collected while the subject's eyes open (EO), eyes close (EC), and doing some task (TASK). Data collected from several channels are not uniform throughout, and the channels vary from 20 to 22 for different subjects. Since the number of channels used to collect the data is not the same for all samples, it is not easy to create a matrix of uniform shape. So, we made all subject's channels uniform, by removing the extra two channels which are there in 22 channel samples. After making all channels uniform, we removed the noise from

TABLE 11.1
Labeling and Duration of EEG Signal

Labels When Signals Are Taken	Activities Performed by the Subjects	Duration of Activities Performed by Each Subject (Minutes)
Sleep	Eyes closed (EC)	5
Awake	Eyes open (EO)	5
Doing some task	Task	5

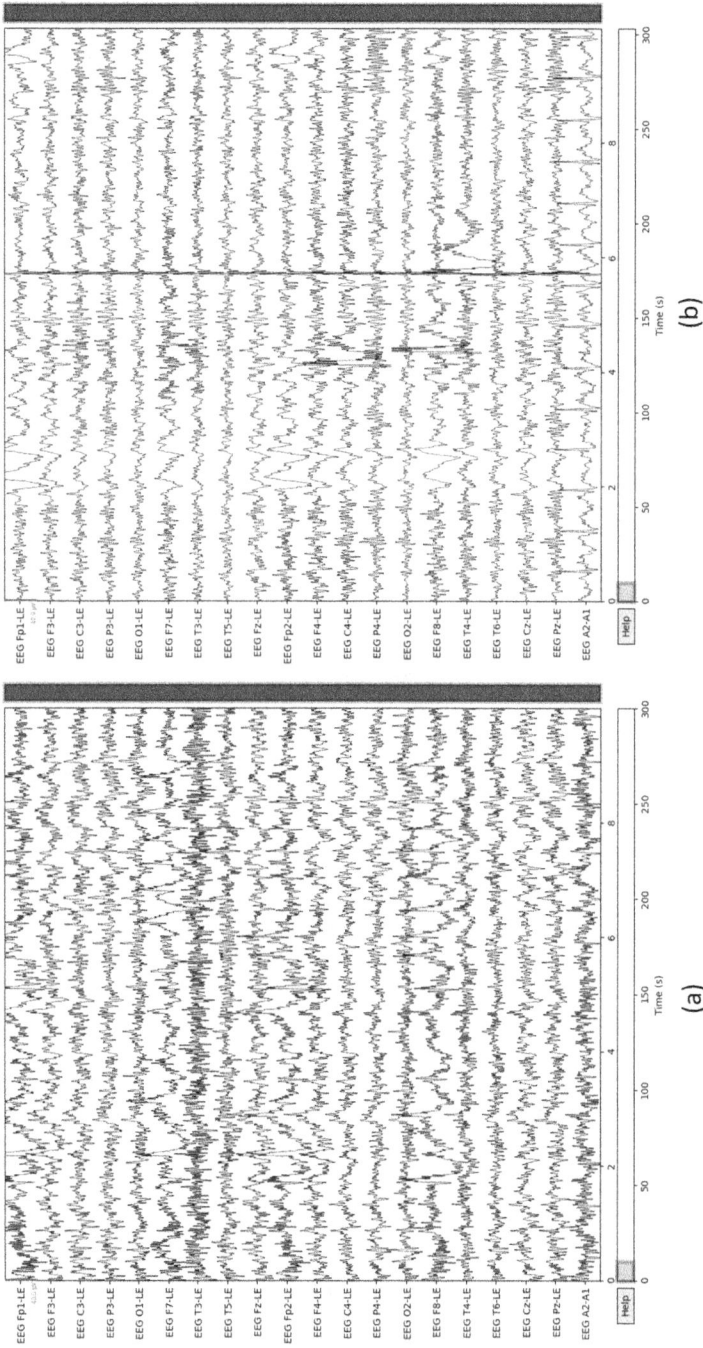

FIGURE 11.1 EEG signals from (a) normal and (b) depressed subjects.

the signal by applying it to the filter. All EEG signals were band-pass filtered with cutoff frequencies at 0.5 and 60 Hz. 0.5 Hz is the low frequency of the signal, and 60 Hz is the high frequency of the signal. Till this, filtering of the data continues, and then we have to convert the data into epochs. MNE-python, a python library, was used to make EEG data into epochs, and "mne.make_fixed_lenth_epochs()" function is used to make epochs of a fixed length of 5 seconds (Figure 11.2).

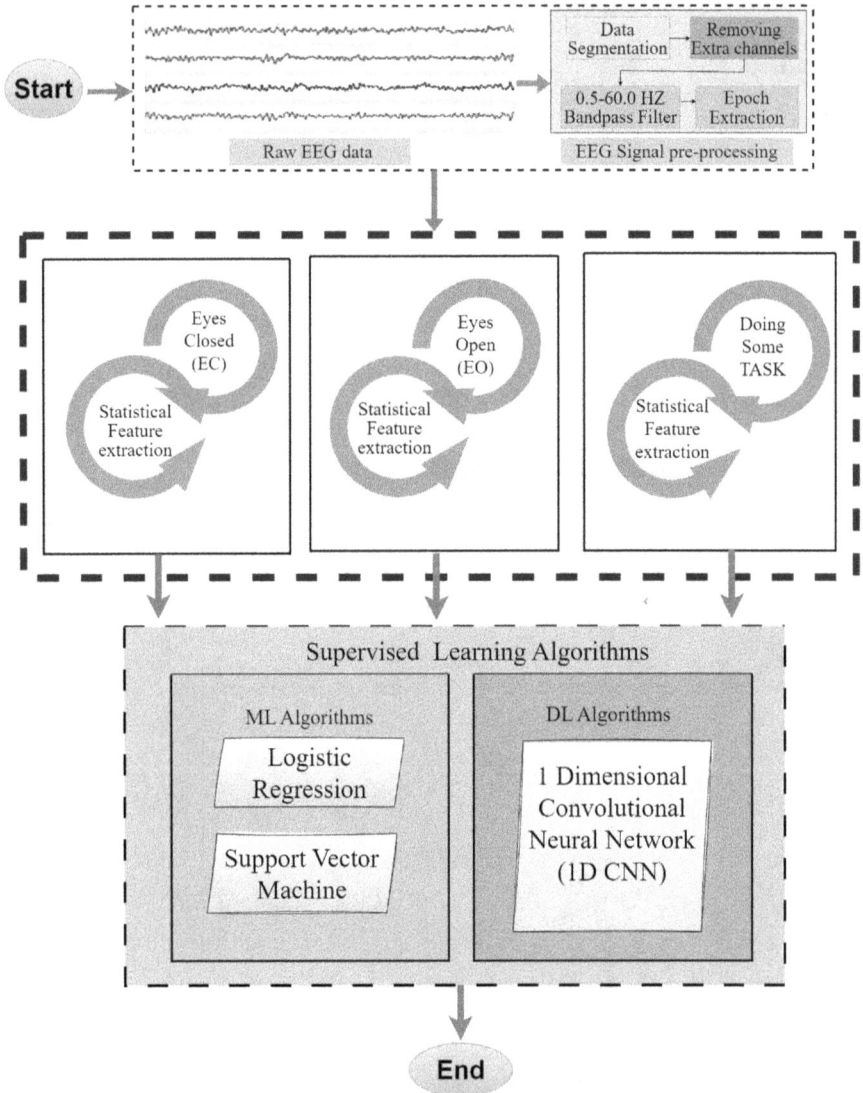

FIGURE 11.2   Experimental flow diagram of the developed models.

11.3.3 FEATURE EXTRACTION

After preprocessing, we have done the feature extraction from the EEG signal before the signal is fed to ML classifiers for classification. Feature extraction is an essential phase in any ML classification algorithm. This step reveals the hidden information about the input signals and their fundamental behaviors. The transformed EEG signals consist of the features that lead the depression recognition. Extracting an optimal feature will reduce the loss of important information in the signals. The 11 statistical features that we have extracted from the data are as follows: mean(x), argmaxim(x), std(x), maxim(x), ptp(x), var(x), minim(x), argminim(x), abs_diff_signals(x), skewness(x), and kurtosis(x).

11.3.4 SUPERVISED LEARNING ALGORITHM

We have built three supervised machine learning classifiers for the classification of diseased patients from a healthy control. The three classifiers we have built are as follows.

11.3.4.1 Logistic Regression

The Logistic Regression (LR) model is the first model that is used in this study to classify signals from depressed patients from healthy ones. Features matrix data are split into the train and test sets on the basis of groups, not on the basis of trials. In this model, the following parameters are considered to train the model. These parameters are considered to train the model after performing hyperparameter tuning operations.

- Clf = LogisticRegression () for classification purpose.
- Gkf = GroupKFold (5) for dividing the number of groups. 'GroupKFold' for splitting the dataset, based on the groups, not on the basis of trials.
- 'GridSearchCV' for Hyper Parameter Tuning.
- Pipeline for scaling with classification purposes.
- StandardScaler is used for scaling the data.
- param_grid = [0.1,0.3,0.5,0.7,1,3,5,7] are some random parameters.
- n_jobs = 12.

11.3.4.2 Support Vector Machine

The second classification model used to develop is SVM. The parameters used to build and train the model are given below:

- Clf = SVC () for classification purposes.
- Gkf = GroupKFold (5) for dividing the number of groups. 'GroupKFold' for splitting the dataset, based on the groups, not on the basis of trials.
- 'GridSearchCV' for Hyper Parameter Tuning.
- Pipeline for scaling with classification purposes.
- StandardScaler is used for scaling the data.
- param_grid = [0.1,0.3,0.5,0.7,1,3,5,7] are some random parameters.
- n_jobs = 12.

11.3.4.3 One-Dimensional Convolutional Neural Network

In this work, we have used various layers like conv1D because the data we are using are one-dimensional, BatchNormalization for normalizing the data to reduce the redundancy, LeakyReLU as an activation function, MaxPool1D for downsample the data, GlobalAveragePooling1D, Dense, Dropout for reducing the dimension and remove irrelevant features, and AveragePooling1D. All these layers were imported from TensorFlow.keras.layers. The detailed description and summary of the CNN we built are shown in Figure 11.3.

11.3.5 PERFORMANCE METRICS

The model's performance evaluation will be done using the performance measure. To measure the performance of the developed models, we used overall accuracy and estimated time as performance metrics. To calculate the accuracy of the model, we require the values of True Positive (TP), True Negative (TN), False Positive (FP), and False Negative (FN). These values will also give the confusion matrix of the model. The counts from the predicted and the actual values can also be found using confusion matrix. Sensitivity and specificity of the model are also determined by using these TP, TN, FP, FN values.

True Positive (TP): When the classification model predicts the same outcome value as the actual value and the value is positive/True, it is called True Positive.

True Negative (TN): When the classification model predicts the same outcome value as the actual value and the value is negative/False, it is called True Negative.

False Positive (FP): When the classification model predicts the outcome value as positive but the actual value negative, it is called False Positive. False Positive is also called a type I error.

False Negative (FN): When the classification model predicts the outcome value as negative but the actual value positive, it is called False Negative. False Negative is also called type II error.

Accuracy: The overall accuracy means the fraction of prediction our model got right from the overall prediction. The overall accuracy of the model is always between 0 and 1. This research will use the following formula to calculate the overall accuracy:

$$\text{Accuracy} = \frac{\text{TP} + \text{TN}}{\text{TP} + \text{TN} + \text{FN} + \text{FP}} \qquad (1)$$

Execution Time: In addition to accuracy, we observed the performance issues in executing the pipeline concerning time. Therefore, we considered the time taken for execution as another metric and calculated the execution time of all nine models. Execution time or run time of a computational program is the amount of time required to execute the complete program without error. The execution time is measured in seconds (execution time is subjected to the number of running processes, the resources available at the time of execution, etc.)

Model :   "sequential"

Layer (type)	Output Shape	Param #
conv1d (Conv1D)	(None, 1278, 5)	305
batch_normalization (BatchN ormalizatio n)	(None, 1278, 5)	20
Leaky_re_lu (LeakyReLU)	(None, 1278, 5)	0
max_pooling1d (MaxPooling1D)	(None, 639, 5)	0
conv1d_1 (Conv1D)	(None, 637, 5)	80
Leaky_re_lu_1 (LeakyReLU)	(None, 637, 5)	0
max_pooling1d_1 (MaxPooling1D)	(None, 318, 5)	0
dropout (Dropout)	(None, 318, 5)	0
conv1d_2 (Conv1D)	(None, 316, 5)	80
Leaky_re_lu_2 (LeakyReLU)	(None, 316, 5)	0
average_pooling1d (AveragePoo ling1D)	(None, 158, 5)	0
dropout_1 (Dropout)	(None, 158, 5)	0
conv1d_3 (Conv1D)	(None, 1586, 5)	80
Leaky_re_lu_3 (LeakyReLU)	(None, 1586, 5)	0
average_pooling1d_1 (AveragePooling1D)	(None, 78, 5)	0
conv1d_4 (Conv1D)	(None, 76, 5)	80
Leaky_re_lu_4 (LeakyReLU)	(None, 76, 5)	0
global_average_pooling1d (GlobalAveragePooling1D)	(None, 5)	0
dense (Dense)	(None, 1)	6

Total params: 651
Trainable params: 641
Non-trainable params: 10

FIGURE 11.3   Architecture of 1DCNN.

11.3.6 EXPERIMENT SETTINGS

This section consists of information about the experimental setup used to compute the various techniques to classify depression using the supervised learning classification algorithm. The experiments were run on the Google Colab [17]. Google Colab provides a machine with 12 GB RAM for basic free version. Detailed specifications of hardware used in this study are mentioned in Table 11.2.

When it comes to software usage, the programming language used is python version 3.7.6 with an anaconda environment of version 1.7.2. The IDE used for the coding part is the Jupiter Notebook, a web-based IDLE. Various software, modules, libraries, and packages of python which are employed to classify depression with the help of supervised learning classifiers are given in Table 11.3 with the used version.

11.4 RESULTS AND DISCUSSIONS

In this section, we will explain the results we got when we conducted the experiments, the models we developed, and the performance of the developed models. We have run a total of nine experiments with the available three types of EEG MDD data (EC, EO, and TASK) by three ML classifiers. We developed a total of three supervised learning models from both ML and DL. Logistic regression and SVM are two machine learning models, and we built 1DCNN, a DL model, for classification.

TABLE 11.2
Hardware Specification of Computation Machine

S. No.	Components	Specifications
1	CPU Model Name	Intel(R) Xeon(R)
2	Available RAM	12 GB (upgradable to 26.75 GB)
3	CPU Family	Haswell
4	CPU Frequency	2.30 GHz
5	No. CPU Cores	2
6	Disk Space	25 GB

TABLE 11.3
Software Specification of Computation Machine

S. No.	Software/Package/Module/Library Used	Version
1	Operating System	Windows 10
2	Python	3.7.6
3	NumPy [18]	1.21.6
4	Pandas [19]	1.0.1
5	Anaconda	1.7.2
6	Scikit-Learn [20]	0.22.1
7	Matplotlib [21]	3.1.3
8	MNE [22]	1.3.0
9	Tensorflow [23]	2.7.0

11.4.1 Results of Logistic Regression

LR is the first supervised machine learning (ML) framework that we built, and the results we obtained are as follows: the highest classification accuracy was obtained by Logistic Regression algorithms, with an accuracy of 73.60% and execution time of 95 seconds for Eyes Closed (EC) dataset; accuracy of 80.50% and execution time of 76 seconds for Eyes Open (EO) dataset; and accuracy of 88.40% and execution time of 104 seconds for TASK EEG dataset.

11.4.2 Results of Support Vector Machine

Support Vector Machine (SVM) is the second model we built for all three different types of signals of the MDD dataset. The results we achieved are an accuracy value of 71.90% and execution time of 180 seconds for Eyes Closed (EC) signals; accuracy of 83.80% and execution time of 187 seconds for Eyes Open (EO) signals; and accuracy of 89.30% and execution time of 521 seconds for TASK EEG data.

11.4.3 Results of Convolutional Neural Network

CNN is the third supervised model which is a deep learning model framework, and the results we got are with the highest classification accuracy of 82.00% and the time taken for execution is 173 seconds in Eyes Closed (EC) EEG signals dataset; accuracy of 83.30% and the execution time of 199 seconds in Eyes Open (EO) dataset; and accuracy of 90.21% and execution time of 756 seconds of TASK dataset.

The evaluation parameter values we got are shown in Figures 11.4 and 11.5. In Figure 11.4, we show the time taken for the execution of each model, and in Figure 11.5, the accuracies we got for each ML and DL model are shown. In Table 11.4, we compare the accuracy of the proposed model with existing state-of-the-art methods.

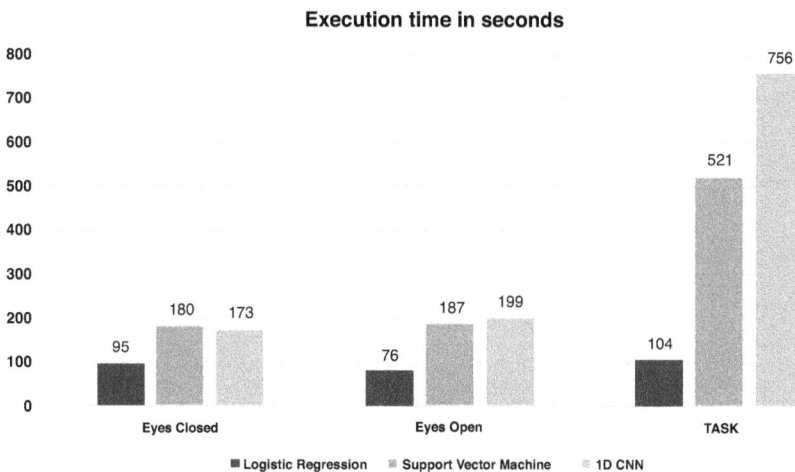

Execution time in seconds

FIGURE 11.4 Execution time of nine developed supervised learning models.

Accuracies of the medel on different signals

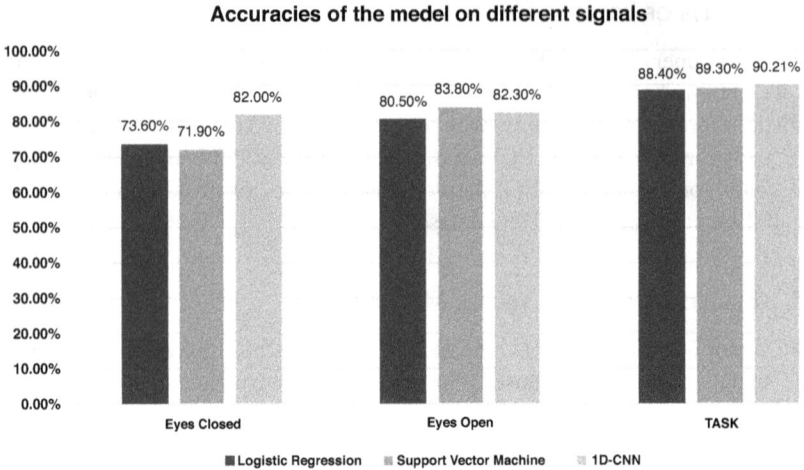

FIGURE 11.5   Accuracy of the nine developed supervised learning models.

11.5 CONCLUSION AND FUTURE WORK

In this section, we give the final remarks of the study we did by creating various ML and DL models for classifying the MDD patient's EEG signal from HC. Through the experiments we have conducted, we have found that the signals which are collected when the subject is doing any task are giving better results than the signals which are collected when the signal is collected from subject when eyes closed and eyes open. Deep learning model 1DCNN gave better results than the traditional machine learning models.

This work can also be extended in various directions in the future by applying various time domain and frequency domain feature extraction methods like Entropy, ANOVA, Auto Regression (AR), ARIMA, SARIMA, wavelet methods like continuous wavelet transform (CWT), discrete wavelet transform (DWT), and wavelet scattering transform (WST). Another direction the study can be extended in is by using the transfer learning. In transfer learning, pre-trained networks' knowledge will be transferred to the model. In this way, a transfer learning model built with a small dataset can also give good results.

TABLE 11.4

Comparison of the Proposed Model with the State-of-the-Art Methods

Authors	Datasets	No. of Electrodes	No. of Participants	Classification Algorithms	Accuracy (%)
Wajid Mumtaz Syed Saad Azhar Ali Mohd Azhar Mohd Yasin Aamir Saeed Malik [16]	The MDD patients met the diagnostic criteria for MDD according to the Diagnostic and Statistical Manual-IV (DSM-IV)	19-Chennai Electrodes	34 MDD patients and 30 healthy subjects	SVM Logistic Regression Naïve Bayesian	98 91.7 93.6
Manish Sharma, Achuth P. V, Dipankar Deb, Subha D Puthankattil, U. Rajendra Acharya [15]	From the Department of Psychiatry, Government Medical College, Kozhikode, Kerala, India	The bipolar channels FP2-T4 (right half) and FP1-T3 (left half)	15 depressed patients and 15 normal subjects	Computer-aided depression diagnosis system using newly designed bandwidth-duration localized (BDL)	99.58
Ding Xinfanga, Yue Xinxin, Zheng Rui, Bi Cheng, Li Dai, Yao Guizhong [12]	A free gazing task was utilized in the present study, following Kellough et al. (2008)	5-Channel Electrodes	144 MDD patients and 204 normal subjects	Logistic Regression	79.63
Wajid Mumtaz, Likun Xia, Syed Saad Azhar Ali, Mohd Azhar Mohd Yasin, Muhammad Hussain, Aamir Saeed Malik [11]	The EEG data acquisition involved vigilance-controlled monitoring during the recordings	19-Chennel Electrodes	33 MDD patients and 30 normal subjects	SVM Logistic Regression Naïve Bayesian	98.4 97.6 96.8
Maie Bachmann, Laura Päeske, Kaia Kalev, Katrin Aarma, Andres Lehtmets, Pille Ööpik, Jaanus Lass, Hiie Hinrikus [13]	The EEG signals were recorded using the Neuroscan Synamps2 acquisition system (Compumedics, NC, USA)	30-Channel Electrodes	13 medication-free depressive outpatients and 13 healthy controls	Single Measure (linear) Single Measure (nonlinear) Combination of two linear measures(li) Combination of two linear measures(non-li)	81 77 88 85
Proposed System	https://figshare.com/articles/EEG_Data_New/4244171	20-Channel Electrodes	34 MDD patients and 30 normal subjects	Logistic Regression Support Vector Machine 1DCNN	EC=73.60 EO=80.50 TASK=88.40 EC=71.90 EO=83.80 TASK=89.30 EC=82.00 EO=82.30 TASK=90.21

ACKNOWLEDGMENTS

1. The author is extremely grateful to University Grants Committee (UGC) for providing the Junior Research Fellowship (JRF) under Maulana Azad National Fellowship for Minorities (MANFJRF), with the award reference number: NO.F.82-27/2019 (SA III).
2. We also acknowledge the Institute of Eminence (IoE) scheme at Banaras Hindu University (BHU) for supporting us.

LIST OF ABBREVIATION

ML	Machine Learning
DL	Deep Learning
AI	Artificial Intelligence
EEG	Electroencephalogram
MDD	Major Depressive Disorder
HC	Healthy Control
EC	Eyes Close
EO	Eyes Open
TASK	Doing Some Task
SF	Statistical Features
SVM	Support Vector Machine
LR	Logistic Regression
1DCNN	One-Dimensional Convolutional Neural Network
NB	Naïve Bayesian
FE	Feature Extraction
MEG	Magneto Encephalogram
MRI	Magnetic Resonance Imaging
BCI	Brain–Computer Interface
TP	True Positive
TN	True Negative
FP	False Positive
FN	False Negative

REFERENCES

1. https://www.who.int/news-room/fact-sheets/detail/mental-disorders#:~:text=A%20mental%20disorder%20is%20characterized,different%20types%20of%20mental%20disorders (Accessed on 14 July 2022).
2. Farnsworth, B. (2018). What is EEG (Electroencephalography) and How Does it Work?. *Imotions. https://imotions.com/blog/what-is-eeg, 8.*
3. Aslan, Z., & Akin, M. (2022). A deep learning approach in automated detection of schizophrenia using scalogram images of EEG signals. *Physical and Engineering Sciences in Medicine, 45*(1), 83–96.
4. Alpaydin, E. (2016). *Machine Learning: The New AI.* MIT Press.
5. Subhash, S., Srivatsa, P. N., Siddesh, S., Ullas, A., & Santhosh, B. (2020, July). Artificial intelligence-based voice assistant. In *2020* Fourth world conference on smart trends in systems, security and sustainability *(WorldS4)* (pp. 593–596). IEEE.

6. Gosala, B., Chowdhuri, S. R., Singh, J., Gupta, M., & Mishra, A. (2021). Automatic classification of UML class diagrams using deep learning technique: convolutional neural network. *Applied Sciences, 11*(9), 4267.

7. Esteva, A., Chou, K., Yeung, S., Naik, N., Madani, A., Mottaghi, A., & Socher, R. (2021). Deep learning-enabled medical computer vision. *NPJ Digital Medicine, 4*(1), 5.

8. Hamet, P., & Tremblay, J. (2017). Artificial intelligence in medicine. *Metabolism, 69,* S36–S40.

9. Gosala, B., Kapgate, P. D., Jain, P., Chaurasia, R. N., & Gupta, M. (2023). Wavelet transforms for feature engineering in EEG data processing: an application on Schizophrenia. *Biomedical Signal Processing and Control, 85,* 104810.

10. Al-Fahoum, A. S., & Al-Fraihat, A. A. (2014). Methods of EEG signal features extraction using linear analysis in frequency and time-frequency domains. *International Scholarly Research Notices, 2014,* 730218.

11. Mumtaz, W., Xia, L., Ali, S. S. A., Yasin, M. A. M., Hussain, M., & Malik, A. S. (2017). Electroencephalogram (EEG)-based computer-aided technique to diagnose major depressive disorder (MDD). *Biomed Signal Process Control, 31,* 108–115.

12. Ding, X., Yue, X., Zheng, R., Bi, C., Li, D., & Yao, G. (2019). Classifying major depression patients and healthy controls using EEG, eye tracking and galvanic skin response data. *Journal of Affective Disorders, 251,* 156–161.

13. Bachmann, M., Päeske, L., Kalev, K., Aarma, K., Lehtmets, A., Ööpik, P., & Hinrikus, H. (2018). Methods for classifying depression in single channel EEG using linear and nonlinear signal analysis. *Computer Methods and Programs in Biomedicine, 155,* 11–17.

14. Akar, S. A., Kara, S., Agambayev, S., & Bilgiç, V. (2015). Nonlinear analysis of EEGs of patients with major depression during different emotional states. *Computers in Biology and Medicine, 67,* 49–60.

15. Sharma, M., Achuth, P. V., Deb, D., Puthankattil, S. D., & Acharya, U. R. (2018). An automated diagnosis of depression using three-channel bandwidth-duration localized wavelet filter bank with EEG signals. *Cognitive Systems Research, 52,* 508–520.

16. Mumtaz, W., Ali, S. S. A., Yasin, M. A. M., & Malik, A. S. (2018). A machine learning framework involving EEG-based functional connectivity to diagnose major depressive disorder (MDD). *Medical & Biological Engineering & Computing, 56,* 233–246.

17. https://colab.research.google.com/ (accessed on 13 December 2022).

18. NumPy documentation – NumPy v1.24 Manual (accessed on 25 November 2022).

19. Pandas – Python Data Analysis Library (pydata.org) (accessed on 20 November 2022).

20. Scikit-learn: machine learning in Python – scikit-learn 1.2.1 documentation (accessed on 25 October 2022).

21. Matplotlib – Visualization with Python (accessed on 17 November 2022).

22. MNE – MNE 1.3.0 documentation (accessed on 7 October 2022).

23. TensorFlow Core|Machine Learning for Beginners and Experts (accessed on 25 October 2022).

12 Dimensionality Reduction for IoMT Devices Using PCA

Rajiv Pandey, Radhika Awasthi,
Archana Sahai, and Pratibha Maurya
Amity University

12.1 INTRODUCTION

There's a lot of buzz right now about the Internet of Things (also known as IoT) and its impact on everything from how we travel and buy to how manufacturers manage inventories. The term "Internet of Things" describes the countless physical objects that have Internet connectivity and actively collect and share data globally. With the development of affordable computer chips and the widespread usage of wireless networks, everything can now be made into an IoT component, from the tiniest pill to the biggest airplane. An IoT network in the healthcare sector allows medical professionals to remotely control devices via software applications and connected infrastructure, all while collecting patient data at key touchpoints, thus giving birth to "Internet of Medical Things" (or IoMT).

The IoMT is a notion that is gaining attention as communication and data storage capability of medical technology increase. It refers to a network of medical equipment that is linked to the Internet. The goal is to seamlessly integrate healthcare with workplace and residential settings. This will improve patient safety, healthcare delivery, and corporate efficiency. Furthermore, novel medical therapies can be developed by harnessing data collected from millions of connected medical devices. As healthcare advances, it becomes more prevalent to connect medical devices to the Internet. Doctors now have immediate access to information about patients' illnesses and medical histories. It also facilitates communication between doctors, nurses, and hospitals. Additionally, Internet accessibility enables medical professionals to educate patients on the proper usage of their medical equipment. Thus, the Internet offers a plethora of options for everyone to improve their daily lives. One of the key advantages of IoMT at the moment is that it addresses the growing concern about healthcare systems that cater to the needs of patients. IoMT devices have enormous potential to assist in dealing with rising healthcare expenditures. Such devices can assist in tracking vital heart performance, monitoring glucose and other bodily systems such as sleep levels, and other activities to sustain chronic issues, if any. Wearables such as fitness trackers, blood glucose monitors, and other medical gadgets such as pain relief devices have also taken the healthcare industry by storm.

DOI: 10.1201/9781003359951-14

IoMT dramatically lowers medical expenditures while enhancing people's quality of life, which benefits health in general. Its significant components include wireless sensors, which are used to monitor a patient's health status remotely, and communication technology, used to transmit the information to healthcare professionals. The first step in creating a smart healthcare system is to realize the potential of the already available technology to give people the best services while also making their lives better.

Artificial intelligence (AI) has piqued the interest of researchers and the biomedical industry due to its ability to analyze enormous volumes of data, deliver accurate results, and control procedures to achieve the best possible result. Reliable machine-algorithm-coordinated outcomes are used to consider a number of aspects, including fairness, explainability, accountability, reliability, and acceptance. Machine learning (ML) is the most extensively used AI technology for generating predictions based on patterns [13]. The pharmaceutical and healthcare sectors have proven the value of AI-based models by increasing the effectiveness of therapeutic drug manufacture, real-time monitoring of health, and prognostic forecasting. AI is already being used in several stages of drug development, from pharmaceutical design to drug screening, and has already demonstrated its potential. In 2020, the "Alphafold" DL model successfully predicted the molecular makeup of a protein from its amino acid sequence, solving a difficulty that had existed for 50 years [17]. A list of specific applications of different AI algorithms and learning techniques in medical literature was published in Ref. [10]. This review suggests that supervised learning may be utilized to learn about healthcare because the outcomes are more therapeutically applicable. AI-based solutions are used to process data and identify abnormalities for specific individuals because the amount of data is large and comes from numerous sources. Similar to this, data gathered from patient-specific health monitoring equipment in hospitals can identify potential emergencies and notify medical personnel. Healthcare system analysis is already being used in some nations, such as Norway and Denmark, to identify treatment errors and ineffective workflow processes. AI is decreasing the workload in the medical field as a consequence, preventing incorrect diagnoses, and saving patients' time and money by cutting out unneeded sessions. Similarly, DL models for breast cancer [2] and pancreatic cancer [18] analyses have also been established. The system may learn normal and abnormal judgments using ML and deep learning techniques, utilizing the data produced by health experts and workers as well as patient input.

For many pregnant women in India, the moment of birth is oftentimes terrifying. Maternal mortality (or MMR) refers to the death of a woman during or after pregnancy, including post-abortion or post-birth periods. Statistically speaking, MMR in the nation decreased from 130 per lakh live births in 2014–2016 to 122 in 2015–2017, and further decreased by 9 points to 113 in 2016–2018. India's MMR dropped to 103 in 2017–2019, compared to a global MMR of 211 [4]. However, the adoption of life-saving medical interventions and measures continues to be low due to expertise, policy, and resource gaps. UNICEF reports that nearly two-thirds of all maternal fatalities are caused by serious complications, which include severe bleeding, infections, high blood pressure throughout pregnancy, difficulties after delivery, and failed abortions [27]. The majority of maternal deaths can be avoided because there are well-established medical strategies to treat or prevent complications. All women need to have access to top-notch care throughout their pregnancies, labors, and postpartum periods. This dire circumstance inspires us to propose a framework, and the concept of minimizing this misery emerges. In these situations, a prompt and affordable diagnosis is crucial so that the doctors can take proactive steps to lessen the severity of the underlying reason. To achieve this, our team of authors have previously suggested a framework for IoMT-based medical equipment. The framework for the equipment is termed 'Prenatal Healthcare System of Remote Mother and Fetal Surveillance via IoMT' [22], and it employs a handheld ultrasound device and a small wearable worn by the mother. By gathering information about the mother's and the fetus's health and wellbeing, the gadgets will serve as a surveillance system for them. The metrics will be sent to a server, which will use the relevant AI/ML module to analyze and forecast the mother's and the fetus's gestational state. Medical professionals caring for specific mothers and fetuses will have access to the central system. The central system will alert attending doctors to take the necessary measures upon detection of any irregularity. However, keeping in mind the importance of accuracy prediction in the equipment, the features seem to be more than what is essentially required for a precise analysis. Thus, we require a tool that minimizes the number of variables in a dataset while retaining the maximum amount of data.

The aim of this chapter is to employ Principal Component Analysis (PCA) [11], a dimensionality reduction technique in ML that is used to reduce data dimensionality for feature extraction. It is a popular unsupervised learning method that has been utilized in many different applications, including data analysis, data compression, and lowering the dimension of data. PCA analysis aids in the reduction or elimination of comparable data that do not even slightly aid in decision-making in the line of comparison. For the analysis, we have created a dataset of parameters based on the fetus' gestational age, which will help in analyzing its real-time health. Our main focus is to apply PCA over the dataset in order to reduce the dimensions and extract the most important features for the input of the prenatal device. The prenatal healthcare system will ultimately become less complex thanks to the essential elements.

The chapter is structured as follows. Section 12.3 delves into the Internet of Things in depth, while Section 12.4 focuses on the IoMT and its components. Section 12.5 discusses the experimental setup, by providing an overview of the datasets used in the recommended device. It also demonstrates the application of real-time PCA to the dataset, thus producing the minimized and most notable features. Section 12.6 concludes the chapter by summarizing the arguments and examining the implications of the problem at hand.

12.2   INTERNET OF THINGS

The IoT) is a network of real-world objects, or "things," that are outfitted with sensors, electronics, networks, software, and other parts to gather and exchange data [26]. IoT enables the Internet connectivity of everyday objects like toasters and relatively basic gadgets like PCs, cellphones, and tablets. The 'Thing' in IoT can be any gadget with built-in sensors [6] that can gather and transport data across a network without human interaction. The object's embedded technology enables interaction with internal states and the outside environment, which aids in decision-making. By enhancing parts of our lives with the power of data collecting, AI algorithms, and networks, IoT makes almost everything "smart." A person with a diabetes monitor implant, an animal with tracking devices, etc. can also be the thing in an IoT system.

IoT employs innovative problem-solving strategies and Internet-connected smart devices to tackle a variety of global economic, political, and public/private sector issues. It is becoming more and more significant, permeating every aspect of our everyday life. The Internet of Things (IoT) is an innovation in technology that incorporates a variety of smart systems, frameworks, intelligent devices, and sensors. In terms of storage, monitoring, and processing speed, it additionally gains from quantum and nanotechnology in manners that were previously unimaginable. Numerous studies have been carried out to show the general applicability and usefulness of IoT innovations. These studies are accessible in the form of press reports, academic articles, and printed materials both online and offline.

12.2.1 IoT Architecture

IoT system hardware consists of sensors, servers, a routing device, remote dashboard devices, and control devices. These devices manage essential tasks and activities such as system activation, action requirements, security, communication, and surveillance in order to support certain goals and actions. The IoT system hardware could be divided into four main categories, namely, building blocks, sensors, wearables, and basic devices. Let us first delve into the basic understanding of each of the subparts.

12.2.1.1 Sensors

A sensor is the key element in all IoT applications. It is a real-world apparatus that determines and measures specific physical values and converts them into signals that may be sent into processing or control systems to be used as inputs for analysis. The sensor connects the IoT gadget to the outside world or a human being. It senses the changes and transmits data to the cloud for analysis. They continuously collect data from their surroundings and relay it to the next layer, which includes pressure sensors, temperature sensors, and light intensity detectors. Energy, power management, radio frequency, and sensing modules are the components that make up sensors. The RF modules regulate communication via signal handling, Wi-Fi, Bluetooth, radio transceiver, duplexer, and BAW (Figure 12.1).

12.2.1.2 Building Blocks

The 'Thing': The item you intend to control, monitor, or closely watch is referred to as a "thing" in IoT. The "thing" is completely integrated into a smart device in many IoT goods. Consider items like a smart TV or an automatic car as examples. These items manage and watch over themselves. There are occasionally numerous more instances where the object is used as a standalone device and is connected to a different product to give it smart features.

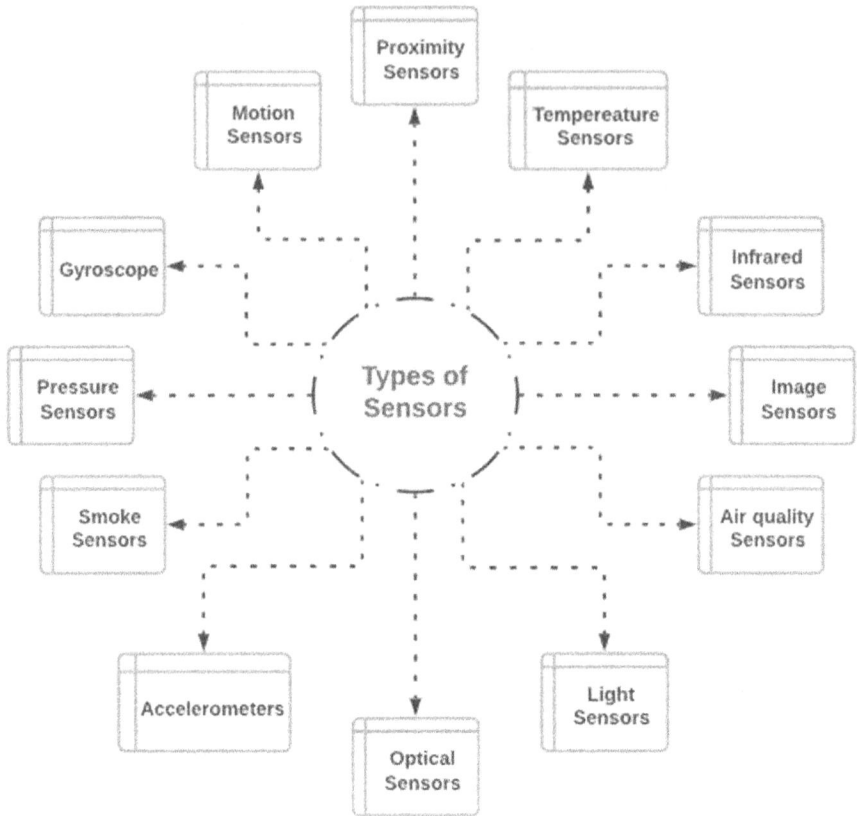

FIGURE 12.1   Types of sensors.

Data acquisition module: The main objective of the data acquisition module is to gather physical signals from the monitored item and convert them into electrical signals that a computer can modify or comprehend. This component is made up of all the sensors that are utilized to gather real-world signals such as temperature, pressure, density, motion, light, and vibration. Depending on the application, a certain type and number of sensors may be needed. It also contains the technology required to transform the incoming sensor signal into digital data so that the computer can utilize it. Input signal conditioning, noise reduction, analog-to-digital conversion, interpretation, and scaling are included in this.

Data processing module: The data processing module is the third component of an IoT device. This is the actual "computer" and the central processing component that handles data processing, local analytics, local data storage, and other computing tasks.

Communication module: The communications module is the final building component of IoT hardware. This section facilitates connectivity with your cloud platform as well as systems from third parties, either locally or remotely.

12.2.1.3   Wearables

Wearable smart devices are becoming increasingly popular in our daily lives. They collect and evaluate data, and in certain circumstances, they may communicate with the user and make wise judgments. Wearable devices could be programmed to connect to other devices in the home. Maybe you like a certain chair to watch TV in a certain lighting level. When the TV is on, your wearable device can help control the lighting in the space using the connected LED lights. Even automatic glare-blocking from window light might be supported in an intelligent home. The LCD TV screen's illumination could also be changed, and all settings could be customized for energy conservation and the best possible viewing experience. Once the general plan has been determined via a smart phone interface, all of these interactions may be carried out automatically and directly across devices. IoT-enabled medical wearable technology gives users the knowledge they need to improve their health results. Smart jewelry includes items like rings, wristbands, watches, and pins. These frequently use a smartphone app for display and interaction.

For the goal of monitoring and transmitting biological data, body-mounted sensors are affixed to the body. Fitness trackers are another kind of wearables that continuously measure physical activity and vital indicators. These devices are mostly in the shape of bracelets, straps, or headbands.

12.2.1.4   Basic Devices

The basic devices including desktop computers, mobile phones, and tablets continue to be crucial components of the IoT system. The desktop gives the user a very high level of authority over the machine and its settings. Users may access the primary system components and utilize the tablet as a remote control.

Remote capability and some basic configuration customizations are possible on a cellphone. Standard network hardware like switches and routers are among the other important connected devices.

12.2.2   Applications of IoT

IoT has a comprehensive vision to benefit numerous industries, including the public and private sectors of government, business, healthcare, and transportation. Different academics have offered different explanations of the IoT with regard to specific topics and problems. IoT's power and promise are seen in a variety of application fields. According to 2020 report on IoT spending by IDC, the worldwide spending on IoT will return to double-digit growth in 2021, with a CAGR of 11.3%. As everything from smart sensors to smart home appliances, and from smart factories to connected healthcare devices, the extension of the IoT market reflects that of a booming business sector [7]. Some of the most significant IoT applications [15] include a wide range of sectors, including manufacturing, medical care, agriculture, smart cities, security, and emergencies, among many others.

Smart cities: In today's world, everyone prefers smart cities. Owing to IoT, smart cities will keep growing their influence by utilizing cutting-edge technologies that will use data from IoT devices to connect entities. IoT provides new opportunities for smarter cities, like automatic parking, intelligent lighting, smart agriculture, garbage generation, walkable neighborhoods, and smart houses, to better utilize infrastructure, safeguard resident safety, and support efficient resource management. Intelligent transport systems [9], smart buildings, traffic jams, garbage management, smart lighting, smart parking, and urban mapping are a few IoT application areas for creating smart cities. This might entail placing sound-monitoring equipment in sensitive urban locations, keeping an eye on the number of people walking and driving, the vibrations of buildings and bridges, the material conditions of such structures, and the accessibility of parking spaces inside the city. Figure 12.2 shows how traffic congestion can be monitored, managed, and lessened in smart cities by utilizing IoT and AI capabilities.

Healthcare: The frequency of health monitoring has altered as a result of smartwatches and fitness equipment. Individuals can periodically check on their own health. In addition, patients who arrive at hospitals via ambulance now have their health reports diagnosed by the time they get there, allowing the hospital to begin their treatment right away. The information received from various healthcare applications is currently collated, analyzed, and used to discover a cure for various diseases. IoT services in healthcare can be used for both single condition and cluster condition management. Some of these applications include, for example, the ability to track and monitor health progress by healthcare professionals remotely, improve self-management of chronic conditions, help in the early detection of abnormalities, fast-track symptom identification and clinical diagnoses, deliver early intervention, and improve adherence to prescribed treatments [12].

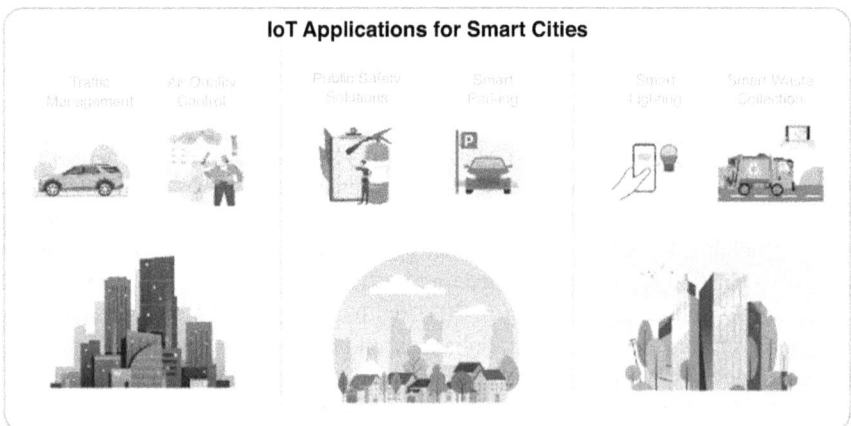

FIGURE 12.2 Smart city applications of IoT.

Retail and logistics: Retail businesses may manage inventory, enhance customer service, streamline the supply chain, and cut costs by using IoT apps. For instance, RFID-based data can be collected via smart shelves equipped with weight sensors, and the data can then be sent to an IoT platform to automatically check inventory and give notifications when supplies are low. Many IoT applications have positive effects on logistics [25] and transportation networks. Thanks to IoT sensor data, cars, trucks, ships, and trains that transport inventory can be rerouted based on the weather, the availability of available vehicles, or the availability of drivers. Our self-driving cars are equipped with sensors, and other technologies like parking assistance that can show us where parking spots are available and traffic lights that can detect traffic and alter automatically also have sensors.

Agriculture: IoT boosts agricultural productivity while also successfully enhancing the quality of the produced goods, reducing labor costs, boosting farmer income, and truly realizing agriculture's modernity and intelligence. Thanks to IoT-enabled agriculture, farmers can monitor their crops and environmental conditions in real time. They are able to swiftly come to conclusions, foresee issues before they happen, and make informed decisions on how to avoid them. IoT solutions for agriculture also incorporate automation, including demand-based irrigation, fertilization, and robot harvesting. IoT offers capabilities including local and remote data acquiring, in-cloud intelligent information, user interface, and agriculture operation automation in addition to the communication infrastructure to connect every smart object from sensors, vehicles, to the user's mobile device over the Internet [23].

Recent advancements in IoT are attracting the attention of researchers and programmers from every corner of the world. IoT researchers and developers are working together to advance the field and increase society's benefits from it. IoT's future appears optimistic. It has already begun to alter how we live our lives, how we work, where we live in cities or in our homes, when we travel, and how we communicate with others throughout the world. Industrial Internet development will probably be fueled by IoT technologies, improved network flexibility, AI, and the ability to automate, deploy, and secure a variety of use cases on a large scale.

12.3   INTERNET OF MEDICAL THINGS (IoMT)

IoMT, sometimes referred to as IoT in the field of healthcare, enables wireless and distant devices to securely communicate via the Internet to enable rapid and flexible analysis of medical data. Every medical device will be connected to the Internet and supervised by healthcare experts in the future by healthcare systems known as IoMTs.

As it develops, this provides for quicker and less expensive medical care. Figure 12.3 illustrates an example of an IoMT where patient vitals are collected using electronic sensors and sent to the IoMT applications over the Internet. Following the

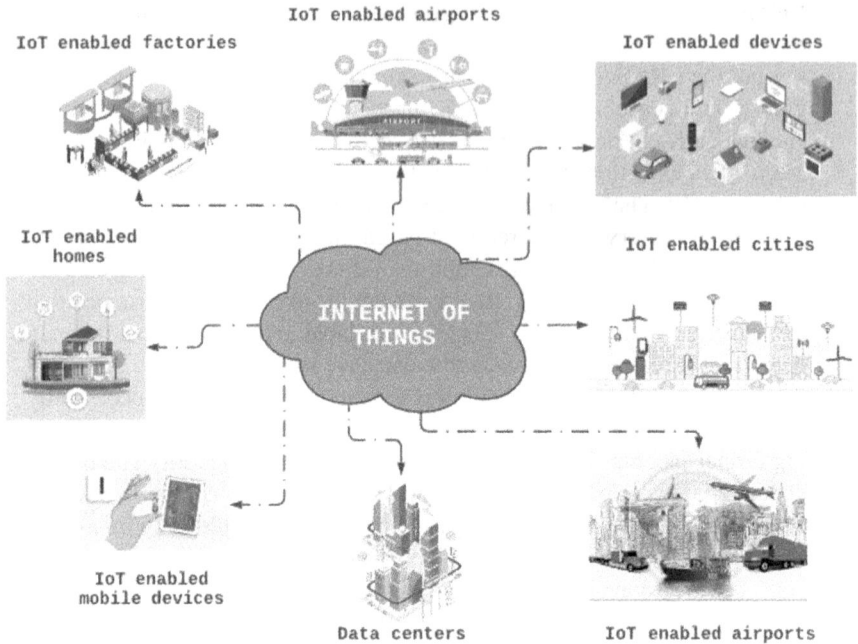

FIGURE 12.3   Internet of Medical Things.

flow of information to the medical staff and healthcare professionals, a response is subsequently forwarded to the required patients [24].

12.3.1 ARCHITECTURE OF IoMT

Three layers make up the IoMTs architecture: one for objects, one for fog, and one for clouds. With this design [20], healthcare professionals can communicate directly over the router connecting the Thing and Fog layers and via the regional fog layer processing servers. The following is a description of each layer:

> **The things layer**: Patient monitoring equipment, sensors, actuators, medical records, pharmaceutical controls, diet plan generators, etc. are all included in this layer. This layer has direct touch with the ecosystem's consumers. At this layer, information is gathered from components like wearables, patient monitoring, and remote care. The gadgets used here must be securely placed to guarantee the integrity of the data collected. To provide useful information, the data are further processed at the cloud and fog layers.
> **The fog layer**: This layer functions in between the things and cloud layers. For a sparsely distributed fog networking infrastructure, it comprises local servers and gateway devices. The bottom layer devices make use of the local computing power to provide consumers with real-time responses. These servers are also used to manage and monitor the security and integrity of

the system. The data must be sent from these servers to the cloud layer for further processing through its gateway components.

The cloud layer: The cloud layer includes data storage and computing resources for data analysis and the creation of decision-making systems that depend on such analysis. It also provides a wide range to incorporate substantial medical and healthcare systems to easily manage their daily operations. The cloud resources of this layer will store the data produced by the medical infrastructure, allowing for further analytical work as needed (Figure 12.4).

12.3.2 TECHNOLOGIES INTEGRATED WITH IoMT

Traditional healthcare monitoring is drastically changing, and customers now have access to integrated cloud products and automated detection tools thanks to digital healthcare. Patients, physicians, and inhabitants of remote locations may now readily obtain high-quality healthcare services thanks to this digital advancement. The development and integration of new technologies are essential for improvised healthcare. Let us discuss some of the emerging technologies and their role in IoMT.

Artificial intelligence: By using medical, laboratory, and demographic data, AI techniques are being utilized in the healthcare sector to screen, diagnose, and project prognosis for a number of illnesses. In addition, a lot of research utilizing AI techniques was carried out during the pandemic to cater to the desire for early identification and screening. Resource allocation, resource finding, extensive screening, monitoring, and the prediction

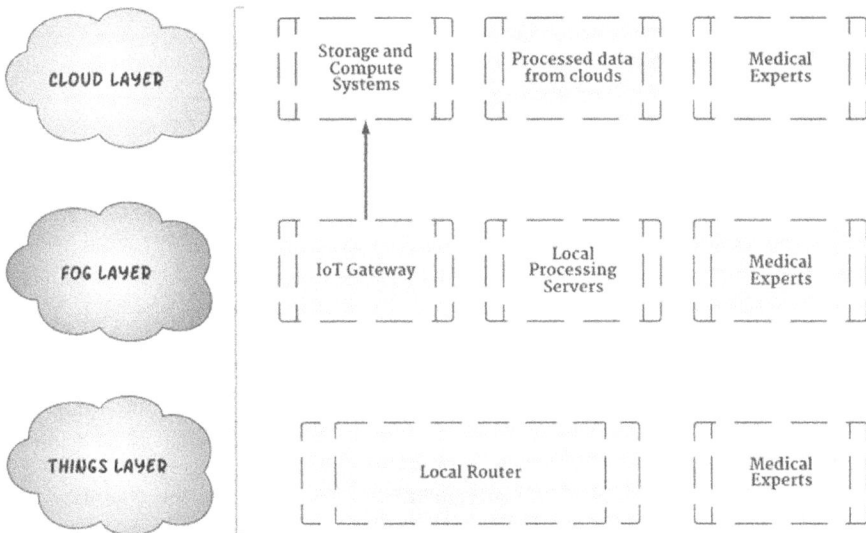

FIGURE 12.4   Layers of IoMT.

of probable interactions with recently suggested drugs are all tasks that AI may assist with. With AI-based solutions, the many components of the health sector can be changed. This will include AI methods for building classifiers that automatically record healthcare data, schedule patient appointments, and decide on lab testing, treatment plans, drugs, and surgical procedures, among other things. These classifiers could support decision-making processes by being trained further. For other classifiers that cannot be recorded digitally, NLP (or Natural Language Processing) provides ways for extracting information from such unstructured data points in the infrastructure [1]. Also, ML makes predictions about the future based on the past. To forecast future circumstances, it uses supervised, unsupervised, or reinforced learning (Figure 12.5).

Blockchain: A significant level of data sharing between medical equipment and healthcare professionals is essential in smart healthcare systems. This frequently results in data fragmentation, which can leave gaps in transferred information, make it difficult to comprehend, and impede the healing process. To solve this problem, blockchain technology was developed to connect the data repositories that are already existent in the network. The blockchain is a growing collection of records (blocks), connected to one another by the use of a hashing algorithm. To link the records together and make them immune to changes, each record includes a cryptographic hash of the one before it. The use of blockchain-based healthcare solutions

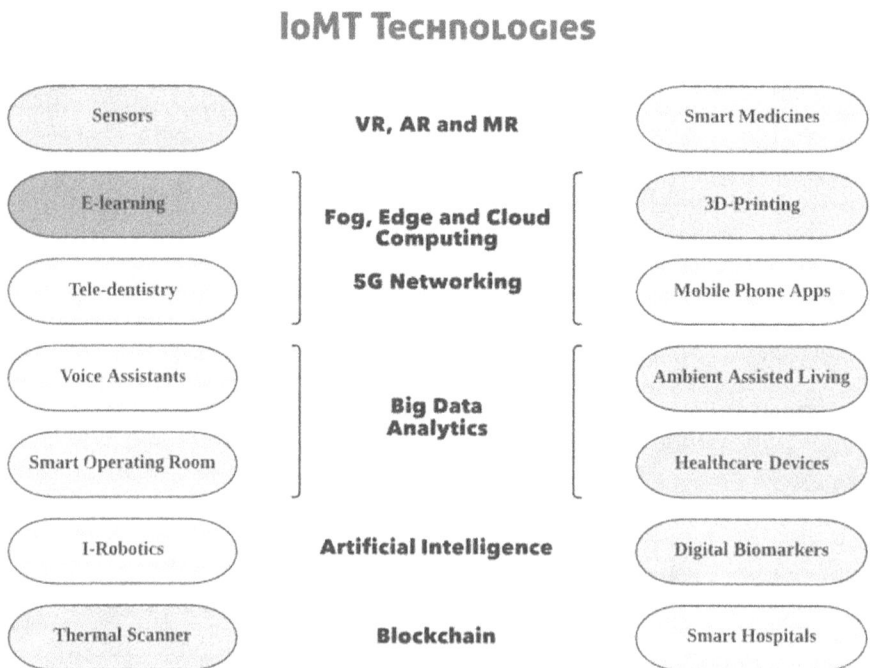

IoMT Technologies

Sensors	VR, AR and MR	Smart Medicines
E-learning	Fog, Edge and Cloud Computing	3D-Printing
Tele-dentistry	5G Networking	Mobile Phone Apps
Voice Assistants	Big Data Analytics	Ambient Assisted Living
Smart Operating Room		Healthcare Devices
I-Robotics	Artificial Intelligence	Digital Biomarkers
Thermal Scanner	Blockchain	Smart Hospitals

FIGURE 12.5 Integrated technologies with IoMT.

necessitates the division of infrastructure into smaller components. The IoMT framework's relevant devices can then be integrated with these modules. Vangipuram et al. developed 'CoviChain,' which uses a blockchain implementation and edge infrastructure to allow the secure transfer of COVID-19-infected people's data to the hospital system [28]. Currently, clinical use trials are being conducted globally for the usage of blockchain in EHR systems in hospitals [3].

Big data visualization and analytics: For accurate decision-making, the massive amounts of data generated by IoMT devices must be properly processed. Big Data Visualization and Analytics (BDVA) is a practical response to this problem, which is defined by attributes like volume, diversity, velocity, veracity, validity, volatility, etc. The use of internal networks by Cloud-IoMT-based big data analytics can assist in monitoring patient health conditions in real time across the globe, which eases workloads and pressure on medical professionals, improves diagnostic precision, lowers healthcare costs, and results in patient treatment satisfaction.

Software-defined networking: Data plane and control plane are the two components that make up the network in IoMTs. The data plane routes traffic to its destination, whereas the control plane handles the operations that allow the data plane to make routing decisions. A standardized method of communication between the data plane and the control plane is offered by software-defined networking (SDN). OpenFlow, Open vSwitch Database Management Protocol, and OpenFlow Configuration Protocol are a few examples of common SDN protocols. Ref. [24] discusses IoMTs with SDN support, in which IoMT devices are linked to e-healthcare applications via the SDN control plane. Data from IoMT devices are gathered by the SDN control plane and provided to an e-healthcare application, such as one for patient diagnostics or patient safety.

Parallel computing: Distributed computing methods using concepts like grid computing, cloud computing, fog computing, and edge computing are built on parallel processing techniques. IoMT and its applications that involve real-time interactions are an emerging source of big data, so it is crucial to identify and separate the data that must be maintained locally from the data that must be shared across cloud servers. Data processing architecture of IoMT systems has switched from centralized cloud computing to distributed fog computing. A hierarchy of layers is created by fog computing between the physical parts and the cloud server (core level). By storing less data across more cloud servers, it minimizes network and Internet latency. By speeding up data streaming and processing real-time IoMT data, edge computing reduces latencies and gives smart devices immediate responsiveness. Ref. [14] proposes an IoMT-based computational approach that detects brain tumors with respect to its grade.

5G Networking: The future of the present healthcare system is 5G and IoMT, where every medical equipment will be more securely connected to the Internet and monitored by healthcare providers. This provides a quicker and less expensive alternative to the widespread, current technology. IoMT applications with 5G support offer online replicable information on demand and in real time. High-speed data transfer rates, ultra-low latency or delay in the data transmission-response system, connectivity and capacity, and high bandwidth and durability per unit area are some of the key characteristics of 5G technology. The 5G network has the capacity to serve thousands of medical gadgets at once, including sensors, mobile devices, medical equipment, video cameras, and VR/AR. Ref. [5] created a 5G-enabled fluorescence sensor for the quantitative detection of COVID-19 virus spike protein and nucleo-capsid protein.

12.3.3 Categories of IoMT Devices

On-body segment: Consumer health wearables and wearables with a medical or clinical grade are the two main categories of the on-body market.

- Consumer-grade fitness or wellness gadgets like activity trackers, bracelets, wristbands, sports watches, and smart clothing are examples of consumer health wearables. While many of these gadgets are not subject to health authorities' regulation, they may be recommended by professionals for particular health uses.
- Clinical-grade wearables comprise platforms and regulated devices that have been granted certification or approval for usage by one or more regulatory or health agencies. The majority of these gadgets are used in conjunction with professional guidance or a prescription from a doctor. For example, an intelligent belt created by Active Protective may detect when an elderly person falls and immediately engage hip protection features to prevent damage.

Home segment: Segment for in-home care includes telehealth virtual visits, remote patient monitoring, and personal emergency response systems (PERS).

- For elderly people who are housebound or have restricted mobility, a PERS mixes wearable technology/relay devices with a live healthcare contact center service to boost independence. Users can swiftly communicate and access emergency medical assistance with the help of the package.
- RPM includes every home surveillance sensor and equipment used for the management of chronic illnesses, which includes ongoing physiological parameter tracking to support sustained care in the patient's home in a bid to slow disease advancement and critical home surveillance for ongoing observation of patients who have been discharged to accelerate recovery and avoid readmission, along with other types of home monitoring.
- Virtual consultations that assist patients in managing their diseases and obtaining prescriptions or suggested treatment regimens are included in telehealth virtual visits.

Community segment:

- Passenger mobility services enable automobiles carrying passengers to track health indicators while moving.
- Emergency response intel is meant to help first responders, paramedics, and hospital emergency department staff.
- Kiosks are physical structures that usually have touchpad computers and may be used to sell things or provide services like access to medical experts.
- Healthcare supplies used by a physician at the point of care are used outside the home or in more conventional healthcare settings, like a medical camp.
- Transporting and delivering medical supplies, medications, surgical instruments, medical equipment, and other items that healthcare professionals require fall under the category of logistics.

In-clinic segment: The IoMT devices in this section are utilized for administrative and clinical tasks (either in the clinic, in the telehealth model, or at the point of care). Point-of-care devices in this context differ from those in the community segment in one crucial way: the care provider can be located remotely while a device is used by qualified personnel, as opposed to the care provider physically using a device. A cloud-based examination tool like Rijuven's Clinic in a Backpack, which allows professionals to evaluate patients at any point of care, is an example.

In-hospital segment: IoMT-enabled devices and a larger range of solutions in various management domains make up this segment.

- Asset management keeps an eye on and keeps track of expensive fixed assets around the hospital, like wheelchairs and infusion pumps.
- People management gauges the effectiveness and productivity of employees.
- By avoiding bottlenecks and optimizing the patient's experience, patient flow management enhances facility operations. One example is the tracking of patient arrival times from an operating room to post-care to a wardroom.
- To cut inventory expenses and boost employee productivity, inventory management streamlines the ordering, storage, and usage of consumables, medications, medical devices, and hospital supplies.
- Monitoring of the environment (such as temperature and humidity) and energy ensures the best conditions in patient areas and storage rooms while keeping an eye on electricity use.

12.4   EXPERIMENTATION

12.4.1   DATASET AND FEATURES

We have generated a dataset for the fetus's prenatal device. The dataset consists of the fetal ultrasound parameters measured in relation to its gestational age. Fetal ultrasonography measures might include the estimated fetal weight (EFW), the crown-rump length (CRL), the biparietal diameter (BPD), the femur length (FL), the head circumference (HC), the occipitofrontal diameter (OFD), the abdominal circumference (AC), and the humerus length (HL). To be more precise, we have included four critical aspects which are briefly outlined below:

  i. Biparietal Diameter (BPD) – One of the most often seen fetal characteristics is the biparietal diameter (BPD). This is the head's diameter at 13 weeks; it increases from around 2.4 cm at 13 weeks to about 9.5 cm at maturity.
 ii. Head Circumference (HC) – Usually done after 13 weeks of pregnancy, it gauges the fetus's head circumference. In terms of accuracy, HC measurement is on par with BPD measurement in estimating gestational age. Yet, in fetuses with aberrant head shape, HC may be a more reliable assessment of fetal age than BPD.
iii. Abdominal Circumference (AC) – It is the most crucial measurement to take in the later stages of pregnancy. It is more a reflection of fetal weight and size than of age. The AC is measured in the trans axial view of the fetal abdomen. It is assessed at the fetal liver level, with the umbilical part of the left portal vein serving as a landmark.
 iv. Femur Length (FL) – The femur length gauges the body's longest bone and illustrates the fetus's longitudinal growth. Its utility is comparable to the BPD's. Similar to BPD, FL dating should begin as soon as possible. FL's gestational age evaluation is especially effective when head measurement is problematic owing to fetal position.

Tables 12.1 and 12.2 show the curated dataset values as well as the specified range of values for all four parameters. However, we will be applying PCA on an extrapolated version of the dataset shown below. The range of characteristics comprises dimensions typical of a growing fetus at various stages of pregnancy, from 14 weeks to 39 weeks.

12.4.2 PRINCIPAL COMPONENT ANALYSIS

PCA is a dimensionality reduction approach that seeks to keep the most information from the initial dataset as possible while decreasing the number of attributes in a dataset into a smaller collection of features called principal components.

TABLE 12.1

Dataset Defined for Analysis (Only Displaying the First Five Rows) in Relation to Table 12.2

| | Vital Parameters Based on Gestational Age (in mm) | | | | |
Patient	Biparietal Diameter (BPD)	Head Circumference (HC)	Abdominal Circumference (AC)	Femur Length (FL)	Weeks
Fetus 1	47.3	171	145	31.8	19.5
Fetus 2	80	287.3	268.2	58	30
Fetus 3	59.2	212	184.5	42.2	23
Fetus 4	94	333.4	336	72.1	38
Fetus 5	73.1	256.1	228.8	51	27

TABLE 12.2
Range Prescriptions of the Desired Features

Prescribed Range (in mm)

	Biparietal Diameter (BPD)			Head Circumference (HC)			Abdominal Circumference (AC)			Femur Length (FL)		
Weeks	5th percentile	AVG	95th percentile	5th 'tile	AVG	95th 'tile	5th 'tile	AVG	95th 'tile	5th 'tile	AVG	95th 'tile
14	28	31	34	102	110	118	80	90	102	14	17	19
15	31	34	37	111	120	129	88	99	112	17	19	22
16	34	37	40	120	130	140	96	108	122	19	22	25
17	36	40	43	130	141	152	105	118	133	21	24	28
18	39	43	47	141	152	164	114	128	144	24	27	30
19	42	46	50	151	163	176	123	139	156	26	30	33
20	45	49	54	162	175	189	133	149	168	29	32	36
21	48	52	57	173	187	201	143	161	181	32	35	39
22	51	56	61	184	198	214	153	172	193	34	38	42
23	54	59	64	195	210	227	163	183	206	37	41	45
24	57	62	68	206	222	240	174	195	219	39	43	47
25	60	66	71	217	234	252	184	207	233	42	46	50
26	63	69	75	227	245	264	195	219	246	44	48	53
27	66	72	78	238	256	277	205	231	259	47	51	55

(Continued)

TABLE 12.2 (Continued)
Range Prescriptions of the Desired Features

Prescribed Range (in mm)

Parameters

Weeks	Biparietal Diameter (BPD)			Head Circumference (HC)			Abdominal Circumference (AC)			Femur Length (FL)		
	5th percentile	AVG	95th percentile	5th 'tile	AVG	95th 'tile	5th 'tile	AVG	95th 'tile	5th 'tile	AVG	95th 'tile
28	69	75	81	248	267	288	216	243	272	49	53	58
29	72	78	85	257	277	299	226	254	285	51	56	60
30	74	81	88	266	287	309	237	266	298	53	58	63
31	77	83	90	273	296	319	246	277	310	55	60	65
32	79	86	93	282	304	328	256	287	322	57	62	67
33	81	88	96	288	311	336	265	297	334	59	64	69
34	83	90	98	294	317	342	274	307	345	61	66	71
35	85	92	100	299	323	348	282	316	355	63	68	73
36	86	94	102	303	327	353	289	324	364	64	69	74
37	87	95	103	306	330	356	295	332	372	66	71	76
38	88	96	104	308	332	358	302	339	380	67	72	77
39	89	97	105	309	333	359	307	345	387	68	73	78

PCA's widespread use would not have been possible without several essential characteristics that set it apart from other tools—all of which are linked to the replicability dilemma. PCA may be used on any numerical dataset [8], no matter how large or tiny, and it always produces results. It has no parameters and almost no assumptions. It excludes the use of error estimates, effect size calculations, or significance tests. PCA's origins are in multivariate data analysis, yet it also has numerous other applications. One of the biggest effects of using linear algebra, PCA is frequently used as the initial stage in attempts to evaluate massive datasets. Data compression, blind source separation, and signal denoising are a few more typical applications. Big datasets are often dimensionally reduced via PCA using a vector space transformation. Mathematical projection is widely used to understand the original data collection, which may have had a large number of variables, in a smaller set of variables.

In medicine, PCA is used to handle a variety of challenges, including multicollinearity diagnostic tests [29]. In order to predict the severity of COVID-19 and execute an individualized treatment, researchers employ PCA to identify phenotypes. By adding adrenal steroid tests and clinical indicators, PCA approach may potentially be utilized to enhance treatment for a group of young kids with CAH [16]. The application of PCA can be applied to multiple fields like IoT [21], finance, psychology, and even medical sector [19].

As a high attribute indicates a clear separation between classes, PCA considers each attribute's variation to reduce the dimensionality. PCA can be used to eliminate a linearly dependent vector from a dataset or matrix. The features are eliminated by rotating the entire set of data into a new vector space, where it behaves as perfect data. Throughout the procedure, the data's covariance matrix is changed into a sparse matrix. By lowering the inter-feature covariance to zero or extremely low values, this reduces the linear dependence between the features.

The following is a summary of how PCA functions:

i. Normalizing the data – We first standardize the data to get it ready for PCA. Each characteristic will therefore have a mean of 0 and a variation of 1.

ii. Creating the covariance matrix – The covariance matrix is a $p \times p$ symmetric matrix containing entries for all possible pairings of the starting variables (where p represents the total number of dimensions). For instance, a three-dimensional dataset with three variables (x, y, and z) has a covariance matrix that is a 33 matrix with the type given below. A square matrix is used to depict the association between two or more attributes in a multidimensional dataset (Figure 12.6).

iii. Determining the eigenvalues and eigenvectors – To find the primary components of the data, we must compute the linear algebra concepts of eigenvectors and eigenvalues from the covariance matrix. The variance held by each principal component is expressed by its eigenvalues, which are simply the coefficients associated to the eigenvectors. Principal components depict the data directions that explain the greatest amount of variance, or the lines that retain most information from the data. The relationship between variance and information in this case is that the greater the variance carried by

$$\begin{bmatrix} \text{Cov}(\boldsymbol{x},\boldsymbol{x}) & \text{Cov}(\boldsymbol{x},\boldsymbol{y}) & \text{Cov}(\boldsymbol{x},\boldsymbol{z}) \\ \text{Cov}(\boldsymbol{y},\boldsymbol{x}) & \text{Cov}(\boldsymbol{y},\boldsymbol{y}) & \text{Cov}(\boldsymbol{y},\boldsymbol{z}) \\ \text{Cov}(\boldsymbol{z},\boldsymbol{x}) & \text{Cov}(\boldsymbol{z},\boldsymbol{y}) & \text{Cov}(\boldsymbol{z},\boldsymbol{z}) \end{bmatrix}$$

FIGURE 12.6   Covariance matrix.

a line, the greater the dispersion of data points along it, and the greater the dispersion along a line, the more information it has.

iv. Constructing principal components – Principal components are created in a way that the first principal component reflects the most possible variability in the dataset since there are exactly as many principal components as there are variables in the data. We can determine the order of significance of the principal components by sorting the eigenvectors from highest to lowest in terms of their eigenvalues (Figures 12.7 and 12.8).

v. Feature vector – In this step, we develop a feature vector to select the primary components that should be kept (with high significance) and discard those of lesser significance. Simply put, the feature vector is a matrix containing the principal component eigenvectors that we decide to keep as columns. As a result, if we choose to maintain only p eigenvectors out of n, the end result dataset will only have p dimensions, making it the first stage in dimensionality reduction.

vi. Recasting – The last goal is to use the feature vector produced in the previous stage to transform the data from its initial axes to the ones represented by the principal components. To do this, we multiply the transpose of the feature vector by the transpose of the original dataset.

12.4.3 APPLYING PCA TO DATASET

The extrapolated dataset consists of vital parameters of 105 fetuses, obtained in real time from the remote prenatal device. The four parameters are based on the fetus's gestational age in weeks. For this analysis, we are using Project Jupyter's Jupyter Notebook, an open-source web application that we can use to visualize our dataset. By performing PCA on the dataset, we can easily decrease the dimensions of the features. This will lessen the complexity of determining which aspects should be given precedence while using the prenatal gadget.

For the analysis, we would first need to import the necessary Python packages, **NumPy**, **Pandas**, and **Matplotlib**. The NumPy package in Python includes specialized data structures, methods, and other features for computation. For working with tabular data, the Pandas package offers utility methods to read data from many different types of files, such as CSV, Excel spreadsheets, and HTML tables. And lastly, the end PCA plot will be visualized using Matplotlib, the primary plotting package for Python.

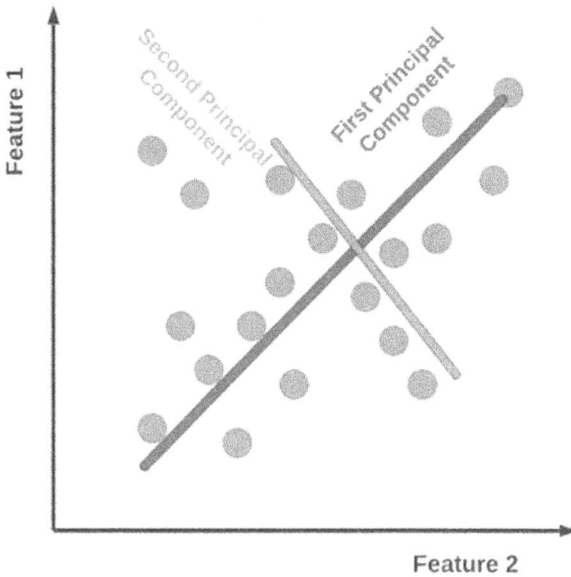

FIGURE 12.7   Finding main components in PCA.

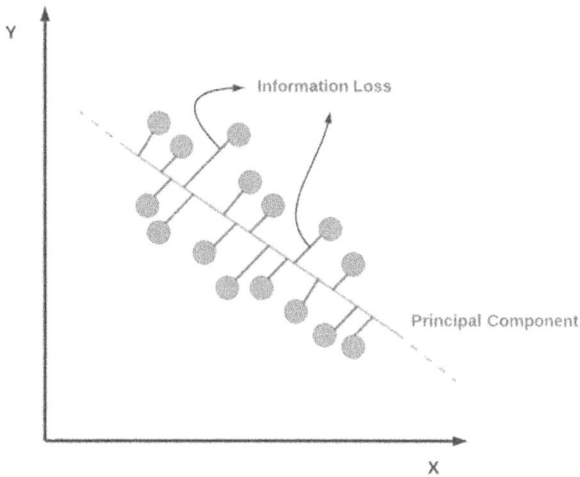

FIGURE 12.8   Identifying PCA and information loss.

a. Importing the required libraries to analyze our data. Here, the '%matplotlib inline' command enables an inline plotting that displays plotted visuals in our Jupyter Notebook (Figure 12.9).

```
In [1].  import matplotlib.pyplot as plt
         import numpy as np
         import pandas as pd
         %matplotlib inline
```

FIGURE 12.9   Importing Python packages.

b. In order to start applying PCA over the dataset, we first need to load it on our Jupyter Notebook (Figure 12.10).

```
In [2]:  dataset = pd.read_csv("pcadataset.csv")
```

FIGURE 12.10 Loading the dataset.

c. Checking if the dataset has the right set of values. The 'head()' function returns the first five rows of our dataset (Figure 12.11).

```
In [3]:  dataset.head()
```

	Biparietal Diameter (BPD)	Head Circumference (HC)	Abdominal Circumference (AC)	Femur length (FL)
0	47.3	171.0	145.0	31.8
1	80.0	287.3	268.2	58.0
2	59.2	212.0	184.5	42.2
3	94.0	333.4	336.0	72.1
4	73.1	256.1	228.8	51.0

FIGURE 12.11 Confirming the dataset values.

d. Splitting the dataset into X and Y components for data analysis (Figure 12.12).

```
In [4]:  X = dataset.iloc[:, 0:3].values
         Y = dataset.iloc[:, 3].values
```

FIGURE 12.12 Distributing the dataset.

e. Checking the current shape of our arrays using the '.shape' function (Figure 12.13).

```
In [6]:  X.shape
Out[6]:  (105, 3)

In [7]:  Y.shape
Out[7]:  (105,)
```

FIGURE 12.13 Checking array size.

f. Importing PCA and reducing the dimensionality of input space from '3' to '2'. The 'fit()' function creates the principal components of our data (Figure 12.14).

```
In [8]:  from sklearn.decomposition import PCA

         pca = PCA(n_components = 2)

In [9]:  pca.fit(X)

Out[9]:  PCA(n_components=2)
```

FIGURE 12.14   Reducing the dimensions of dataset.

    g. Checking eigenvectors of the covariance matrix of the test set (X) (Figure 12.15).

```
In [10]:  pca.components_

Out[10]:  array([[-0.19039362, -0.66070852, -0.72609539],
                 [-0.08955628, -0.72485128,  0.68305951]])
```

FIGURE 12.15   PCA components.

    h. Transforming the X set into the feature space by storing the value of transformed variables into a different variable Z. Finally, verifying that Z has been transformed to '2' dimensions (Figure 12.16).

```
In [11]:  Z = pca.transform(X)

In [12]:  Z.shape

Out[12]:  (105, 2)
```

FIGURE 12.16   Transforming dimensions.

    i. Observing the amount of data preserved after PCA. Using the 'scatter()' function to visualize the result through a scatter plot (Figure 12.17).

```
In [15]:  plt.scatter(Z[:,0], Z[:,1], c=Y)

Out[15]:  <matplotlib.collections.PathCollection at 0x18af2e616a0>
```

FIGURE 12.17   Visualizing the test set results.

12.4.4   RESULT

As seen in Figure 12.18, the dimensions of our dataset have been reduced from three to two using PCA. This experiment provided a high-level overview of the actual analysis that could aid in resolving the dimensionality problem in general. In practice, the parameters are typically high-dimensional and complex, making it difficult to detect trends in data analysis. By using PCA, we may conserve the most critical data, reduce the size of the dataset, improve visualization, and make it easier to discover correlations between elements.

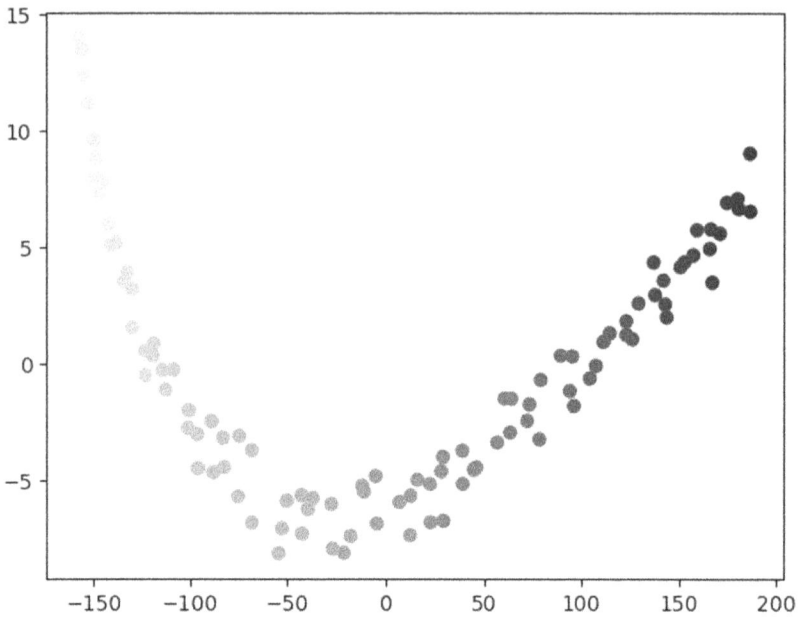

FIGURE 12.18   Scatter plot of reduced dimensions.

12.5   CONCLUSION

This chapter proposed using PCA across datasets to reduce the number of variables in the table of multivariate data in order to identify trends, leaps, clusters, and outliers. Large amounts of data require a lot of effort to manage and comprehend, which could produce unpredictable effects. By limiting the number of characteristics and concentrating only on the most important ones, PCA allows us to get around the problem of data overfitting. This chapter was analyzed on the basis of one of our own research papers that examined a monitoring device for fetus's prenatal care. With regard to the recommended gadget, the effect of PCA on lowering data dimensions is quite beneficial and could eventually result in a superior healthcare tracker. The prenatal device might notify the medical staff of any unusual observations in the fetus's health with reliable analysis. PCA could facilitate IoMT in producing advanced features that could benefit the medical sector immensely. Combining the two can result in cutting-edge medical gadgets that exchange highly sensitive and precise data which could result in an evolved health surveillance. Notably, PCA also functions as a tool for enhanced data visualization in addition to offering a streamlined model of a dataset. In addition to IoT, PCA may be used to discover patterns in a variety of fields, including economics, psychology, and data mining. There appears to be a lot of room for exploration of different areas where PCA might serve as a means of improvisation.

REFERENCES

[1] Ahmed, Z., & Mohamed, K. (2020). Artificial intelligence with multi-functional machine learning platform development for better healthcare and precision medicine. *Database, 2020.* https://doi.org/10.1093/database/baaa010

[2] Amethiya, Y., & Pipariya, P. (2022). Comparative analysis of breast cancer detection using machine learning and biosensors. *Intelligent Medicine, 2,* 69–81. https://doi.org/10.1016/j.imed.2021.08.004

[3] Dilawar, N., & Rizwan, M. (2019). Blockchain: securing internet of medical things (IoMT). *(IJACSA) International Journal of Advanced Computer Science and Applications, 10.*

[4] Government of India. (2022). *Significant Decline in Maternal Mortality in India. Government of India, Ministry of Women and Child Development.* Retrieved from https://pib.gov.in/FeaturesDeatils.aspx?NoteId=151238&ModuleId%20=%202

[5] Guo, J., & Chen, S. (2021). 5G-enabled ultra-sensitive fluorescence sensor for proactive prognosis of COVID-19. *Biosensors and Bioelectronics, 181,* 113160 https://doi.org/10.1016/j.bios.2021.113160

[6] Intuz. (n.d.). *An All-Inclusive Guide On The Top IoT Sensors In The Market.* Retrieved from Intuz: https://www.intuz.com/guide-on-top-iot-sensor-types

[7] IoT Technology in India. (n.d.). Retrieved from NASSCOM: https://community.nasscom.in/communities/emerging-tech/iot-ai/iot-technology-in-india.html

[8] Jaadi, Z. (n.d.). *A Step-by-Step Explanation of Principal Component Analysis (PCA).* Retrieved from Builtin: https://builtin.com/data-science/step-step-explanation-principal-component-analysis

[9] Jain, R. (2018). A congestion control system based on VANET for small length roads. *Annals of Emerging Technologies in Computing, 2.* https://doi.org/10.33166/AETiC.2018.01.003.

[10] Jiang, F., Jiang, Y., Zhi, H., Dong, Y., Li, H., Ma, S., & Wang, Y. (2017). Artificial intelligence in healthcare: past, present and future. *Stroke and Vascular Neurology, 2,* 230–243. https://doi.org/10.1136/svn-2017-000101

[11] Keboola. (2022, April 2). *A Guide to Principal Component Analysis (PCA) for Machine Learning.* Retrieved from Keboola: https://www.keboola.com/blog/pca-machine-learning

[12] Kelly, J. T., & Campbell, K. L. (2020). The Internet of Things: impact and implications for health care delivery. *Journal of Medical Internet Research, 22*(11), e20135. https://doi.org/10.2196/20135

[13] Keshari, S. (2019, May 29). *Introduction to Machine Learning and its Techniques.* Retrieved from Medium: https://medium.com/@eshukeshari/introduction-to-machine-learning-and-its-techniques-a383053539f2

[14] Khan, S. R. (2020). IoMT-based computational approach for detecting brain tumor. *Future Generation Computer Systems, 109,* 360–367. https://doi.org/10.1016/j.future.2020.03.054

[15] Kumar, S. T. (2019). Internet of Things is a revolutionary approach for future technology enhancement: a review. *Journal of Big Data, 6,* 111. https://doi.org/10.1186/s40537-019-0268-2

[16] Ljubicic Marie Lindhardt, M. A. (2021). The application of principal component analysis on clinical and biochemical parameters exemplified in children with congenital adrenal hyperplasia. *Frontiers in Endocrinology, 12.* https://doi.org/10.3389/fendo.2021.652888

[17] Manickam, P. M. (2022). Artificial intelligence (AI) and Internet of Medical Things (IoMT) assisted biomedical systems for intelligent healthcare. *Biosensors, 12*(8). https://doi.org/10.3390/bios12080562

[18] Muhammad, W., & Gregory, R. H. (2019). Pancreatic cancer prediction through an arti-ficial neural network. *Frontiers in Artifical Intelligence, 2*, 2. https://doi.org/10.3389/frai.2019.00002

[19] Naeem Ahmed Qureshi, V. S. (2017). Application of principal component analysis (PCA) to medical data. *Indian Journal of Science and Technology, 10*(20), 1–9. https://doi.org/10.17485/ijst/2017/v10i20/91294

[20] Naresh, V. S., & Pericheria, S. S. (2020). Internet of things in healthcare: architecture, applications, challenges, and solutions. *Computer Systems Science and Engineering, 35*, 411–421. https://doi.org/10.32604/csse.2020.35.411

[21] Nguyen, D. H. (2018). A PCA-based method for IoT network traffic anomaly detection. *20th International Conference on Advanced Communication Technology (ICACT)* (p. 1). IEEE. https://doi.org/10.23919/ICACT.2018.8323765

[22] Pandey, R. E. (2022). *Prenatal Healthcare Framework Using IoMT Data Analytics.* IGI Global. https://doi.org/10.4018/978-1-6684-3533-5

[23] Perwej, D. Y. (2019). The Internet of Things (IoT) and its application domains. *International Journal of Computer Applications, 182*, 36–49. https://doi.org/10.5120/ijca2019918763

[24] Sharma, S. R. (2022). Internet of Medical Things (IoMT): overview, emerging tech-nologies, and case studies. *IETE Technical Review, 39*, 775–778. https://doi.org/10.1080/02564602.2021.1927863

[25] Song, Y., & Richard, F. (2021). Applications of the Internet of Things (IoT) in smart logistics: a comprehensive survey. *IEEE Internet of Things Journal, 8*, 4250–4274. https://doi.org/10.1109/JIOT.2020.3034385

[26] Tibco. (n.d.). *What is the Internet of Things (IoT)?* Retrieved from Tibco: https://www.tibco.com/reference-center/what-is-the-internet-of-things-iot

[27] UNICEF. (n.d.). *Maternal Health - India.* Retrieved from https://www.unicef.org/india/what-we-do/maternal-health

[28] Vangipuram, S. L. T., & Mohanty, S. P. (2021). CoviChain: a blockchain based framework for nonrepudiable contact tracing in healthcare cyber-physical systems during pandemic outbreaks. *SN Computer Science, 2*, 346. https://doi.org/10.1007/s42979-021-00746-x

[29] Zhang, Z., & Castello, A. (2017). Principal components analysis in clinical studies. *Annals of Translational Medicine, 5*(17), 351. https://doi.org/10.21037/atm.2017.07.12

13 Face Mask Detection System

Rajiv Pandey, Radhika Awasthi, and Archana Sahai
Amity University

13.1 INTRODUCTION

Since the first known case of coronavirus was identified in Wuhan, China, the disease has spread worldwide, causing the COVID-19 pandemic. The common symptoms of this disease are fever, cough, headache, breathing difficulties, loss of smell, and loss of taste. It spreads when an infected person is in close contact with others. When the infected person coughs, sneezes, or breathes, the virus particles spread. If the other person inhales the same air containing the virus or comes in direct contact with it through the eyes, mouth, or nose, then the virus is said to have infected the other person. The coronavirus is airborne, due to which it remains in the contaminated air for a longer period, thus increasing the spread of virus to a greater number of people. Even after handling things or surfaces that have been in contact with the virus, people may get the disease by touching their lips, eyes, or nose. This outbreak has caused millions of casualties and anomalous threats to the world. It is spreading rapidly causing loss in economies, lack of food and public health crisis, unemployment, lack of medical facilities, disruption in work environment, and a catastrophic threat to human species. Therefore, at times like these, safety becomes the utmost concern for us. Taking precautionary measures for our safety is necessary for our survival. Official bodies have set up compulsory measures like wearing masks and maintaining social distancing. History has proved in the past that using face masks has helped to stop the spread. It is a custom that dates to China's Manchurian pandemic of 1910–1911. Face masks have been used by individuals in the past to survive numerous pandemics. Additionally, several studies have definitely demonstrated that the use of face masks may reduce the transmission of the infection. An automatic face mask detector is required since it has been discovered that countries with higher face mask usage rates have fewer COVID-19 outbreaks overall [1].

We employed a face mask detector in our study, along with auto-focus [2] and human–computer interaction [3] access. Using a mask detection approach, colour two-dimensional principal component analysis (2DPCA)-convolutional neural network (C2D-CNN) [4], for instance, is employed to identify a person wearing a mask or covering their face. In effect, we are identifying human faces in digital data, such as an image or a video, using a face detection algorithm [5]. It may also be viewed as an example of object-class identification. Although face recognition has traditionally been a challenging task for computer vision, it is growing easier as technology develops [6]. In the 1960s, there existed semi-automated facial recognition technology.

DOI: 10.1201/9781003359951-15

The first semi-automatic facial recognition technology was introduced in 1988 and used to take surveillance images and draw attention at the Super Bowl event in 2001. This technology is widely used nowadays for a variety of purposes, and it is crucial to the computer vision and IT industries. It is used for almost every activity, from straightforward camera detection to a variety of research applications.

In this study, we use deep learning techniques for picture classification across both datasets and face identification of both masked and unmasked subjects. Only those wearing masks on their faces may enter using this notification system, which may be compatible with any CCTV or sensor. In this study, we utilised two datasets—one with mask-wearing subjects and the other without—and we went through a number of steps, including preprocessing of the data, face identification, and picture classification using convolution neural networks and the Alex Net architecture. This study may be used in various places including public spaces like airports, healthcare facilities, and other places, to facilitate the monitoring of visitors arriving or exiting by security staff.

13.2 METHODOLOGY

13.2.1 DATABASE USED

The Real-World Masked Face Dataset (RMFD) [7], the biggest masked face dataset in existence, is the first dataset that has been proposed for identifying people wearing masks. It is available on a GitHub repository. Another dataset that we employed is Celeb Faces Attributes (CelebA) [8]. The Kaggle dataset includes approximately 200,000 images of celebrities with 40 binary feature classifications, as shown in Figures 13.1 and 13.2.

13.2.1.1 Face Detection

Here, we have utilised C2D CNN (Color 2-Dimensional Convolutional Neural Networks) [9, 4, 10] and Principal component Analysis [11, 12] for face recognition. In this approach, we estimate the result or make a choice based on a blend of characteristics learned from original pixels and the depiction learned by CNN.

13.2.2 CONVOLUTIONAL NEURAL NETWORK

Convolution is a method for combining two signals to produce a new one. As an image is a two-dimensional signal, convolution is one of the most crucial methods in signal processing and is used to transform one picture into a new one with customised attributes. CNN is a deep learning method that can analyse an input image, apply significance to various characteristics and objects within the image, and differentiate between them [13].

Figure 13.3 displays numerous structures in a variety of hues; however, it is unclear how a computer system will interpret or read this picture. Every channel in a standard image, also known as colour space, has a corresponding pixel value since RGB (RED, BLUE, GREEN) is the colour space utilised in computers. $B \times A \times 3$ in general refers to B rows, A columns, and three channels. $48 \times 48 \times 3$ in particular refers to 48 rows and columns, respectively, with three channels. Each pixel has a

FIGURE 13.1 Dataset of images of people without a mask—celebrity face dataset.

value in each of the three colour channels, which the computer interprets to produce a coloured image. We only have two channels for black-and-white photos, and each pixel in an image has a grayscale value, a shade of grey, and white.

A basic CNN architecture comprises four main layers:

i. **Convolution layer**: This layer is the main part of CNN. The initial convolutional layer could be followed by another convolutional layer. A kernel inside of this layer travels through the receptive areas of the image throughout the convolution process to determine if a feature is present.

The kernel runs over the entire image repeatedly. At the conclusion of each cycle, a dot product between the input pixels and the filter is computed. A feature map is the result of joining the dots in a certain pattern. The picture is finally converted to numerical values in this layer so that the CNN can recognise them and draw useful patterns from them.

ii. **ReLu layer**: Rectified linear activation function, or ReLU, is an ordered linear function that generates zero if the input is negative and the output is zero otherwise. Since a model that employs it trains more rapidly and often

FIGURE 13.2   Dataset of images of people with a mask.

outperforms other models, it has evolved to become the standard activation mechanism for many diverse kinds of neural networks.

iii. **Pooling layer**: The pooling layer applies a kernel to the input image similarly to the convolutional layer. Nevertheless, in contrast to the convolutional layer, the pooling layer reduces the number of input parameters and also causes some information loss. On the upside, this layer makes CNN more effective while reducing complexity [14].

iv. **Fully connected layer**: This layer is where image classification based on the features gathered from the underlying layers takes place. Fully linked here refers to the connection of each activation unit of the subsequent layer to each input of the preceding layer.

Consider a picture of form 10×10 (shown in Figure 13.4) and a kernel/filter of shape 3×3 (shown in Figure 13.5) with $I1$–$I64$ pixel intensity range and $F1$–$F9$ pixel intensity range, to comprehend convolution.

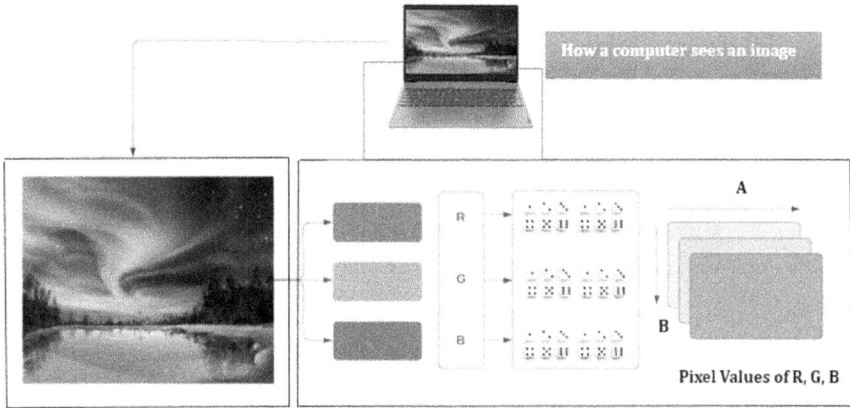

FIGURE 13.3   Computer image recognition.

I_1	I_2	I_3	I_4	I_5	I_6	I_7	I_8
I_9	I_{10}	I_{11}	I_{12}	I_{13}	I_{14}	I_{15}	I_{16}
I_{17}	I_{18}	I_{19}	I_{20}	I_{21}	I_{22}	I_{23}	I_{24}
I_{25}	I_{26}	I_{27}	I_{28}	I_{29}	I_{30}	I_{31}	I_{32}
I_{33}	I_{34}	I_{35}	I_{36}	I_{37}	I_{38}	I_{39}	I_{40}
I_{41}	I_{42}	I_{43}	I_{44}	I_{45}	I_{46}	I_{47}	I_{48}
I_{49}	I_{50}	I_{51}	I_{52}	I_{53}	I_{54}	I_{55}	I_{56}
I_{57}	I_{58}	I_{59}	I_{60}	I_{61}	I_{62}	I_{63}	I_{64}

FIGURE 13.4   Picture.

F_1	F_2	F_3
F_4	F_5	F_6
F_7	F_8	F_9

FIGURE 13.5   Kernel.

Convolution will be conducted by sliding the filter from top left to bottom right across the image. The equation below gives a mathematical explanation:

$$\sum_{k=1}^{m} \sum_{l=1}^{n} I\left(i+k-1,\ j+1-1\right) K\left(k,l\right)$$

where $j = 1$ to $(N-n+1)$ and $i = 1$ to $(M-m+1)$, respectively.

A CNN architecture can be explained as follows:

- A convolutional tool, through a procedure called feature extraction, isolates and identifies the distinctive aspects of a picture for analysis. There are several sets of convolutional/pooling layers that make up the feature extraction network.
- A fully connected layer uses the outputs of the convolutional process to predict the class of the image using the information obtained in previous phases.

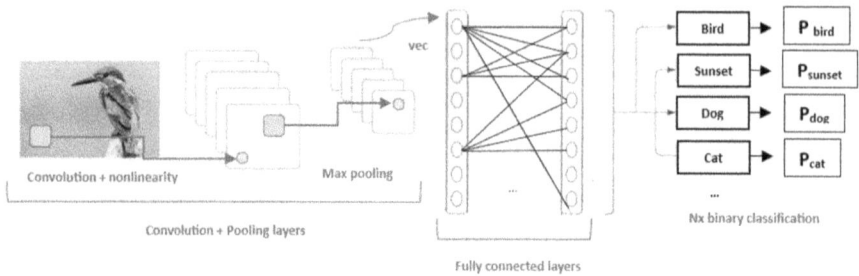

FIGURE 13.6 Convolutional neural network.

The objective of this CNN feature retrieval model is to extract as few features from a dataset as possible. The features of an initial set of features are combined into a single new feature to generate new features [15].

The convolution technique is repeated across all levels to build a network-like structure known as a convolutional neural network, as seen in Figure 13.6. This graphic depicts the structure of this network's CNN (Convolutional Neural Network). In order to produce SoftMax output, this network processes 11 layers worth of coloured face training photos as input. It also uses the backpropagation algorithm to train the network. The parameters that the model will learn from the algorithm also include the loss function. To produce projection vectors, the same picture that was earlier used as the input for the previous CNN is now run through PCA.

13.2.2.1 Principal Component Analysis

PCA is an approach to dimensionality reduction that attempts to retain as much information from the original dataset as feasible while reducing the number of features in a dataset into a fewer number of features called principal components.

A linearly dependent vector can be removed from a dataset or matrix using the PCA. The features are removed by reorienting the complete collection of data into a brand-new, rotated vector space where the data act as ideal data. The covariance matrix of the data is transformed into a sparse matrix during the entire process. This makes features less linearly dependent by bringing the inter-feature covariance to zero or extremely low values.

A short summarisation of the working of PCA is as follows:

    i. **Normalise the data**: We first standardise the data to get it ready for PCA. Each characteristic will thus have a mean of 0 and a variation of 1.
    ii. **Building the covariance matrix**: We utilise a square matrix to depict the association between two or more attributes in a multidimensional dataset.
    iii. **Find the eigenvectors and eigenvalues**: Eigenvalues/unit vectors and eigenvectors are further calculated. Eigenvalues are scalars that multiply the covariance matrix's eigenvector.
    iv. To get the primary component axis, rank the eigenvectors from highest to lowest.

v. Generate a feature vector to select which primary components to keep.

vi. Finally, data should be recast along the axis of the primary components (Figures 13.7 and 13.8).

13.2.2.2 Mask Detection

In this study, we used a unique convolutional architecture to perform mask recognition. It is essentially a classification technique that works best with RGB pictures.

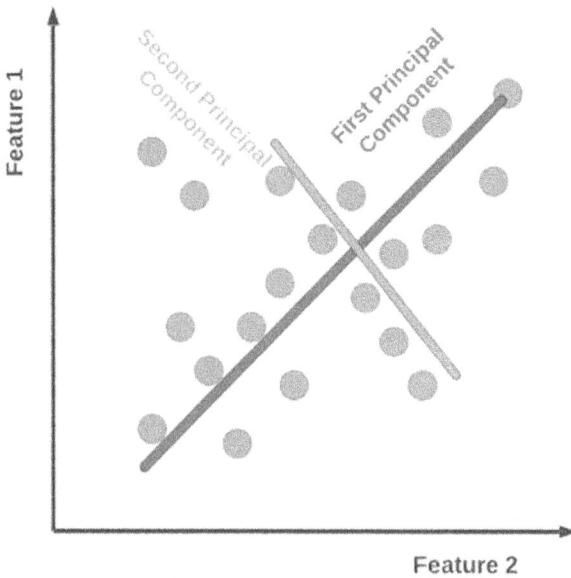

FIGURE 13.7 Finding principal components in PCA.

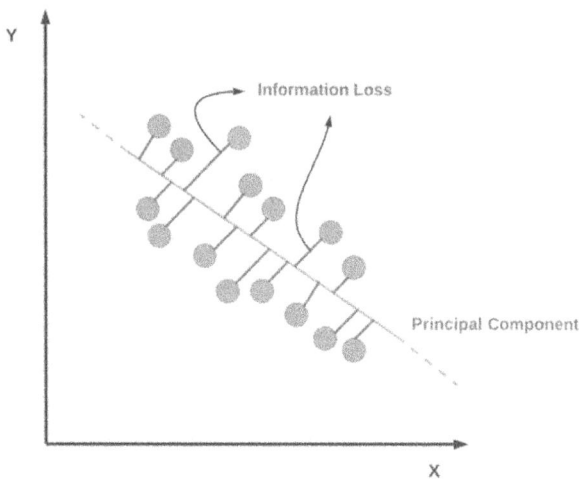

FIGURE 13.8 Identifying information loss and principal component.

Eight layers make up its architecture; the first five are convolutional layers, and the next layers are max-pooling layers, followed by completely linked layers. It mostly makes use of the ReLU activation function (linear function that outputs the input directly if positive or 0 otherwise), which outperforms tanh and sigmoid. Since the data's principal components equal the number of variables, the first principal component is designed to account for the greatest amount of variation in the dataset. The same formula is used to compute the second principal component as the first, but it accounts for the next-largest variance since it is uncorrelated with the first principal component. Until *p* primary components—the initial number of variables—have been calculated altogether, this process continues.

13.2.3 TOPOLOGY AND RESULTS

13.2.3.1 Data Preprocessing

Preparing any raw data for understanding is part of the mining method known as data preprocessing. Actual data may from time to time be unreliable, inconsistent, or lacking in particular tendencies. Since we are employing an image data collection with a large number of unrelated photographs or with form or dimension issues, we are preprocessing our data. We used the celebrity dataset and the RMFD (masked image dataset) in this investigation. We reorganised the datasets to process them, as shown in Figures 13.9–13.11.

13.2.3.2 Detected Faces from the Image

For facial recognition, we used a deep C2D-CNN algorithm. Decision fusion is used to blend the properties of the original pixels with the picture representation

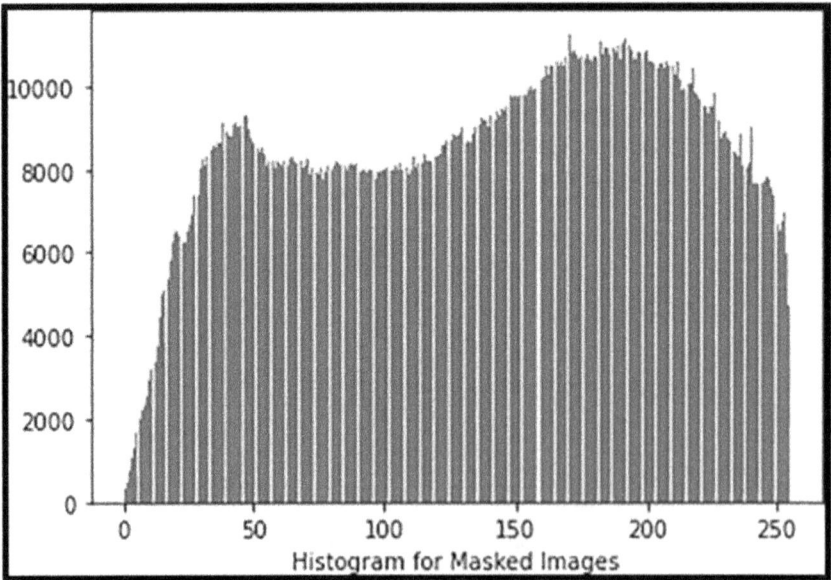

FIGURE 13.9 Histogram of average values of images with a mask.

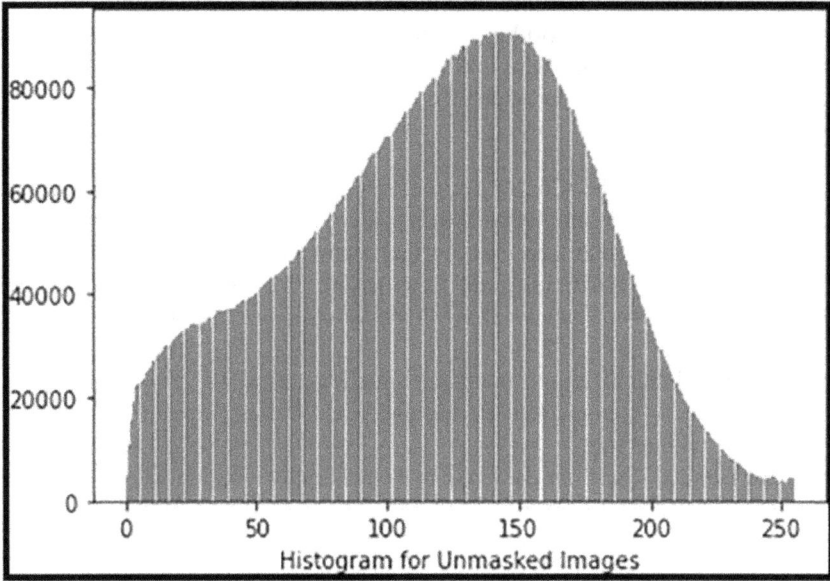

FIGURE 13.10   Histogram of average values of images without mask.

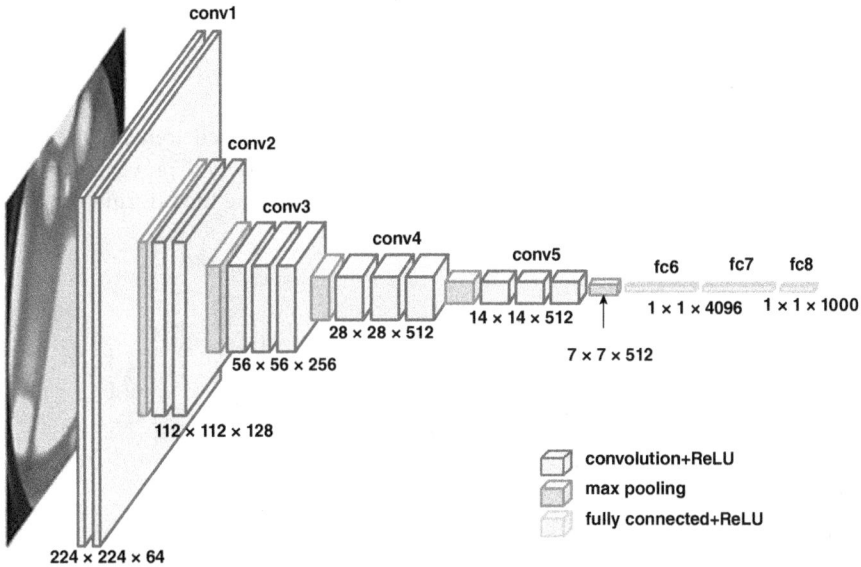

FIGURE 13.11   Detection of masks on faces.

recognised by CNN in order to enhance our face recognition model. The addition of a normalisation layer is the first of several procedures that this network goes through in order to reduce training time. To prevent gradient diffusion, a layered activation function will be employed. Last but not least, to preserve feature information as much as possible, probabilistic max-pooling would be used. We are using this method because the HOG algorithm, which focuses on forms, is unable to recognise individuals wearing masks. We are utilising the C2D-CNN technique because it is pixel-oriented and supports our approach.

13.2.3.3 AlexNet

In this research, we use the AlexNet architecture to categorise the data. AlexNet was the first convolutional neural network to employ GPU to improve performance. In 2012, it won the 'Imagenet large-scale visual recognition' competition. The model was proposed by Alex Krizhevsky and his colleagues in their 2012 research publication, *Imagenet Classification with Deep Convolution Neural Network*.

- Five convolutional layers, three max-pooling layers, two standardisation layers, two fully connected layers, and one SoftMax layer make up the AlexNet architecture.
- Convolutional filters and ReLU, a non-linear activation function, make up each convolutional layer.
- The pooling layers are used to carry out max-pooling. Additionally, the presence of completely linked layers causes the input size to be fixed.

This type of neural network employs convolutions and works well with RGB images. The basic description of the AlexNet architecture is shown in Figures 13.12 and 13.13.

13.3 CONCLUSION

As we are all too aware, the COVID-19 epidemic has caused tremendous harm across the world. Therefore, it becomes everyone's national duty to prevent its spread and to instil the values of conformity in society. By analysing the information from

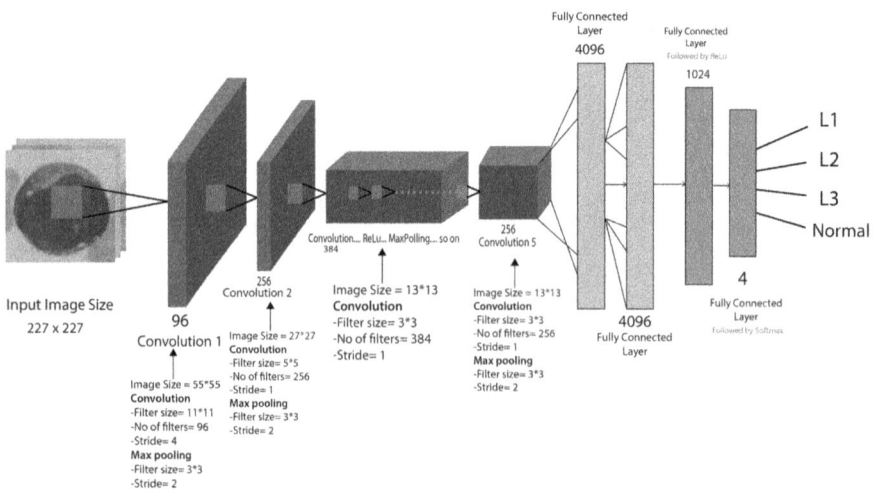

FIGURE 13.12 AlexNet architecture.

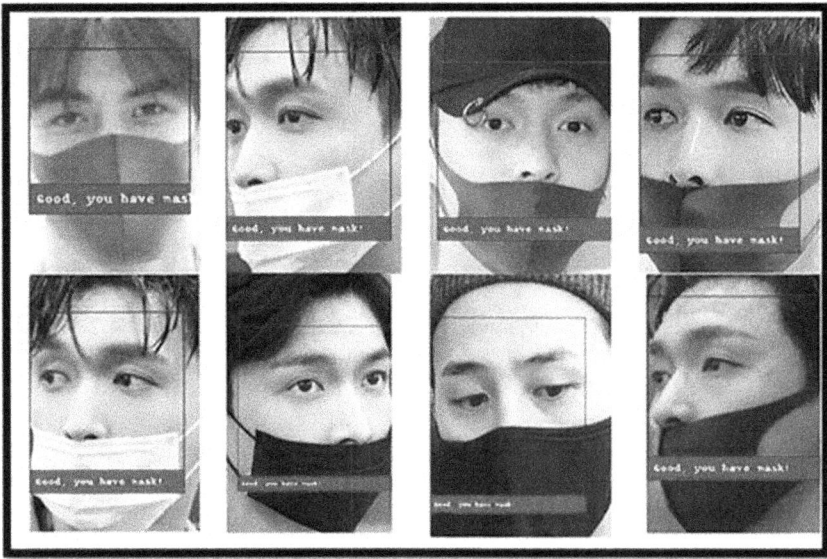

FIGURE 13.13   Result.

public surveillance cameras that use mask recognition to seek people who are not wearing masks, a lot may be discovered about the development of the disease. We will have to exercise more caution when dealing with large groups of people, and masks will start to become customary as a result of the corona virus's entrenched place in the world. We may infer from the data that the machine learning- and deep learning-based topology is better at producing accurate results and is more successful at containing the COVID-19 pandemic. The suggested technology might be improved for future projects by merging it with automated heat-sensing systems. Additionally, the system might be expanded such that it monitors the use of social distance in congested spaces. For biometric applications, face landmark detection might be incorporated as a feature. Furthermore, because innovative approaches are so adaptable, their structures might be improved to produce better outcomes more quickly. As seen in Figure 13.14, the application of deep learning techniques has increased.

Future research in this area might be conducted using a variety of different methodologies, taking advantage of their immense value. Images with inadequate light might be eliminated, enhancing dataset quality. However, the technology may be combined with a model to assess if enough physical distance is kept between individuals. Additionally, it might be used to a pattern that distinguishes a person's kind of mask. Machine learning algorithms may also be utilised to research new feature extraction techniques [16].

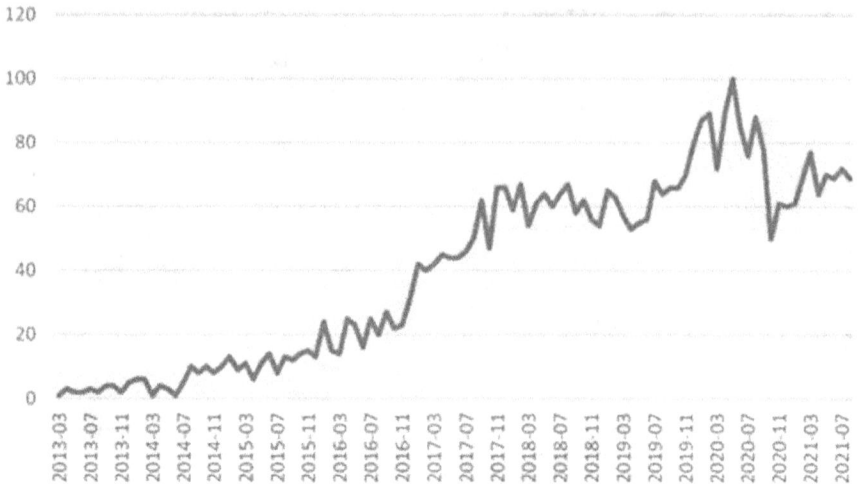

FIGURE 13.14   Increase in deep learning.

ACKNOWLEDGEMENTS

I would like to extend my sincere appreciation to Professor Archana Sahai for overseeing my study and for her patient direction, passionate support, and helpful criticism of my work. Additionally, I would like to extend my thanks to the AIIT department of Amity University Lucknow for providing me the opportunity of working on this project.

REFERENCES

[1] Vibhuti, J. N. (2022). Face mask detection in COVID-19: a strategic review. *Multimedia Tools and Applications*, 40013-40042. https://doi.org/10.1007/s11042-022-12999-6
[2] Gomathi, S., & Rashi, K. (2020). Pattern analysis: predicting COVID-19 pandemic in India using AutoML. *World Journal of Engineering, ahead-of-print*. Retrieved from https://www.emerald.com/insight/content/doi/10.1108/WJE-09-2020-0450/full/html
[3] Gupta, S., & Sreenivasu, S. V. N. (2021). Novel face mask detection technique using machine learning to control COVID'19 pandemic. *Materials Today Proceedings*, *80*(3), 3714–3718. https://doi.org/10.1016/j.matpr.2021.07.368
[4] Huang, C., & Ai, H. (2007). High-performance rotation invariant multi-view face detection. *IEEE TPAMI*, *29*(4), 671–686. Retrieved from https://pubmed.ncbi.nlm.nih.gov/17299224/
[5] Soni, M., & Joshi, S. (2020). Detection of COVID-19 cases through X-ray images using hybrid deep neural network. *World Journal of Engineering*, *19*(1), 33–39.
[6] Anushka. (2020). *OpenCV- a library with sight*. Retrieved from Infinite Potential: https://www.infinitepotential.in/opencv-a-library-with-sight/
[7] Li, J. Q. (2018). Robust face recognition using the deep C2D-CNN model based on decision-level fusion. *Sensors*, *18*, 2080. Retrieved from https://www.ncbi.nlm.nih.gov/pmc/articles/PMC6068932/
[8] Keboola. (2022, April 2). *A Guide to Principal Component Analysis (PCA) for Machine Learning*. Retrieved from Keboola: https://www.keboola.com/blog/pca-machine-learning

[9] Banik, A., & Shrivastava, A. (n.d.). Design, modelling, and analysis of novel solar PV system using MATLAB. *Materials Today: Proceedings, 51*(1), 756–763.

[10] Jun, B., Choi, I. (2013). Local transform features and hybridization for accurate face and human detection. In *IEEE TPAMI* (pp. 1423–1436). IEEE. Retrieved from https://pubmed.ncbi.nlm.nih.gov/23599056/

[11] Bhattacharya, S. E. (2021). Deep learning and medical image processing for coronavirus (COVID-19) pandemic: a survey. *Sustainable Cities and Society, 65*, 102589. Retrieved from https://www.ncbi.nlm.nih.gov/pmc/articles/PMC7642729/

[12] Das, T. K., & Banik, A. (2020). Energy-efficient cooling scheme of power transformer: an innovative approach using solar and waste heat energy technology. In *Advances in Thermal Engineering, Manufacturing, and Production Management. ICTEMA 2020*. IEEE.

[13] Ferguson, M., & Ronay, R. (2017). Automatic localization of casting defects with convolutional neural networks. In *2017 IEEE International Conference on Big Data (Big Data)* (pp. 1726–1735). IEEE. https://doi.org/10.1109/BigData.2017.8258115

[14] Shafique, S., & Tehsin, S. (2018). Acute lymphoblastic leukemia detection and classification of its subtypes using pretrained deep convolutional neural networks. *Technology in Cancer Research & Treatment, 1*(17), 1533033818802789. https://doi.org/10.1177/1533033818802789

[15] Yin, P. X. (2018). Linear feature transform and enhancement of classification on deep neural network. *Journal of Scientific Computing, 76*(3), 1396–1406. https://doi.org/10.1007/S10915-018-0666-1

[16] Keshari, S. (2019). *Introduction to Machine Learning and its Techniques*. Retrieved from Medium: https://medium.com/@eshukeshari/introduction-to-machine-learning-and-its-techniques-a383053539f2

Section 3

IoMT

Data Analytics and Use Cases

14 An IoT-Based Real-Time ECG Monitoring Platform for Multiple Patients

Vishnu S
Amrita Vishwa Vidyapeetham

Lova Raju K
VFSTR

Shanmugam M
Pondicherry University

Sharma S
Besoins Technologies

14.1 INTRODUCTION

Diseases that are long-lasting and incurable but can be controlled through continuous monitoring can be stated as chronic diseases. As humans are more concerned about their health, advanced technologies are implemented to monitor the variations in physiological signals in real time. According to the WHO Reports on Non-communicable Diseases, world statistics on the death rate are shown in Figure 14.1. 45% of deaths due to non-communicable diseases are because of heart diseases, 22% is due to respiratory diseases, 12% is due to cancer, 3% is due to diabetes, and 18% are due to other chronic diseases. The lack of acute care for the patient is one of the reasons behind the increased death rate.

Continuous health examination is an inevitable measure for patients who are suffering from chronic diseases [1] such as asthma, COPD, diabetes, arrhythmia, and cardiovascular diseases [2]. These patients don't require an active medication from the hospital, but continuous monitoring of vital signs is mandatory. Moreover, the long stay in the hospital will be unpleasant for the patients as well as for the bystander. This long stay in the hospital can be avoided if any alternate setup is made available in the patient's home to continuously monitor the vital signs.

The recent works in the fields of smart homes [3, 4], smart cities [5, 6], and smart healthcare [7, 8] show that how rapid the technology is redefining the quality of

DOI: 10.1201/9781003359951-17

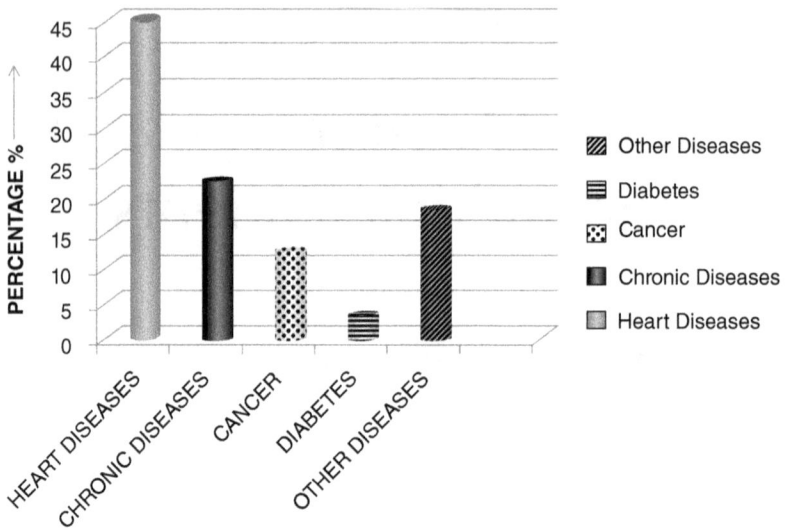

FIGURE 14.1 Statistics of death due to chronic diseases.

human lifestyle. The wireless sensor–based monitoring system [9] will be a solution to provide continuous monitoring to the patients if their conditions allow them to stay at the home itself.

Several systems have been developed for the remote monitoring of patient's health, which includes systems using Bluetooth [10, 11], ZigBee [12, 13], and Wi-Fi [14, 15] technologies for sending the data from the sensor module to PC [16] or personal digital assistant (PDA) [17]. High cost for the PC or PDA, complexity in usage, and accessibility are the main drawbacks of these systems.

In this chapter, an IoT-based remote-monitoring system is developed for patients who are suffering from cardiovascular diseases. The work is focused on the measurement of electrocardiogram (ECG) signals because it shows the reflection according to the heart's electrical activity and is useful to analyze various arrhythmias [18]. Patients are considered as nodes in the network, and continuous ECG signals of the patients are updated to the ECG-data server by establishing a secure connection via a home router. The doctor or the clinician can access the patient's profile to monitor the patient's vital signal status and initiate the necessary actions according to the health condition. With this system, the patient can get expert medical help irrespective of his location, and it is very much beneficial for patients who want to lead a normal life in their homes, even though they are under a continuous monitoring process.

The rest of the chapter is organized as follows. Section 14.2 presents the literature survey, Section 14.3 explains the system model which includes the home unit and ECG-data server, and Section 14.4 describes the evaluation of PWN in terms of current consumption. Finally, Section 14.5 gives the conclusion and Section 14.6 provides the current and future developments.

14.2   LITERATURE SURVEY

Recording of ECG signals, the transmission of recorded data to the hospital server, and the analysis of the received data are the challenging broad research areas in the remote patient monitoring system [19, 20]. Some of the existing systems are explained in this literature. A wearable mobile ECG detector based on Bluetooth is presented in Ref. [16]. The limited range of Bluetooth and the inability to send messages if any call is connected are the drawbacks of this system. A remote-monitoring system based on ZigBee protocol is presented in Refs. [12, 13], and the maximum range this system can achieve is 50 m. The work proposed in Ref. [21] introduced a real-time ECG analysis system, which is powered with a RISC-based ARM processor. The intention is to give warnings to the patients based on their physiological sign status. The work in Ref. [22] developed a GUI to represent the physiological parameters. A system for the early-stage prediction of cardiovascular diseases is developed in Ref. [23]. The limitation of this work is the usage of a PC or a Laptop in the acquisition module. A wearable breathing detection system is presented in Ref. [24]. The complexity of the algorithm used is the drawback of this system, as it consumes more power. To address the power consumption problem, another work is proposed to monitor the ECG and EEG signals with a CMOS-based analog front end [25]. High 'common mode rejection ratio' and reduced input referred noise are the highlights of this work. Since power consumption is less, even button cells can power the device. A wearable ECG sensor to detect arrhythmia is presented in Refs. [17,26]. ECG signals are sampled and sent to a PDA for the extraction of signals. The cost required for the handheld devices and usage of GPRS are the drawbacks of these systems. A mobile phone–based emergency alert system for heart patients is presented in Refs. [11,27]. The work proposed in Ref. [28] presents a system that uses a mobile device for the extraction and analysis of the vital signs of the patients. Compactness and mobility are the highlights of this work. The system utilizes the computing ability of mobile phones. However, it is only for self-diagnosis purposes.

The major drawbacks of smartphone-based health monitoring systems are as follows:

- Interruption due to other services like calls and messages.
- A 24/7 monitoring of the signal is not possible because of the limited battery backup of smartphones.

The work proposed in Ref. [29] presents a system to measure the physiological signals of multiple patients. The system is free from the usage of a PC/laptop in the home unit. The need for an intermediate node for establishing a connection with the home router and the wired connections in the home unit are the limitations of this system.

The main advantages of the proposed system over the existing systems explained previously in this chapter are as follows:

- Using a dedicated microcontroller cum network processor for the PWN development;
- Provides real-time ECG signals of the patient under observation;
- Easiness in configuration;
- Security is preserved by limiting access through login credentials.

14.3 SYSTEM MODEL

The network architecture of the proposed system is shown in Figure 14.2. The system model has two sections: the home unit and the ECG-data server. The home unit consists of PWN (patient wireless node) and WAP (wireless access point). In the ECG-data server, each patient will have a unique identity and profile with his details. The ECG signals from the PWN will be updated periodically to the respective patient's profile once the connection is established. A clinician or doctor who is having access to the patient's profile can view the ECG signal and can take action accordingly.

14.3.1 HOME UNIT

The home unit consists of PWN which senses the ECG signals of the patients via the electrodes attached to their chest and transmits the collected ECG signal samples to the ECG-data server via the home Wi-Fi router. The block diagram of PWN is shown in Figure 14.3. It consists of an ECG sensor (electrodes), a signal conditioning circuit, and a single-chip microcontroller with an integrated Wi-Fi module (NodeMCU). The ECG signals obtained from the patient's body will be in a range of 1 mV peak to peak. A minimum of 1 V is required for the proper processing of the signal by the microcontroller. So, signal amplification and noise cancellation are achieved by using the AD8232 IC.

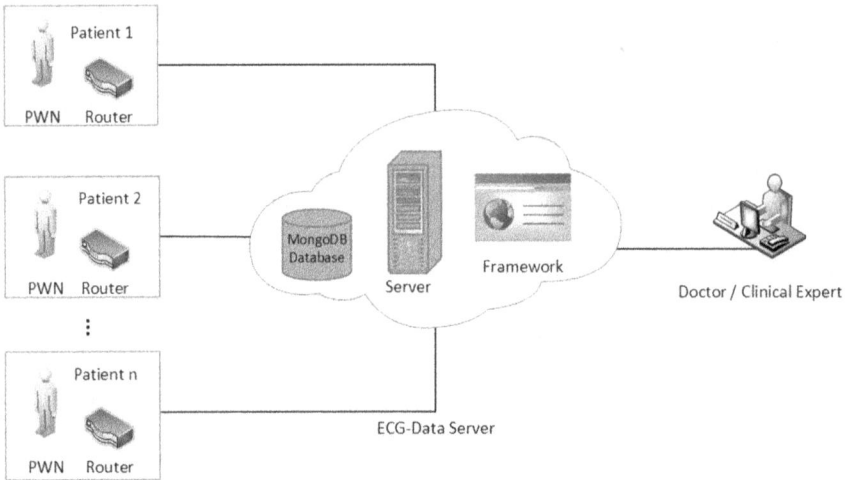

FIGURE 14.2 Network architecture of the proposed system.

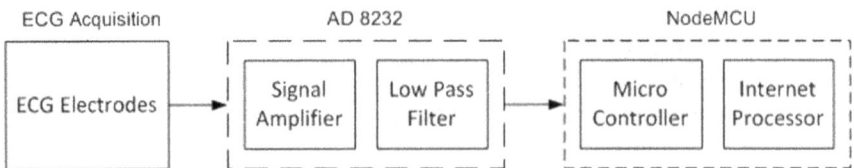

FIGURE 14.3 Block diagram of PWN.

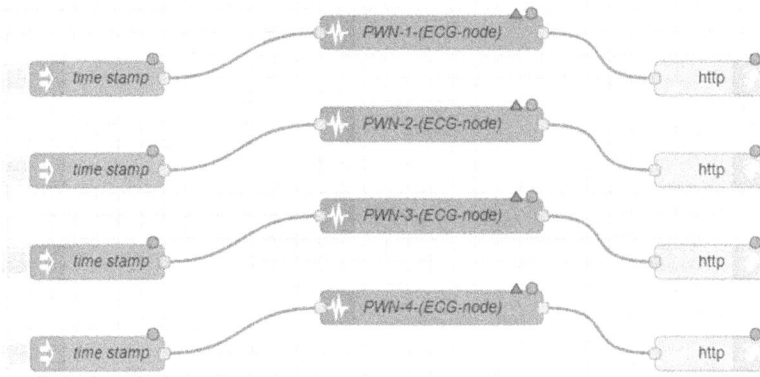

FIGURE 14.4   Node-RED deployment of PWNs.

The home router is establishing the Internet link between the PWN and the ECG-data server. In the proposed system, the WAP is the common router available in the home itself. TP-link-WE740N is used as the home router, which has a data speed of 150 Mbps and is compatible with 802 11b/g/n devices. The proposed system is to monitor multiple patients, and in this work, four nodes are developed and the system is able to monitor four patients in real time. Node-RED programming tool is used to develop the PWN's firmware, and an instance of the same is shown in Figure 14.4.

14.3.2   ECG-DATA SERVER

The real-time ECG signals of the patients can be accessed by the clinicians or doctors who have credentials to log in to the remote-monitoring webpage. They can add, remove, or update the patients' details in the database and access the existing patient's profile for ECG signal analysis in real time. The 'mlabs' database as a service feature with MongoDB is used to develop the patient's ECG database. The front end is developed with HTML and CSS, and the back end is powered with Python, Flask, and pymongo. The ECG-data server is made to run on the local host address 127.0.0.1 with a port number of 8000. Once the doctor's login is authenticated by the server, the home screen of the webpage with the options to Add Patient, Patient's History, and Diagnose Patient will appear as shown in Figure 14.5.

When a new patient comes under the division of remote health monitoring, the admin will add the patient's details such as name, gender, and date of birth. Once the details are entered, the web application will generate a random unique patient identifier (Patient ID) that corresponds to the newly added patient as shown in Figure 14.6. This identifier will be used for the authentication of the patient whenever his PWN requests establishment of a connection to the ECG-data server.

The doctor can log in to the web portal to access the profiles of patients under his treatment and can analyze the ECG signals to find any abnormalities. The registered patients with their IDs corresponding to the four PWNs are shown in Figure 14.7.

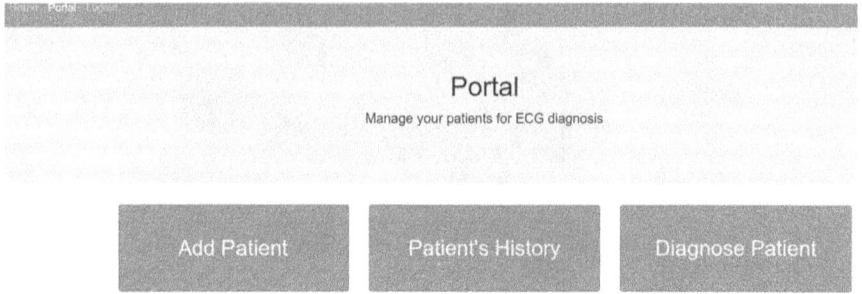

FIGURE 14.5 Home screen of the webpage.

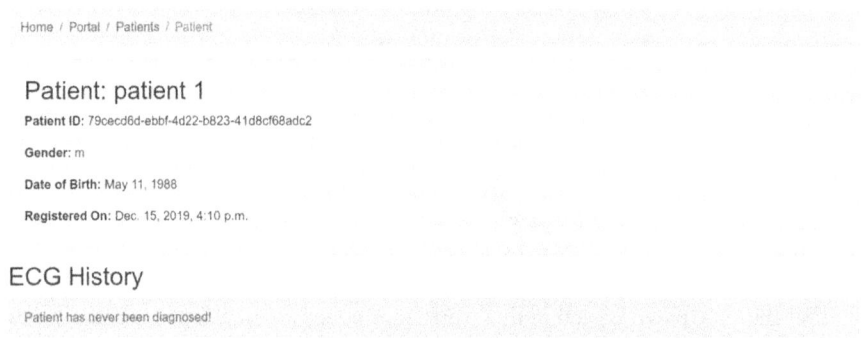

Home / Portal / Patients / Patient

Patient: patient 1

Patient ID: 79cecd6d-ebbf-4d22-b823-41d8cf68adc2

Gender: m

Date of Birth: May 11, 1988

Registered On: Dec. 15, 2019, 4:10 p.m.

ECG History

Patient has never been diagnosed!

FIGURE 14.6 Patient details stored with a unique Patient ID.

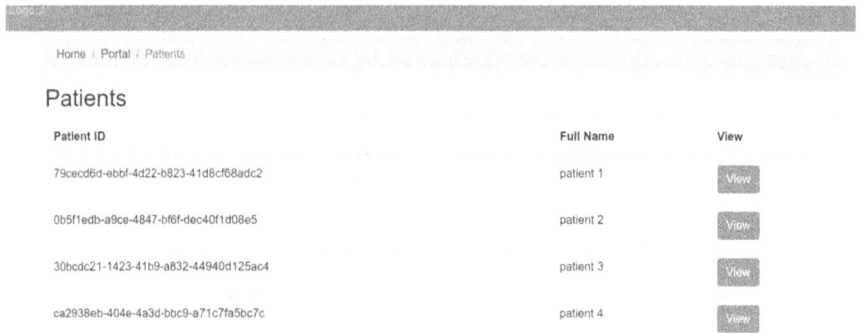

Patients

Patient ID	Full Name	View
79cecd6d-ebbf-4d22-b823-41d8cf68adc2	patient 1	View
0b5f1edb-a9ce-4847-bf6f-dec40f1d08e5	patient 2	View
30bcdc21-1423-41b9-a832-44940d125ac4	patient 3	View
ca2938eb-404e-4a3d-bbc9-a71c7fa5bc7c	patient 4	View

Home / Portal / Patients

FIGURE 14.7 Registered patients' profile.

Each patient's ECG signal can be viewed by clicking the view icon corresponding to the respective patients. The visualization of the ECG signal of patient 1 is shown in Figure 14.8.

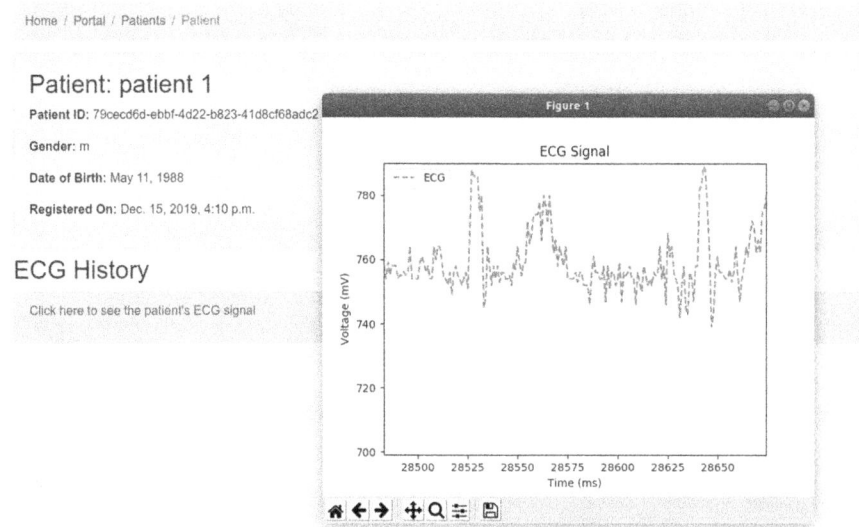

FIGURE 14.8   Plotted ECG signal of patient 1.

14.4   EVALUATION OF PWN

In this work, a three-lead ECG electrode setup is used to measure the ECG signal in real time. The complete functionality of the PWN is based upon the NodeMCU (ESP8266). The PWNs are powered with batteries, and the selection of the right battery is a critical part of the node design. The NodeMCU requires a 3.3 V power supply for its operation. An alkaline battery of size AA with a capacity of 2,400 mAh is chosen to power the NodeMCU. The voltage rating of a single AA battery is 1.5 V, so three batteries are connected in series to provide enough voltage supply to the NodeMCU. Calculations are carried out to determine the life time of the battery, i.e., up to how many days the PWN can transmit the signal once a new set of batteries are inserted. The calculations are performed under the following typical conditions:

- Normal supply voltage required for NodeMCU is 3.3 V.
- Current consumption for normal operation is 70 mA.
- Current consumption in deep sleep mode is 20 µA.
- Assume the module takes 5 seconds to wake up from sleep mode and to establish the connection with the server.
- The data will be sent to the server at a rate of 3 times in an hour, and each time, it will send the data for 60 seconds.
- 3 AA batteries of 2,400 mAh capacity are used to power the NodeMCU.

Equations (14.1) and (14.2) are used to estimate the power consumption per hour and the lifetime of the battery, respectively.

$$\text{Power consumtion per hour} = N \times (t_w + t_d) \times I_{\text{active}}$$
$$= 3 \times (5\text{ s} + 65\text{ s}) \times 70\text{ mA} \qquad (14.1)$$
$$= 13,650\text{ mAs}$$

$$\text{The life time of the battery} = B_{\text{Capacity}} \div \text{Power consumtion per hour}$$
$$= (2,400 \times 3,600)\text{mAs} \div 13,650\text{ mAs} \qquad (14.2)$$
$$= 632.96\text{ hours} \cong 26\text{ days}$$

where

B_{Capacity}	Battery capacity in mAs
I_{active}	Active mode current consumption
t_w	Wake time from deep sleep mode to establish connectivity
t_d	Data transmission time
N	Number of times the PWN sending data in an hour

14.5 CONCLUSION

A reliable and efficient real-time patient monitoring system can play a vital role in increasing the comfort of patients suffering from chronic diseases. In this chapter, a real-time ECG monitoring system is designed and implemented to provide high-quality medical services to patients irrespective of their locations. The system is free from the conventional usage of smartphones to establish the connection between the web servers and sensor nodes. The privacy of the patients is also considered in this work by limiting access to the patient's profile with login credentials. This work will be a solution to overcome the limitations of currently used devices such as Holter monitors and event monitors.

14.6 CURRENT AND FUTURE DEVELOPMENTS

The work presented in this paper needs further improvements in security features and power consumption to increase the feasibility and usability of the system. Also, the proposed work is using the home router to establish Internet connectivity with the server which results in lower mobility for the patients under monitoring. The mobility can be further increased by integrating 4G/5G modules in the PWN design.

REFERENCES

[1] Jimenez-Fernandezr, S., de Toledo, P., & del Pozo, F. (2013). Usability and interoperability in wireless sensor networks for patient telemonitoring in chronic disease management, *IEEE Transactions on Biomedical Engineering*, *60*(12), 33–40.

[2] Muhammad, Y., Temesghen, T., Hani, S., Baker, M., & OzgurSinanoglu, M. I. (2008). Ultra-low power, secure IoT platform for predicting cardiovascular diseases. In *IEEE Transactions On Circuits And Systems* (Vol. *64*). IEEE.

[3] Demiris, G., Hensel, B. K., et al. (2008). Technologies for an aging society: a systematic review of 'smart home' applications. *Yearbook of Medical Informatics*, *3*, 33–40,

[4] Cook, D., Schmitter-Edgecombe, M., Crandall, A., Sanders, C., & Thomas, B. (2009). Collecting and disseminating smart home sensor data in the CASAS project. In *Proceedings of the CHI Workshop on Developing Shared Home Behavior Datasets to Advance HCI and Ubiquitous Computing Research* (pp. 1–7). IEEE.

[5] Gubbi, J., Buyya, R., Marusic, S., & Palaniswami, M. (2013). Internet of Things (IoT): a vision, architectural elements, and future directions. *Future Generation Computer Systems*, *29*(7), 1645–1660.

[6] Petrolo, R., Loscrı, V., & Mitton, N. (2015). Towards a smart city based on cloud of things, a survey on the smart city vision and paradigms. *Transactions on Emerging Telecommunications Technologies*, *28*(1), 1–11.

[7] Doukas, C., & Maglogiannis, I. (2012). Bringing IoT and cloud computing towards pervasive healthcare. In *Proceedings of IEEE International Conference on Innovative Mobile and Internet Services in Ubiquitous Computing (IMIS)* (pp. 922–926). IEEE.

[8] Amendola, S., Lodato, R., Manzari, S., Occhiuzzi, C., & Marrocco, G. (2014). RFID technology for IoT-based personal healthcare in smart spaces. *IEEE Internet of Things*, *1*(2), 144–152.

[9] Shah, T., Yavari, A., Mitra, K., SagunaSaguna, P. P. J., & Rajiv Ranjan, F. R. (2016). Remote health care cyber-physical system: quality of service (QoS) challenges and opportunities. *IET Cyber-Physical Systems, Theory Applications*, *1*(1), 40–48.

[10] Noureddine, B., & Fethi, B. R., (2010). Bluetooth portable device for ECG and patient motion monitoring. *Nature & Technology*, 19.

[11] Saravanan, S. (2014). *Remote Patient Monitoring in Telemedicine Using Computer Communication Network Through Bluetooth, Wi-Fi, Internet Android Mobile*, IJARCCE.

[12] Deshmukh, N. R., Nagekar, N. S., & Harde, A. M. (2014). Wireless ECG monitoring on computer using ZigBee technology. *Discovery*, *18*(52), 58–63.

[13] Ramu, R. & Kumar, S. (2014). Real time monitoring of ECG using ZigBee technology. *IJEAT*, *28*(7), 316–320.

[14] Deshmukh, R. S. (2013). Wi-Fi based vital signs monitoring and tracking system for medical parameters. *IJETT*, *4*(5), 1935-1938.

[15] Ahammed, S. S., & Binu, C. P. (2013). *Design of Wi-Fi Based Mobile ECG Monitoring System on Concerto Platform*. Elsevier.

[16] Jun, D., & Zhu, H. H. (2004). Mobile ECG detector through GPRS/Internet. In *Proceedings of 17th IEEE Symposium On Computer Based Medical Systems*. IEEE.

[17] Zhang, F., & Lian, Y. (2009). QRS detection based on multiscale mathematical morphology for wearable ECG devices in body area networks. *IEEE Transactions on Biomedical Circuits and Systems*, *3*(4), 220–228.

[18] Ting, L., & Yong, J. Y. (2016). Wearable medical monitoring systems based on wireless networks: a review. *IEEE Sensors Journal*, *16*(23), 8186–8199.

[19] Vishnu, S., JinoRamson, S. R., LovaRaju, K., & Anagnostopoulos, T. (2019). Chapter 11 simple-link sensor network-based remote monitoring of multiple patients. *Intelligent Data Analysis for Biomedical Applications, 2019,* 237–252.

[20] Sahil, M. A., Abbas, H., Saleem, K., Yang, X., Derhab, A., Orgun, M. A., Iqball, W., Rashid, I., & Yaseen, A. (2018). Privacy preservation in e-healthcare environments: state of the art and future directions. *IEEE Access, 6,* 464–478.

[21] Cai, K., & Liang, X. (2010). Development of remote monitoring cardiac patient monitoring system using GPRS. In *International Conference on Biomedical Engineering and Computer Science (ICBECS), April, 2010.* IEEE.

[22] Rajput, M., Pai, S., & Mhapankar, U. (2013). Wireless transmission of biomedical parameters using GSM technology", *IJESE, 1*(9), 83–85.

[23] Sukanesh, R., Veluchamy, S., & Karthikeyan, M. (2014). A portable wireless ECG monitoring system using GSM technique with real time detection of beat abnormalities, *International Journal of Engineering Research, 3*(2), 108–111.

[24] Corbishley, P., & Villegas, E. R. (2008). Breathing detection: towards a miniaturized, wearable, battery operated monitoring system. *IEEE Transactions on Biomedical Engineering, 55*(1), 196–204.

[25] Ng, K. A., & Chan, P. K. (2005). A CMOS analog front-end IC for portable EEG/ECG monitoring applications. *IEEE Transactions on Circuits and Systems, 52*(11), 2335–2347.

[26] Fensli, R., Gunnarson, E., & Gunderson, T. A wearable ECG-recording system for continuous arrhythmia monitoring in a wireless tele-home-care situation. In *18th IEEE Symposium on Computer-Based Medical Systems (CBMS'05), 2005.* IEEE.

[27] Shebi, A. S, Pillai, B. C., & Rajesh, R. (2013). Design and implementation of Wi-Fi based imaging system on concerto platform'. In *Annual International Conference on Emerging Research Areas.* IEEE.

[28] Mahmud, M. S., HonggangWang, A. M., & Esfar-E-Alam, A. (2017). Wireless health monitoring system using mobile phone accessories. *IEEE Internet of Things Journal, 4,* 1009–2018.

[29] Dilmaghani, R. S., Hossei, B., Choobkar, M. G. S., & Charles, W. (2011). Wireless sensor networks for monitoring physiological signals of multiple patients. *IEEE Transactions on Biomedical Circuits and Systems, 5*(4), 347–356.

15 Study of Anomaly Detection in Clinical Laboratory Data Using Internet of Medical Things

Richa Singh
KIET Group of Institutions

Nidhi Srivastava
Amity University

15.1 INTRODUCTION

A major portion of medical decision-making is driven by laboratory data, and the number of laboratory tests ordered annually is rising [1]. Laboratory testing must be quick and precise in order to provide the proper illness diagnosis, prognosis and therapy. Errors are unavoidable because of the enormous volume of data and complexity in collecting samples from the bedside to a test result. These mistakes result in further testing, higher expenditures, patient concern and, in the worst-case scenario, patient morbidity or fatality. The digital transformation of health data has amazing potential to harness this enormous quantity of information to enhance the identification of abnormal findings in the clinical laboratory, a process that is already challenging given the rising test volume. This review discusses a variety of "anomaly detecting" methods. Any outcome that considerably deviates from what is expected is considered an abnormality, broadly speaking. According to statistics, any observation that deviates sufficiently from the predicted distribution may be classified as an anomaly. The concept of variance is at the core of the definition. Laboratory testing is impacted by the tremendous variation in human health; analytes may vary depending on the diet, posture, or time of day of the patient; reference intervals fluctuate with age and sex; and disease states create predictable variations in findings as they proceed. All of this is a natural function of physiology. "Biological variance" is a term that is frequently used to describe variation that emerges from variations in individual physiologies. Clinicians utilise changes in

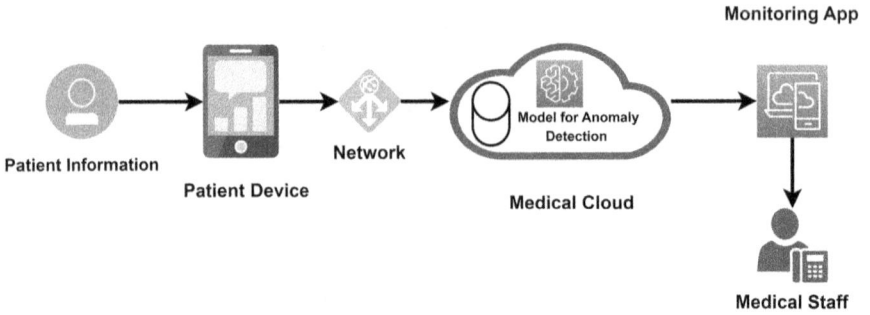

FIGURE 15.1   Anomaly detection using Internet of Medical Things.

test findings to infer changes in physiology when making difficult medical decisions for their patients. "Laboratory mistakes" are acts or processes that lead to results that differ from the underlying biology. Anomalies caused by these variables must be correctly and quickly discovered before they leave the laboratory and reach the patient's medical record, where they may result in inaccurate diagnoses or wrong therapies. This deviation from the biological reality is referred to as "statistical error". Clinical laboratories have created several strategies to reduce the reporting of aberrant test findings in order to address this problem, which is undoubtedly not new [2]. But the digital revolution in healthcare has made it possible to combine, synthesise and use vast volumes of data to create new processes for enhancing this procedure. Figure 15.1 shows the basic concept of Internet of Medical Things (IoMT) for anomaly detection [3].

15.2 LITERATURE SURVEY

Although there are many ways to define an anomaly in clinical laboratory data, D.M. Hawkin's description, "an observation which deviates so much from the other observations as to provoke suspicions that it was created by a distinct process," may be the most accurate one [4]. To comprehend the abnormalities in the context of the clinical laboratory, one must comprehend how the observations are made. The workflow for producing test results is a complicated system with plenty of moving pieces, often known as the overall testing process (TTP) shown in Figure 15.2. In this, each TTP step has its different value known as error which might take the form of random noise or a systematic bias, in terms of statistics [5].

The biological truth and the whole amount of the errors from all the procedure steps make up the final laboratory result. The final result might be categorised as an anomaly if it contains an excessive amount of this inaccuracy in one direction or the other. The difference between "error" in the statistical sense and laboratory mistakes must be made. The term "laboratory errors" often refers to uncommon TTP acts or mistakes that, if substantial enough, can contribute considerable quantities of statistical inaccuracy into a result and result in anomalies. Technical anomalies are those that arise from mistakes in the laboratory, as opposed to

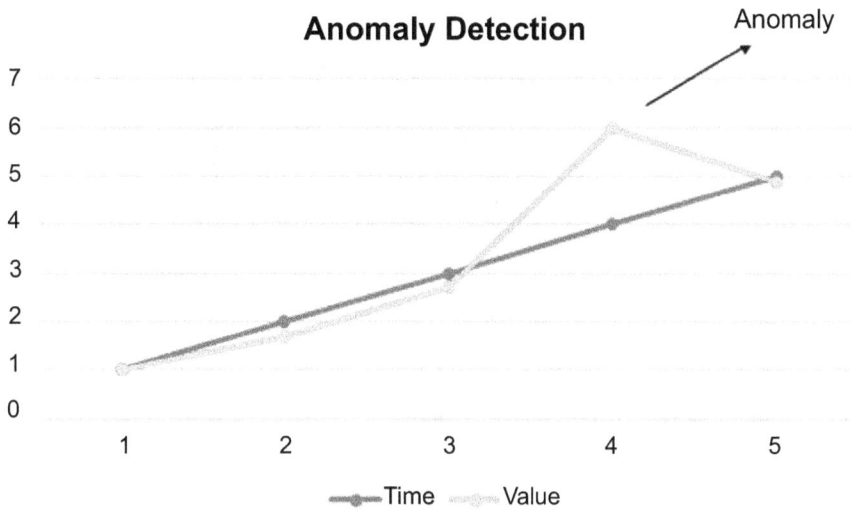

FIGURE 15.2   Unusual pattern from original pattern is anomaly.

biological anomalies, which arise from biological variation within a population (Figure 15.2). Any anomaly, regardless of its cause, must frequently be swiftly and precisely identified in order to prevent a medical mistake that might endanger the patient. The TTP error rate has been estimated via several studies, with varying approaches yielding wildly divergent results. Carraro and Plebani presented the most often referenced set of findings in 1997, and follow-up research was conducted in 2007 [6, 7]. A crucial point to emphasise is that these estimations were the product of human interpretations of the data, necessitating the involvement of nurses and doctors as the screening step for alerting the laboratory to potential anomalies.

Writers have calculated error rates in their own labs using a wider range of potential mistake sources. An extensive analysis of these initiatives is provided by Mrazek et al. [8]. The specimen, the analyser, the patient and the population are among the frequent causes of the abnormalities that the algorithms are designed to identify; therefore, understanding these sources is necessary before algorithm assessment can start. The anomalies that arise from the various factors that affect analytical performance can be linked to two main causes of error: low accuracy or poor precision.

The term "bias" describes an error in a measurement that frequently overestimates or undervalues the actual outcome, such as calibration drift. The Levey–Jennings plot [9] is the most generally used method for analyser-based anomaly identification, and the Westgard rules [10] are frequently used on it. Internal QC has recently come under more scrutiny, and the use of QC standard materials, which can be costly, inconsistent between batches, and difficult to commutate, has decreased [11–14]. Additionally, there is no clinically approved timetable for the ideal timing of these runs for each analyser. When these controls are used too frequently, it wastes both technician and instrument time and raises expenses. The requirement for repeating samples or a considerable number of analytical mistakes might result from not running them often enough. One of the primary reasons for using "live" detection methods is the lag between the failure occurrence and its detection. Online techniques are those that continuously synthesise and evaluate data as they are generated in real time. This necessitates the use of actual patient data in the clinical laboratory as opposed to calibration materials and standard QC regimens. "Patient-based, real-time quality control" (PBRTQC) is the name given to this approach to the live identification of analyser flaws using actual patient data [15].

PBRTQC operates by examining a condensed window of recent outcomes, generating a summary statistic (such as the mean, median, or positive rate) and establishing acceptable bounds for the statistic's short-term variance relative to its mean. There are several ways to define these data and associated alarm thresholds. Running sum-of-positives [18, 19] and average-of-normals [16, 17] are two straightforward, straightforward statistics that are effective at detecting changes in analyser bias. The performance of bias detection can be enhanced by more complex operations such as exponential weighting [20, 21], Bull's method [22] and simulated annealing [23, 24]. Algorithms that integrate measures of variation such as shifting standard deviation and sum-of-outliers can be used to detect changes in imprecision [24].

PBRTQC has a number of benefits over traditional QC. It carries out online detection, which reduces the quantity of samples that must be recalled in the event that an analyser requires recalibration that eliminates the problem of commutability because the samples are from actual patients. It adds sensitivity to recurrent preanalytical errors, and its numerous tuning parameters allow laboratories to significantly control the trade-off between sensitivity and specificity [25, 26]. Nevertheless, PBRTQC has its own limits and is not intended to take the place of traditional internal QC.

The future of analyser-based anomaly detection certainly rests as much on how it can reduce operating expenses and streamline laboratory processes as it does on how well it can detect anomalies. Regular traditional QC demands a lot of material, technician and director time, as well as monitoring to maintain a high sensitivity. Algorithms that can include linear time series data in a repeatable and understandable manner are necessary for the transition from a proactive, schedule-based approach to a collection of algorithms focused on detecting and forecasting drift. When determining if recalibration is required, these algorithms must be able to respond quickly to changes in the input data and generate prediction labels that a medical director may include into their own heuristic. This theory might be used in a number of ways, such as reinforcement learning, hidden variable decompositions and autoregressive integrated moving averages [27–31]. The detailed description of various machine learning algorithms on the basis of its category used for anomaly detection in clinical data is shown in Table 15.1.

TABLE 15.1

Machine Learning Algorithms for Anomaly Detection in Clinical Laboratory Data

Reference	Categories	Machine Learning Algorithm	Advantages	Gap
[36]	Supervised technique	Logistic regression	Easy to use. Simple to use.	Classification is challenging when there are nonlinearly separable classes.
[36]	Supervised technique	Support vector machine	Improved efficiency in datasets with a small number of classes and several instances per class. Scalable. Decreased storage needs.	It might be difficult to determine which kernel function is best.
[32]	Unsupervised technique	K-Nearest neighbour	Easy to use. Simple to use.	A challenge to locate the ideal k. When the number of the k variable, the number of data points, or the number of classes rises, the computing performance falls.
[36]	Supervised technique	Decision tree	Easy to use. Performance is the same for parameters that are separated linearly and nonlinearly.	Susceptible to being overfitted. Unstable (minor data variations may lead to the creation of Decision Trees (DTs) that are very different from one another).
[36]	Supervised technique	Naïve Bayes	Both binary and multi-class classifications are possible to utilise. Easy to use. It only takes a few examples to train.	Low classification accuracy may be caused by the assumption of feature independence. Problem with "zero frequency". A class will be given a chance of zero if it does not arise during training.
[32]	Unsupervised technique	Random cut Forest	Able to withstand overfitting. Features are chosen automatically. Less inputs are necessary.	Fast only when there are few trees present. May call for big datasets.

15.3 TYPES OF ANOMALIES IN CLINICAL LABORATORY

15.3.1 DETECTION OF SPECIMEN-BASED ANOMALIES

While specimen-based approaches may study inconsistent data across many analytes and use traditional relationships between results to help in the discovery of those likely to be anomalies, analyser-based anomaly detection methods concentrate solely on a single analyte (Figure 15.3).

Anomalies based on samples are frequently the consequence of preanalytical mistakes, such as haemolysis, contamination, incorrect labelling, incorrect tube type and transit delays. The approach to "wrong-blood-in-tube" mistakes that might lead to disastrous transfusion responses has been revolutionised by the use of radio-frequency identification, multiple user label verification and the check sample [1, 32]. Spectrophotometric screening indices for haemolysis, icterus and lipoma are another technique that does not rely on laboratory test findings directly (HIL). For specimens with these interferences, their widespread usage has substantially enhanced the process. However, there are no accepted standards for handling these samples, and thus each laboratory must create its own methods. These index values can be correlated to the amount of each inference in a specimen and calibrated with respect to its clinical impact on the reported laboratory results using mathematically derived and clinically validated threshold optimisations, such as those provided by Ho et al. [33] and Mays et al. [34]. Clinical practitioners are generally aware of these modifications; therefore, this study will instead focus on fresh, data-driven methods for finding more cunning origins of specimen-based aberrations. The final chemical findings could be challenging to distinguish from their uncontaminated equivalents depending on the level of contamination.

A decision tree approach was shown by Baron et al. to distinguish between actual and fake glucose values [35]. Doctor first utilised a similar method to foretell aberrant outcomes in liver function tests, demonstrating the effectiveness of the naïve Bayes classifier in simulating and diagnosing wrong-blood-in-tube problems using just the data themselves [36, 37].

Benirschke and Gniadek recognised this as a great target for management when they described how a logistic regression model can successfully discriminate Proof of Concept (POC) samples with pseudohyperkalemia from haemolysis [38].

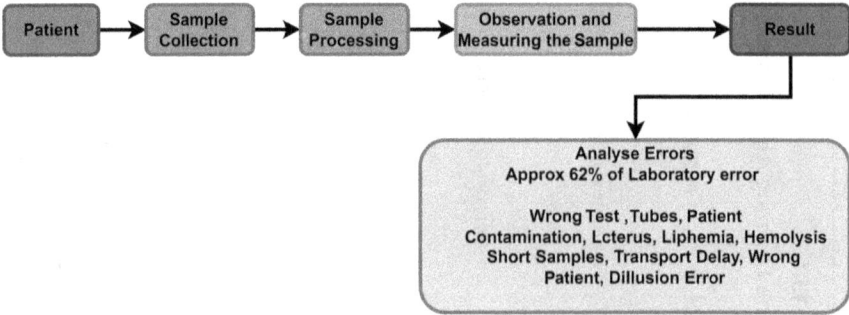

FIGURE 15.3   Specimen-based anomaly detection.

In today's clinical laboratory, specimen-based anomaly detection is especially important due to the substantial load of preanalytical mistakes on the TTP. As a result, several writers have created solutions for specific situations that use the previously briefly stated patterns between analytes to identify abnormalities. These methods all share the same flaw, though, in that they can't properly train and test their models without access to a sizable dataset of labelled normal outcomes and abnormalities.

Unsupervised or semisupervised methods may be used in future specimen-based anomaly detection techniques to lessen the need for labelled datasets. Creating a representation of the latent structure contained in a data collection and then transforming the data to bring comparable points closer together is a common feature of such methods (Figure 15.4). These adjustments, also known as "embeddings," provide the original data-enhanced representations that highlight the characteristics that distinguish one class from the others. This makes it simpler to carry out tasks with great precision and accuracy, such as labelling things as "normal" or "anomaly." Application of semisupervised architectures (e.g., positive and unlabelled learning, self-supervised modelling) to use subsets of the data that do have trustworthy labels as the inputs for models, which can then be expanded to new data that are still unlabelled, is one theoretical approach with significant promise. Given that labelling is a problem that affects all disciplines equally, this is an area that is fast developing and will probably enter the clinical laboratory space sooner rather than later. Although relabelled data are uncommon and expert label assignment takes time, the advantages of supervised techniques make them a high-value target for innovation. High-value anomaly types might be simulated using generative models to act as a stand-in for labelling specialists, particularly if those anomalies originate from a source with

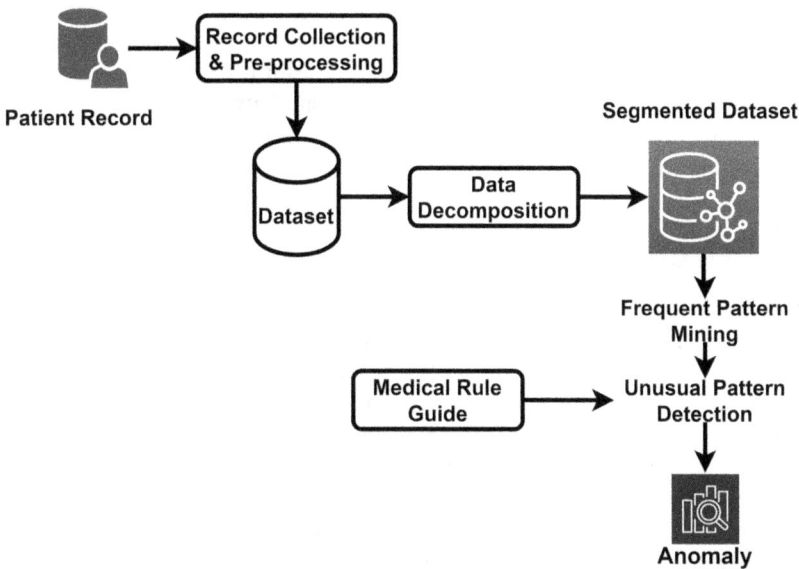

FIGURE 15.4   Detection of anomaly in patient.

a predictable impact on the tests they influence (e.g., haemolysis, IV fluid contamination). In lieu of relabelled datasets, supervised machine learning algorithms might be given the output of generative models such as latent Dirichlet allocation [39], Gaussian mixture models [40], Bayesian networks, or, more recently, generative adversarial networks [41] and variational autoencoders [42]. These methods would need to be thoroughly validated before being used on actual patient outcomes, but they provide a more workable option than the manual labelling that would otherwise be necessary.

15.4 DETECTION OF ANOMALIES IN PATIENTS

Ladenson, who originally used the phrase "delta check," initially explained this idea in 1975 [43]. By including data on the rate of change, percentage of change, difference of differences and reference change value, this method has been enhanced and iterated upon [44–50]. The thresholds have been enhanced by the addition of several analytes and statistical procedure optimisation [51–56]. The thresholds have been enhanced by the addition of several analytes [56] and statistical procedure optimisation. However, with the widespread use of barcoding and other process enhancements, specimen mix-ups have significantly decreased, prompting some writers to wonder what the delta check will be used for in the clinical laboratory in the future [57, 58]. Although there are a lot of biological variations across individuals, there is far less variation from day to day within the same patient. By taking advantage of this discrepancy, a patient's historical outcomes may be used as their own internal control from which no individual analyst should vary noticeably without a clinical explanation (Figure 15.4) [59, 60].

Optimal resource use is a common defence used against delta checks. The fundamental idea is still valid though, and more recent machine learning approaches have shown enhanced positive and negative predictive values which provide compelling justifications for their adoption as best practises [61–63]. These strategies have shown a lot of promise for enhancing the harmony between pursuing a policy of zero-tolerance for specimens that have been incorrectly labelled and avoiding the overwhelming workflow with findings that call for technician intervention.

15.5 DETECTION OF POPULATION-BASED ANOMALIES

The strategies that have been covered thus far have mostly concentrated on technical abnormalities and how laboratory findings are produced. But a significant amount of the variation in test results is biological [64]. Extreme laboratory findings that signify genuine biological abnormalities ought to be accompanied with a clinical explanation, although these clinical reasons aren't often immediately apparent. Laboratorians must comprehend the significance of clinical context in order to be persuaded that an anomalous value is the consequence of a real biological extreme and not a laboratory error. For instance, a creatinine test of 1.3 mg/dL in a male bodybuilder would not be noteworthy, but in a pregnant woman, it might be the first indication of life-threatening pre-eclampsia. In a mathematical sense,

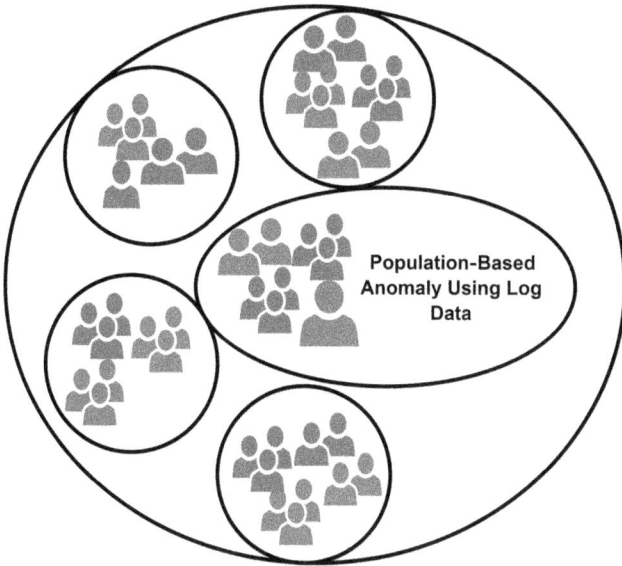

FIGURE 15.5   Population-based anomaly.

when physicians use clinical context, they are redefining the population from which the result was produced. After the population is changed, the provider's internal heuristics determine how likely and significant the outcome is to be in the modified population as shown in Figure 15.5. These mental arguments are simple for skilled providers to execute, but incorporating this decision-making process into an automated system is far more challenging undertaking. The creation of reference ranges is one area where the process of redefining the normal population is essential. The initial demographically unique populations with an emphasis on age, sex and race were utilised to deviate from global reference ranges [65–67]. Since then, as labs have studied the contextualisation of test data and reference ranges in the era of "personalised medicine" [68–75], reflection within the industry has produced the following unresolved issues, i.e., the suitable ranges for making clinical decisions and those range who promote patient care based on evidence and who doesn't promote.

If we divide reference ranges based on things like colour, age and sex, how can we assure fairness in healthcare delivery? As laboratories provide novel descriptions of what is normal or abnormal for a given analyst, based on patients and clinical situation, but still correct identification of anomalies is very difficult. These questions will become increasingly important. The practical dilemma of how to create and use clinically informed criteria for test data continues even while these discussions go on. The vast quantity of metadata in the electronic medical record is ready to be mined, but choosing which analysis to run first should include a multidisciplinary effort. When attempting the difficult process of generating clinically informed reference ranges, it will be crucial to enlist the assistance of the providers who must act on these laboratory results.

15.6 TYPES OF ALGORITHMS FOR ANOMALY DETECTION

The laboratory's most important function is to support illness diagnosis and prognosis. There are basically three algorithms, i.e., classification, regression and spatial sampling density in which regression or classification are the fundamental mathematical operations that might be used to describe these behaviours.

> **Classification**: Assigning an observation to a set of related observations is referred to as classification. The majority of categorisation in medicine is binary referring to just two potential states. Despite having a variety of labels, such as positive or negative, normal or aberrant, present or absent, true or untrue, these states are equally statistical. Any medical categorisation issue aims to provide an observation a name that may be utilised to enhance clinical judgement.
>
> **Regression**: Deals with quantifying a value. It demands continuous scale data analysis rather than the categorical labelling used in classification. It frequently entails creating a set of independent variables that may be used to predict an outcome variable. These independent variables are known as "predictors" in statistical literature or "features" in the context of machine learning. The "outlier score" that regression algorithms offer in anomaly detection might be a numerical indicator of how unusual an observation can be. Results from a regression technique offer a level of specificity that classification approaches do not but it also takes a solid base of knowledge to be able to interpret a continuous result in a way that is clinically relevant. After defining normal, may use a variety of decision boundaries to divide the anomalies. Anomaly detection methods greatly differ in how they approach this decision-making process.
>
> **Spatial sampling density**: There are various techniques for spatial sampling density such as k-nearest neighbours, local outlier factor [76, 77], isolation forests [78] or distance measure. Apart from this, principal component analysis, support vector clustering [79], manifold learning [80] and autoencoders [79] can create a whole new space that captures more significant aspects inside the latent structure of the data.

15.7 SMART E-HEALTHCARE AND IoMT

The digital revolution in the medical industry began sooner than in other industries, it has advanced more slowly [81]. Medical care has drawn personal, societal and even national attention as a result of the rapid advancements in science, technology and the economy. The conventional medical approach contains issues including the cost of therapy, the difficulty in getting a doctor's appointment and the concealment of medical information. The IoMT is the central component of the medical industry's digital revolution and the concentrated manifestation of IoT technology in the medical sector. The IoMT combines the Internet of Things with medical equipment (IoT). Future healthcare systems will be based on IoMTs where every medical gadget will be connected to the Internet and supervised by medical specialists. As it develops, this promises good speed and less expensive health treatments.

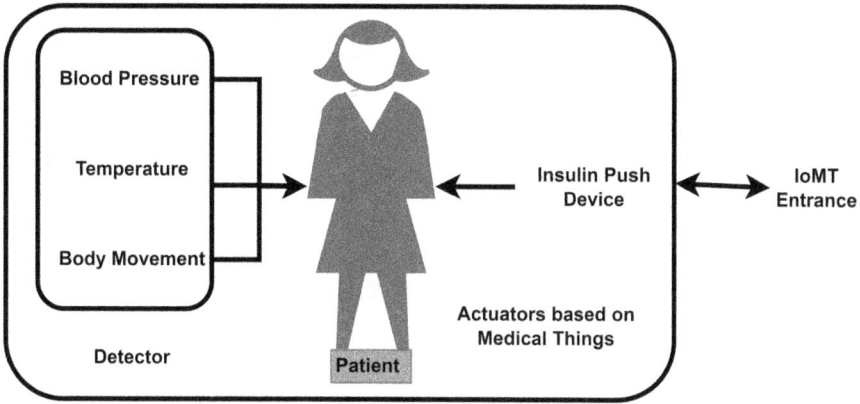

FIGURE 15.6   IoMTs architecture in patient environment.

15.7.1 ARCHITECTURE

The IoMTs architecture (Figure 15.6) have a patient environment which include bio-sensors as well as IoMT actuators.

An actuator is basically a machine component which is used for controlling and moving any system. Bio-sensors includes various sensors such as blood pressure sensor which is responsible for blood pressure of the patient, body temperature sensor gives the information about temperature and motion sensor indicates that the body parts are in moving condition or not. Monitoring of patients is part of the objects layer, medical records, technology, sensors, actuators, pharmacies controls and a diet plan creator etc. This layer has direct touch with the ecosystem's consumers. At this layer, information is gathered from components like wearables, patient monitoring and remote care.

15.7.2 INTELLIGENT E-HEALTHCARE SYSTEM

Hospitals that use intelligent automated and optimised modules (perhaps based on AI/ML) on the Information and communication Technology (ICT) infrastructure to enhance patient care processes and provide new capabilities are referred to as "smart hospitals." Smart hospitals have several uses including telemedicine, telehealth and remote robot surgery [82]. Figure 15.7 is an illustration of a smart healthcare system in which data from multiple sources is first gathered (e.g., physically or remotely) and delivered to an EHR (Electronic Health Records) system. If the staff collects data like medical notes offline and on paper, it may be categorised as unstructured data.

If the data is gathered from the devices and sensors in an organised format using specified data fields for users to fill, it will be simple to process in other systems like CRM (Customer Relationship Management) System. The CRM makes available the resources for data analysis and data assignment to the ecosystem's predetermined targets. These processed data produce additional triggers for patients and medical staff across the ecosystem.

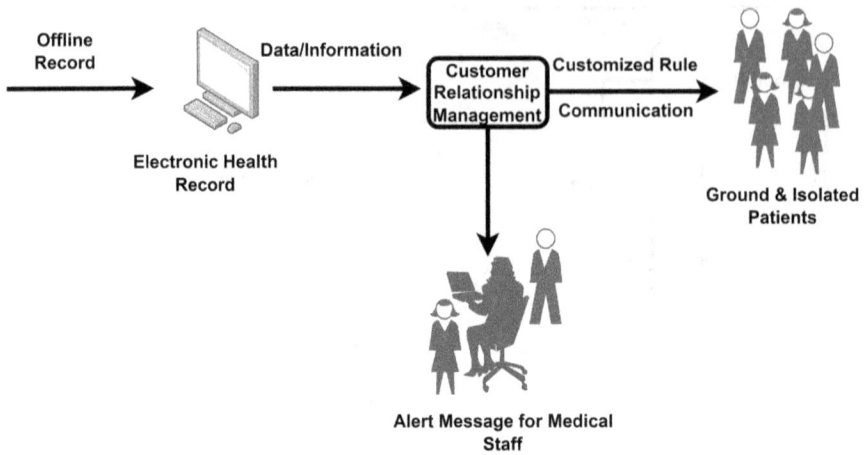

FIGURE 15.7 Illustration of smart electronic healthcare system.

Hospitals and medical professionals communicate with the patients outside by sending them personalised health regimens. The software notifies the physicians and other medical personnel about the reminders and other notifications.

15.7.3 IoMT Systems in Electronic Healthcare

Sensors and other devices make up IoT systems which are linked to one another via a network of cloud ecosystems and fast Internet. The unprocessed data collected at these devices/sensors is sent to the ample storage space offered by cloud services.

To acquire further understanding from this data, it is further cleaned and then analysed. This calls for the adoption of new software, hardware and administration tools that will enhance data viewing, analysis, processing and administration. Figure 15.8 depicts the interconnection of various wireless technologies including RFID (Radio-Frequency Identification), NFC (Near-Field Communications), Bluetooth, LTE (Long-Term Evolution) and 5G/6G with a variety of devices including smartphones, monitoring devices, sensors, smart wearables and other medical devices [82]. Basically, there are three layers, i.e., application layer on the top of the diagram that provides services, network that gives information using different networks, and at the last layer, various technologies are used for data transmission such as radio-frequency identification that is involved in the network for transmitting signals to the user. Bluetooth and Wi-Fi are used for small-distance communication.

15.8 COMPONENTS USED IN IoMT

15.8.1 IoMT Monitoring and Data Acquisition (MDA)

IoMT sensor and actuator are linked with the IoMT gateways in which the MDA component runs. The major objective of this MDA component [82] is to monitor and observe the behaviour of IoMT devices. It has various components shown in Figure 15.9.

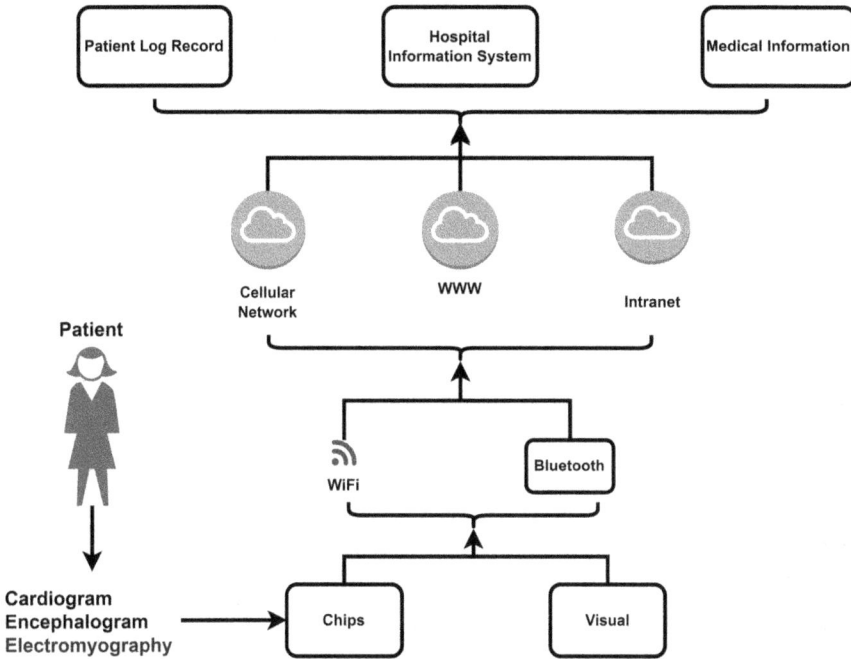

FIGURE 15.8   Interconnection of wireless technologies.

FIGURE 15.9   Monitoring and data acquisition component.

The data selection selects the specific data from real time and generates the report in CSV file format which contains the information related to CPU, memory, storage, bandwidth, etc. The data recoding module records all the data that are collected from the previous step in a log file and monitor the size of the file. During the data collection, it gives the information about the percentage of CPU utilisation, total number of active CPU, internal memory, external storage and bandwidth used by Wi-Fi. Reporting of data is made in which each record contains the maximum log files. It compresses the report to reduce the bandwidth and finally send to the IoMT hatchway to send at the application layer.

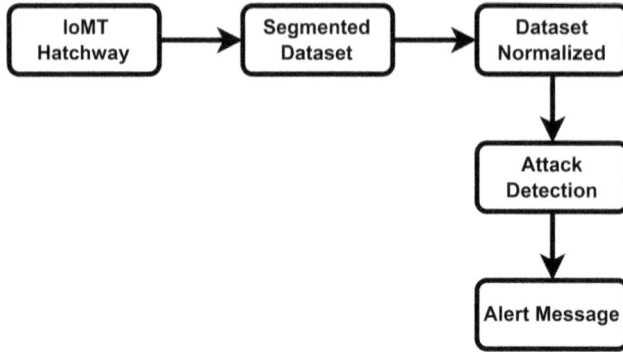

FIGURE 15.10 Central detection component of IoMT devices.

15.8.2 IoMT CENTRAL DETECTION (CD)

The IoMT edge network's gateway hosts the central detection (CD) component shown in Figure 15.10 [82]. During a predetermined monitoring time, keep an eye on the behaviour of the gateway, hosting it and gathering pertinent behaviour data.

The network traffic goes through the gateway and collects pertinent network traffic data, such as source IP address, destination IP address, connection status information and packet content information. The data are coming from the various sources such as gateway, network and IoT device dataset. The "Gateway Data Recording" module logs data in CSV format that was generated by the "Gateway Data Collection" module. The maximum size of the log file determines the number of entries that may be stored in each log file. The "Gateway Dataset Creation" module creates a gateway dataset in CSV format using the log files produced by the "Gateway Data Recording" module. The maximum number of log files for a given gateway dataset determines the number of log files that go into each gateway dataset. The amount of log files that may be stored for a given gateway dataset relies greatly on the needs of the detection engine. The IoMT devices connected to the gateway should get the reports sent by the MDA components executing on those devices. All the reports coming from the hatchway is normalised. The dataset is normalised after feature selection and normalisation. On the basis of this record, if there is any attack possible, then the decision takes place and an alert message is sent to the staff. The normalisation module receives datasets from a gateway, a network, and one or more IoMT devices. It then normalises the features of the datasets. The module generates the following three distinct types of normalised datasets in CSV format: normalised gateway dataset, normalised network dataset and normalised IoMT device dataset. Each dataset is normalised independently from the other datasets.

There are various methods to make all the communication secure [83]. Security is a major concern, and it may lead to false data and thus result in anomaly. Nowadays for anomaly detection, there are various methods and algorithms available. To make every communication secure on the Internet or to send the patient information from one place to another, various security measurements can be implemented. To make patient information reliable and secure in IoMT, emerging technologies such as blockchain and machine learning may play important roles. There is various recent trending stuff mentioned in current environment [84, 85].

15.9   CONCLUSION AND FUTURE SCOPE

Clinical lab anomalies originate from several origins and come in various shapes. While many anomalies are caused by what are known as "laboratory mistakes," not all laboratory oddities are. Anomalies occur from errors; however, not all anomalies are caused by mistakes in the lab. Extreme biological variations may result in anomalies that are illness states. Finding and highlighting biological abnormalities may be just as crucial as finding and avoiding technological ones. By incorporating data-driven strategies made feasible by constantly rising testing quantities, this workflow may be greatly enhanced. For better performance, more recent algorithms have increased the thresholds, added several analyst and even used machine learning methods. Other sectors frequently use data-driven techniques to anomaly detection, creating a variety of alternatives that may be customised for the director of the laboratory's objectives. Overall, the profession must move away from manual verification and towards more automated techniques to anomaly identification due to the steadily rising volume of laboratory testing.

While the digital transformation of healthcare has created a wide range of opportunities for bettering the way to produce, validate and communicate laboratory results, it has also increased the body of knowledge necessary to be successful in the industry. As the paradigm evolves more and more in favour of automation, awareness and comprehension of new data-driven methodologies for anomaly detection will serve as a starting point. In future, researchers can check the result with the help of real-time approaches and mitigate the anomaly using deep learning techniques.

REFERENCES

[1] Spies NC, Farnsworth CW, Jackups Jr, R. Data-driven anomaly detection in laboratory medicine: past, present, and future. *J Appl Lab Med* 2023;8(1):162–79.
[2] Agarwal R. Quality-improvement measures as effective ways of preventing laboratory errors. *Lab Med* 2014;45:e80–8.
[3] https://www.marktechpost.com/2022/01/01/hierarchical-federated-learning-based-anomaly-detection-using-digital-twins-for-internet-of-medical-things-iomt/
[4] Hawkins DM. Identification of outliers. Dordrecht (the Netherlands): Springer Netherlands; 1980.
[5] https://nutanxt.com/making-anomaly-detection-work-for-your-cognitive-ai-solutions/
[6] Plebani M, Carraro P. Mistakes in a stat laboratory: types and frequency. *Clin Chem* 1997;43:1348–51.
[7] Carraro P, Plebani M. Errors in a stat laboratory: types and frequencies 10 years later. *Clin Chem* 2007;53:1338–42.
[8] Mrazek C, Lippi G, Keppel MH, Felder TK, Oberkofler H, Haschke-Becher E, Cadamuro J. Errors within the total laboratory process, from test selection to medical decision-making-a review of causes, consequences, surveillance and solutions. *Biochem Med (Zagreb)* 2020;30:020502.
[9] Levey S, Jennings ER. The use of control charts in the clinical laboratory. *Am J Clin Pathol* 1950;20:1059–66.
[10] Westgard JO, Barry PL, Hunt MR, Groth T. A multi-rule Shewhart chart for quality control in clinical chemistry. *Clin Chem* 1981;27:493–501.
[11] Howanitz PJ, Tetrault GA, Steindel SJ. Clinical laboratory quality control: a costly process now out of control. *Clin Chim Acta Int J Clin Chem* 1997;260:163–74.

[12] Wg M, Erek A, Cunningham TD, Oladipo O, Scott MG, Johnson RE. Commutability limitations influence quality control results with different reagent lots. *Clin Chem* 2011;57:76–83.

[13] Hsu Y-MS, McClellan A, Jackups R, Gronowski AM, Scott MG. Was a recent manufacturer recall of CA-125 reagents necessary? *Clin Chim Acta* 2011;412:1886–7.

[14] Fleming JK, Katayev A. Changing the paradigm of laboratory quality control through implementation of real-time test results monitoring: for patients by patients. *Clin Biochem* 2015;48:508–13.

[15] Badrick T, Bietenbeck A, Cervinski MA, Katayev A, van Rossum HH, Loh TP. Patient-based real-time quality control: review and recommendations. *Clin Chem* 2019;65:962–71.

[16] Hoffmann RG, Waid ME. The "average of normals" method of quality control. *Am J Clin Pathol* 1965;43:134–41.

[17] Cembrowski GS, Chandler EP, Westgard JO. Assessment of "average of normals" quality control procedures and guidelines for implementation. *Am J Clin Pathol* 1984;81: 492–9.

[18] Liu J, Tan CH, Badrick T, Loh TP. Moving sum of number of positive patient results as a quality control tool. *Clin Chem Lab Med* 2017;55:1709–14.

[19] Li T, Cao S, Wang Y, Xiong Y, He Y, Ke P, et al. Moving rate of positive patient results as a quality control tool for high-sensitivity cardiac troponin T assays. *Ann Lab Med* 2021;41:51–9.

[20] Smith FA, Kroft SH. Exponentially adjusted moving mean procedure for quality control: an optimized patient sample control procedure. *Am J Clin Pathol* 1996;105:44–51.

[21] Smith JD, Badrick T, Bowling F. A direct comparison of patient-based real-time quality control techniques: the importance of the analyte distribution. *Ann Clin Biochem Int J Lab Med* 2020;57:206–14.

[22] Bull BS, Elashoff RM, Heilbron DC, Couperus J. A study of various estimators for the derivation of quality control procedures from patient erythrocyte indices. *Am J Clin Pathol* 1974;61:473–81.

[23] Kirkpatrick S, Gelatt CD, Vecchi MP. Optimization by simulated annealing. *Science* 1983;220:671–80.

[24] Ng D, Polito FA, Cervinski MA. Optimization of a moving averages program using a simulated annealing algorithm: the goal is to monitor the process not the patients. *Clin Chem* 2016;62:1361–71.

[25] van Rossum HH, Kemperman H. Optimization and validation of moving average quality control procedures using bias detection curves and moving average validation charts. *Clin Chem Lab Med* 2017;55:218–24.

[26] van Rossum HH. Moving average quality control: principles, practical application and future perspectives. *Clin Chem Lab Med* 2019;57:773–82.

[27] Whittle P. Prediction and regulation by linear least-square methods. 2nd Ed. Minneapolis (MN): University of Minnesota Press; 1983.

[28] Singh, R. "Performance Optimization of Autoencoder Neural Network Based Model for Anomaly Detection in Network Traffic." In *2nd International Conference on Advance Computing and Innovative Technologies in Engineering (ICACITE)* (pp. 598–602). IEEE; 2022.

[29] Singh R, Srivastava N, Kumar A. "Machine Learning Techniques for Anomaly Detection in Network Traffic." In *2021 Sixth International Conference on Image Information Processing (ICIIP)* (vol. 6, pp. 261–6). IEEE; 2021.

[30] Papadimitriou S, Sun J, Faloutsos C. Streaming pattern discovery in multiple time-series. In: Böhm K, Jensen CS, Haas LM, Kersten ML, Larson P- A, Chin Ooi B, editors. In *Proceedings of the 31st International Conference on Very Large Data Bases; Trondheim (Norway)* (pp. 697–708). New York (NY): VLDB Endowment; 2005.

[31] Wiering M, van Otterlo M. *Reinforcement Learning: State-of-the-Art.* Heidelberg (Germany): Springer; 2012.

[32] Schneider T, Jackups R. Transfusion medicine informatics: a review of current practice and a glimpse into the future. *Clin Lab Med* 2021;41:713–25.

[33] Ho CKM, Chen C, Setoh JWS, Yap WWT, Hawkins RCW. Optimization of hemolysis, icterus and lipemia interference thresholds for 35 clinical chemistry assays. *Pract Lab Med* 2021;25:e00232.

[34] Mays JA, Greene DN, Merrill AE, Mathias PC. Evidence-based validation of hemolysis index thresholds by use of retrospective clinical data. *J Appl Lab Med* 2018;3:109–14.

[35] Baron JM, Mermel CH, Lewandrowski KB, Dighe AS. Detection of preanalytic laboratory testing errors using a statistically guided protocol. *Am J Clin Pathol* 2012;138:406–13.

[36] Strylewicz G, Doctor J. Evaluation of an automated method to assist with error detection in the ACCORD central laboratory. *Clin Trials Lond Engl* 2010;7:380–9.

[37] Le QA, Strylewicz G, Doctor JN. Detecting blood laboratory errors using a Bayesian network: an evaluation on liver enzyme tests. *Med Decis Making* 2011;31:325–37.

[38] Benirschke RC, Gniadek TJ. Detection of falsely elevated point-of-care potassium results due to hemolysis using predictive analytics. *Am J Clin Pathol* 2020;154:242–7.

[39] Blei DM, Ng AY, Jordan MI. Latent Dirichlet allocation. *J Mach Learn Res* 2003;3:993–1022.

[40] Reynolds D. *Gaussian mixture models. Encyclopedia of Biometrics.* Boston (MA): Springer US, pp. 659–63; 2009.

[41] Goodfellow IJ, Pouget-Abadie J, Mirza M, Xu B, Warde-Farley D, Ozair S, et al. Generative adversarial networks. Preprint at https://doi.org/10.48550/arXiv.1406.2661 (2014).

[42] Kingma DP, Welling M. Auto-encoding variational Bayes. Preprint at https://arxiv.org/abs/1312.6114 (2014).

[43] Ladenson JH. Patients as their own controls: use of the computer to identify "laboratory error". *Clin Chem* 1975;21:1648-53.

[44] Lacher DA, Connelly DP. Rate and delta checks compared for selected chemistry tests. *Clin Chem* 1988;34:1966–70.

[45] Ko D-H, Park H-I, Hyun J, Kim HS, Park M-J, Shin DH. Utility of reference change values for delta check limits. *Am J Clin Pathol* 2017;148:323–9.

[46] Schifman RB, Talbert M, Souers RJ. Delta check practices and outcomes: a Q-probes study involving 49 health care facilities and 6541 delta check alerts. *Arch Pathol Lab Med* 2017;141:813–23.

[47] Ali AA, Khalid A, Moiz B. Performance evaluation of delta checks for error control in a hematology laboratory. *Int J Lab Hematol* 2021;43:e118–21.

[48] Hong J, Cho E, Kim H, Lee W, Chun S, Min W. Application and optimization of reference change values for delta checks in clinical laboratory. *J Clin Lab Anal* 2020;34:e23550.

[49] Singh R, Srivastav G. "Novel Framework for Anomaly Detection Using Machine Learning Technique on CIC-IDS2017 Dataset." In *2021 International Conference on Technological Advancements and Innovations (ICTAI),* pp. 632–6. IEEE; 2021.

[50] Singh R, Singh, A. Challenges of various load balancing algorithms in distributed environment. *IJITEE* 2018;2018:9–13.

[51] Strathmann FG, Baird GS, Hoffman NG. Simulations of delta check rule performance to detect specimen mislabeling using historical laboratory data. *Clin Chim Acta* 2011;412:1973–7.

[52] Lund F, Petersen PH, Fraser CG, Sölétormos G. Different percentages of false-positive results obtained using five methods for the calculation of reference change values based on simulated normal and ln-normal distributions of data. *Ann Clin Biochem* 2016;53:692–8.

[53] Sampson M, Rehak NN, Sokoll LJ, Ruddel ME, Gerhardt GA, Remaley AT. Time adjusted sensitivity analysis: a new statistical test for the optimization of delta check rules. *J Clin Ligand Assay* 2007;30:44–54.

[54] Iizuka Y, Kume H, Kitamura M. Multivariate delta check method for detecting specimen mix-up. *Clin Chem* 1982;28:2244–8.

[55] Rheem I, Lee KN. The multi-item univariate delta check method: a new approach. *Stud Health Technol Inform* 1998;52:859–63.

[56] Ovens K, Naugler C. How useful are delta checks in the 21st century? A stochastic-dynamic model of specimen mix-up and detection. *J Pathol Inform* 2012;3:5.

[57] Karger AB. To delta check or not to delta check? That is the question. J *Appl Lab Med* 2017;1:457–9.

[58] Rosenbaum MW, Baron JM. Using machine learning-based multianalyte delta checks to detect wrong blood in tube errors. *Am J Clin Pathol* 2018;150:555–66.

[59] https://www.researchgate.net/figure/Framework-for-anomaly-detection-in-healthcare-systems_fig1_263312632

[60] Singh R, Singh A Bhattacharya P. "A machine learning approach for anomaly detection to secure smart grid systems." In *Research Anthology on Smart Grid and Microgrid Development* (pp. 911–23). IGI global; 2022.

[61] Jackson CR, Cervinski MA. Development and characterization of neural network-based multianalyte delta checks. *J Lab Precis Med* 2020;5:10.

[62] Zhou R, Liang Y-F, Cheng H-L, Wang W, Huang D-W, Wang Z, et al. A highly accurate delta check method using deep learning for detection of sample mix-up in the clinical laboratory. *Clin Chem Lab Med* 2021;60:1984–92.

[63] Mitani T, Doi S, Yokota S, Imai T, Ohe K. Highly accurate and explainable detection of specimen mix-up using a machine learning model. *Clin Chem Lab Med* 2020;58:375–83.

[64] Schneider AJ. Some thoughts on normal, or standard, values in clinical medicine. *Pediatrics* 1960;26:973–84.

[65] Siegel M, Reilly EB, Lee SL, Fuerst HT, Seelenfreund M. Epidemiology of systemic lupus erythematosus: time trend and racial differences. *Am J Public Health Nations Health* 1964;54:33–43.

[66] Keyser JW. The concept of the normal range in clinical chemistry. *Postgrad Med J* 1965;41:443–7.

[67] Katayev A, Fleming JK, Luo D, Fisher AH, Sharp TM. Reference intervals data mining. *Am J Clin Pathol* 2015;143:134-42. Defining, establishing, and verifying reference intervals in the clinical laboratory; approved guideline. CLSI document C28-A3. Wayne (PA): Clinical and Laboratory Standards Institute; 2008.

[68] Ichihara K, Boyd JC. An appraisal of statistical procedures used in derivation of reference intervals. *Clin Chem Lab Med* 2010;48:1537–51.

[69] Colantonio DA, Kyriakopoulou L, Chan MK, Daly CH, Brinc D, Venner AA, et al. Closing the gaps in pediatric laboratory reference intervals: a CALIPER database of 40 biochemical markers in a healthy and multiethnic population of children. *Clin Chem* 2012;58:854–68.

[70] College of American Pathologists (CAP). All common checklist. CAP Accreditation Program. Northfield (IL): CAP; 2014.

[71] Katayev A, Balciza C, Seccombe DW. Establishing reference intervals for clinical laboratory test results: is there a better way? *Am J Clin Pathol* 2010;133:180–6.

[72] Poole S, Schroeder LF, Shah N. An unsupervised learning method to identify reference intervals from a clinical database. *J Biomed Inform* 2016;59:276–84.

[73] Martinez-Sanchez L, Marques-Garcia F, Ozarda Y, Blanco A, Brouwer N, Canalias F, et al. Big data and reference intervals: rationale, current practices, harmonization and standardization prerequisites and future perspectives of indirect determination of reference intervals using routine data. *Adv Lab Med Av En Med Lab* 2021;2:9–16.

[74] Association for Computing Machinery, Special Interest Group on Management of Data. In *Proceedings of the 2000 ACM SIGMOD International Conference on Management of Data: Dallas (TX)*. New York (NY): Association for Computing Machinery; 2000.

[75] Breunig MM, Kriegel H-P, Ng RT, Sander J. LOF: identifying density-based local outliers. *ACM SIGMOD Rec* 2000;29: 93–104.

[76] Liu FT, Ting KM, Zhou Z-H, Isolation forest. In: Giannotti F, Gunopulos D, Turini F, Zaniolo C, Ramakrishnan N, Wu X, editors. *Proceedings of the 2008 Eighth IEEE International Conference on Data Mining;* Pisa (Italy). Washington (DC): IEEE Computer Society; 2008. p. 413–22.

[77] Ben-Hur A, Horn D, Siegelmann HT, Vapnik V. Support vector clustering. *J Mach Learn Res* 2002;2:125–37.

[78] Pandey R, Pandey A, Maurya P, Singh GD. Prenatal Healthcare Framework Using IoMT Data Analytics. In *The Internet of Medical Things (IoMT) and Telemedicine Frameworks and Applications* (pp. 76–104). IGI Global; 2023.

[79] Melas-Kyriazi L. The mathematical foundations of manifold learning. Preprint at https://doi.org/10.48550/arXiv.2011.01307 (2020).

[80] Kramer MA. Nonlinear principal component analysis using autoassociative neural networks. *AIChE J* 1991;37:233–43.

[81] Pandey R, Paprzycki M, Srivastava N, Bhalla S, Wasielewska-Michniewska K. *Semantic IoT: Theory and Applications*. Springer Nature, Switzerland; 2021.

[82] Rodrigues JJPC, Segundo DBDR, Junqueira HA, Sabino MH, Prince RMI, Al-Muhtadi J, De Albuquerque, VHC. Enabling technologies for the internet of health things. *IEEE Access* 2018;6:13129–41.

[83] Razdan S, Sharma, S. Internet of Medical Things (IoMT): overview, emerging technologies, and case studies. *IETE Technical Review* 2021;39:775–88.

[84] Sun L, Jiang X, Ren H, Guo Y. Edge-cloud computing and artificial intelligence in internet of medical things: architecture, technology and application. *IEEE Access* 2020;8:101079–92.

[85] Srivastava, N. Analyzing the Suitability of Deep Learning Algorithms in Edge Computing. In *Applications of Artificial Intelligence, Big Data and Internet of Things in Sustainable Development* (pp. 271–85). CRC Press.

16 Computational Intelligence Framework for Improving Quality of Life in Cancer Patients

Perla Sunanda and Dwaram Kavitha
G. Pulla Reddy Engineering College

16.1 INTRODUCTION

16.1.1 BIG DATA

Gartner defined big data [1] as "high-volume, high-velocity and/or high-variety information assets that demand cost-effective, innovative forms of information processing that enable enhanced insight, decision making, and process automation." Big data is often generated by machines, people, and organisations. It is characterised by several Vs. The first three Vs as defined by Gartner are Volume, Velocity, and Variety. Volume represents the huge amount of data generated at every instant of time in this digital world. Variety represents the different forms of data like text, audio, video, images, etc. Velocity is the speed at which data are being generated and moved from one place to the other. As big data community discovered new challenges, more Vs like Veracity, Valence and Value have also been introduced. Accuracy and truthfulness of the data are represented by another 'V' called Veracity. Valence refers to the connectedness. The more connected the data, the higher its valence. Value has a value if its value is valued (Figure 16.1).

16.1.2 DIFFERENCE BETWEEN ARTIFICIAL INTELLIGENCE AND COMPUTATIONAL INTELLIGENCE

Machine intelligence is of two types: (1) Artificial Intelligence (AI) and (2) Computational Intelligence (CI). AI refers to the system's capability to correctly determine external data, learn from such data, and use those knowledge to achieve specific targets through flexible conversion. The CI is defined as the system's ability to learn a specific task from data or empirical observations. CI is a subset of AI. Bezdek et al. [2] determined that CI uses soft computing techniques, while AI is based on hard computing techniques. Fuzzy logic, expert systems, artificial neural networks, genetic algorithms, and machine learning are soft computing techniques.

DOI: 10.1201/9781003359951-19

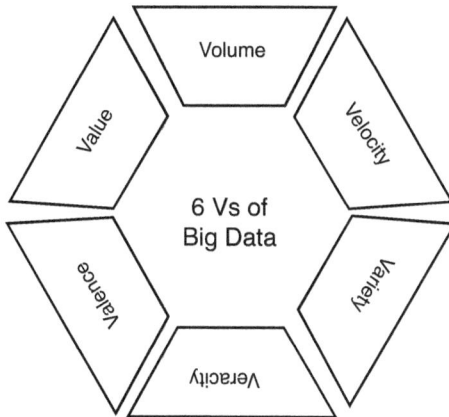

FIGURE 16.1   Characteristics of big data.

16.1.3   FUZZY LOGIC

Fuzzy logic, a computing approach, is built on "degrees of truth" which is different from Boolean logic. Hamon et al. [3] used fuzzy logic to identify and validate gene signatures on breast cancer patients. Juan Carlos Martin et al. [4] used fuzzy logic to propose a methodological framework to calculate synthetic indicator to measure Quality of Life (QoL).

16.1.4   EXPERT SYSTEMS

An expert system is a computer application that uses the expert knowledge in the specific domain. It can pattern the reasoning process of experts and employ their knowledge in solving peculiar tasks. Atanasova Irena [5] proposed an expert system strategy for evaluating QoL using qualitative data.

16.1.5   ARTIFICIAL NEURAL NETWORKS

Artificial neural networks (ANN) are computing systems motivated by the functionality of biological neurons, and they produce nonlinear functional outputs for the weight-controlled inputs. The network studies and analyses the weights of inputs in the course of supervised or unsupervised training. ANNs are determined by their learning and self-organising capabilities. Few research studies used ANNs to assess QoL in different kinds of cancer patients. Takehira et al. [6] modelled an artificial neural network to determine the relationship between QoL scores in the perspectives of patients, nurses and chemists.

16.1.6   GENETIC ALGORITHMS

Genetic algorithms (GA) use heuristic search technique and solve optimisation problems by randomly creating new solutions. They strive to improve the solution quality by repetitions. After creating initial set of solutions, crossover and mutation methods are applied to determine a new set of solutions. The quality of solutions is determined by the extent of solution space. No research studies focussed on improving QoL using GA. However, Oztekin et al. [7] proposed a hybrid genetic algorithm model to predict QoL in lung transplant recipients.

16.1.7 MACHINE LEARNING

Machine learning refers to a system's mechanical ability to learn and progress from its experience without being explicitly programmed. It is a subset of AI. Medical researchers developed new strategies for cancer prognosis and prediction using several machine learning algorithms.

16.2 QUALITY OF LIFE

Health refers to a state of physical, mental and social well-being and not just being free from disease or illness. Well-being refers to physical, mental, social and environmental conditions of an individual to live life with happiness and internal energy. QoL is an individual's perception of their position in the society and the happiness they experience in regard to their goals, expectations, and concerns. Holmes and Dickerson [8] represented QoL based on the responses given by an individual to the physical, mental and social aspects of sickness that affected their personal satisfaction in life and well-being. The physical domain refers to the troublesome physical sensations like pain, fatigue, etc. as experienced by an individual. Emotional domain refers to the psychological performance of an individual like stress, depression, etc. during positive and negative mental statuses. Social domain refers to the socio-personal role and social support perception of an individual on a disease diagnosis. Perception of health is changed from thinking of it as a state to a process. This provided opportunities for understanding, strengthening and learning healthcare. Measurement of healthcare and its effects should assess changes in the disease severity and well-being. This can be measured by monitoring the improvement in the health-related QoL. Today, self-assessments have become the main instruments to measure, monitor and improve human health.

16.2.1 CHARACTERISTICS OF QoL INSTRUMENTS

The QoL instruments can be used to assess the influence of healthcare in the patients, when the complete disease cure is not possible. Any QoL instrument [9] should satisfy three characteristics: reliability, validity and responsiveness. Reliability determines the degree to which an instrument gives consistent results at different times, provided that there is no proof of change. Validity refers to the degree to which an instrument measures what it is supposed to measure. Responsiveness refers to the measurement tool's ability to identify changes over time. Table 16.1 shows the categories and the corresponding measures of the QoL measurement characteristics.

TABLE 16.1

QoL Measurement Characteristics

Characteristics	Categories	Measurements
Reliability	Internal consistency	Cronbach's alpha coefficient
	Inter-rater	Kappa, Intraclass Correlation Coefficient (ICC), Cronbach's alpha coefficient
	Test–retest	Classical test theory, Intraclass Correlation Coefficient (ICC), Model-based, Graphical-based
	Parallel forms (equivalent forms)	Miller's analogy test
Validity	Content	
	Criterion	
	Construct	
Responsiveness	Statistics	Paired t-statistic, effect size statistic, responsiveness statistic

16.2.2 Quality of Life Measurement Instruments

QoL instruments measure the disease impact and its effect, and well-being state after treatment. Development process of a QoL instrument consists of a series of steps as follows:

1. Identification of concepts and development of a conceptual framework
2. Instrument creation
3. Assessment of instrument properties
4. Instrument modification.

In the past two decades, many instruments were developed to measure QoL on different populations. Most popular ones are (1) generic instruments and (2) disease-specific instruments.

16.2.3 Generic versus Disease-Specific Instruments

16.2.3.1 Generic Instruments

Generic instruments are developed to assess QoL in a wide-ranging population with or without chronic illness. There are many generic instruments available with a different number of items, domains and dimensions. Some of the commonly used are as follows:

1. **SF-36**: A 36-item, 8-dimension Short-Form instrument used to measure the generic health status.
2. **SF-12**: A 12-item, 4-dimension Short-Form instrument, the same as SF-36, measures the generic health status.

3. **PAIS-SR**: Psychological Adjustment to Illness Scale-Self Report, a 46-item instrument measures the psychological adjustment across health, vocational, domestic, family, social environments, psychological, and sexual relationship scales.
4. **HADS**: A 14-item Hospital Anxiety and Depression Scale is used for global measurement of anxiety and depression.

Apart from these, there are many other generic instruments such as the Center for Epidemiological Studies-Depression Scale (CES-D), Dispositional Hope Scale (DHS), Goal Interference Scale (GIS), Life Orientation Test-Revised (LOT-R), Nottingham Health Profile (NHP), Posttraumatic Growth Inventory (PTGI) and Posttraumatic Stress Disorder Checklist-Civilian (PCL-C).

Most of these instruments are characterised by internal consistency reliability, convergent and construct validity.

16.2.3.2   Disease-Specific Instruments

Disease-specific instruments are developed focusing on the issues specific to the illness. These instruments measure changes in QoL at different times, before and after treatment, which is not viable with generic instruments.

16.2.3.3   Cancer-Specific Instruments

This is a type of disease-specific instrument that focuses on only cancer disease. Among the number of cancer-specific instruments available in oncology, the most widely used QoL instruments are EORTC and FACT.

There are different types of EORTC and FACT instruments based on the type of cancer they are used for:

1. EORTC QLQ-C30 consists of 30 items divided into three domains. It is used as a common core instrument for all cancer survivors.
2. EORTC QLQ-BR23 consists of 23 items used most commonly for breast cancer survivors to access their QoL.
3. EORTC QLQ-PR25 consists of 25 items for accessing QoL in prostate cancer patients.

These are instrumented by European Organization for Research and Treatment of Cancer (EORTC).

The different forms of FACT (Functional Assessment of Cancer Therapy) instruments are a 27-item FACT-G (FACT-General) used for accessing functional well-being among different cancer survivors, FACT-B (FACT-BREAST) used for breast cancer survivors and FACT-BL (FACT-Bladder), a bladder-cancer-specific instrument used for assessing QoL in terms of functional well-being among bladder cancer patients. A FACT-LUNG is used to assess QoL among lung cancer survivors.

These all instruments are mainly characterised by internal consistency reliability, concurrent and discriminant validity.

Widely used instruments for measuring QoL are EORTC QLQ-C30 (European Organization for Research and Treatment of Cancer Quality-of-Life Questionnaire

Core 30) and FACT-G (Functional Assessment of Cancer Therapy-General). Many research studies measure QoL of cancer patients using EORTC QLQ-C30.

16.2.4 CALCULATION OF QoL SCORES

EORTC QLQ-C30 is a multidimensional instrument, consisting of 30 items: I1, I2, I3 ,…, I30. It is divided into five functional scales: physical, role, emotional, cognitive and social; one overall QoL measure; three symptom scales: fatigue, nausea-vomiting and pain; and six single items: dyspnoea, insomnia, appetite loss, constipation, diarrhoea and financial difficulty. Each item is either a multi-item scale or a single-item measure. Table 16.2 represents the process of calculation of QoL scores using QLQ-C30 [10].

TABLE 16.2
The Process QoL Calculation Using EORTC QLQ-C30

Scale	Items	Raw Score	Item Range	Score	Reference Value Mean	Reference Value Standard Deviation
Functional Scales						
Physical functioning	I_1, I_2, I_3, I_4, I_5	$(I_1 + I_2 + I_3 + I_4 + I_5)/5$	3	$\{1-(\text{RawScore}-1)/3\}*100$	76.7	23.2
Role functioning	I_6, I_7	$(I_6 + I_7)/2$	3		70.5	32.8
Emotional functioning	$I_{21}, I_{22}, I_{23}, I_{24}$	$(I_{21} + I_{22} + I_{23} + I_{24})/4$	3		71.4	24.2
Cognitive functioning	I_{20}, I_{25}	$(I_{20} + I_{25})/2$	3		82.6	21.9
Social functioning	I_{26}, I_{27}	$(I_{26} + I_{27})/2$	3		75.0	29.1
Symptom Scales						
Fatigue	I_{10}, I_{12}, I_{18}	$(I_{10} + I_{12} + I_{18})/3$	3	$\{(\text{RawScore}-1)/3\}*100$	34.6	27.8
Nausea and vomiting	I_{14}, I_{15}	$(I_{14} + I_{15})/2$	3		9.1	19
Pain	I_9, I_{19}	$(I_9 + I_{19})/2$	3		27.0	29.9
Dyspnoea	I_8	I_8	3		21.0	28.4
Insomnia	I_{11}	I_{11}	3		28.9	31.9
Appetite loss	I_{13}	I_{13}	3		21.1	31.3
Constipation	I_{16}	I_{16}	3		17.5	28.4
Diarrhoea	I_{17}	I_{17}	3		9.0	20.3
Financial difficulties	I_{28}	I_{28}	3		16.3	28.1
Global Health Status	I_{29}, I_{30}	$(I_{29}+I_{30})/2$	6	$\{(\text{RawScore}-1)/3\}*100$	61.3	24.2

QLQ-C30 considers only two socio items and no demographic, clinical items. To overcome this and to have an accurate assessment, the following questionnaire (SCD-21) with social and demographic questions may be used. SCD-C21 is another multidimensional, made up of 21 items (3 socio items: caregiver, parental/spouse education, and occupation; 12 clinical items: health of caregiver, family history of the disease, time interval since diagnosis, stage of disease, treatment, type of treatment, medicine usage, food habits, following diet, practicing physical activity, and duration of practicing; and 6 demographic items: gender, education, occupation, residence, marital status and number of children) (Table 16.3).

16.2.5 TYPES OF INTERVENTIONS

Intervention is the act of intervening in a situation. When the QoL score of a cancer patient is below the EORTC reference value, the patient should be considered under the risk group and the right intervention strategy should be employed. Interventions can be categorised into (1) pharmacological and (2) non-pharmacological. Several non-pharmacological interventions are as follows:

1. Physiotherapy
2. Social and physical activity
3. Yoga interventions with meditation
4. Activity or mindfulness
5. Digital intervention: Bradbury et al. developed a digital intervention for cancer survivors using evidence-, theory-, and person-based approaches and multidimensional and multidisciplinary rehabilitation.

This chapter provides a systematic review of the literature selected from the database PubMed. The aim of this systematic review is to address the following overall research question: What is the relationship between intervention method and QoL. In order to explore this issue, we also looked into the following secondary research question: What intervention methods are proposed for the patients with poor QoL? This chapter highlights the selected studies to illustrate the state of the research in the type of intervention methods practiced on the chosen population.

16.3 METHODS AND DESIGN

16.3.1 SEARCH AND SELECTION CRITERIA

We conducted a literature search using the online PubMed database to identify studies in English during the past 5 years of interventions in cancer survivors designed to improve QoL. The keywords used for the selection are 'Quality of Life', 'cancer survivors' and 'intervention'. The selection criteria are to include only those studies with prospective interventions to improve QoL. Table 16.4 shows the flow diagram of study selection. Screening of the articles was completed in two stages: first, articles were screened for the relevance based on the title and the information provided in the abstract, and then evaluated for the inclusion based on the full text of the article.

TABLE 16.3

The Socio Clinical Demographic QoL Questionnaire (SCD QLQ-C21)

1.	**Type of Cancer**			
2.	Sex	☐ Female	☐ Male	
3.	Residence	☐ City	☐ Village	
4.	Education	☐ Uneducated ☐ Elementary- Secondary	☐ High school- diploma	☐ University
5.	Occupation	☐ Have a Job	☐ No Job	
6.	Marital status	☐ Single	☐ Married	☐ Divorced or widowed
7.	Spouse's/Parental education	☐ Uneducated ☐ Elementary- Secondary	☐ High school- diploma	☐ University
8.	Spouse's/Parental Occupation	☐ Unemployed ☐ Employee	☐ Self- employee	☐ Retired
9.	Number of children	☐ No	☐ 1-2 Child/ Children	☐ 3 and more
10.	Caregiver person	☐ Spouse	☐ Mother, sister, other relatives	☐ Others (not even relatives (outsiders))
11.	Health of caregiver	☐ Good	☐ Bad	
12.	Family history of the disease	☐ Grandparents	☐ Parents	☐ None
13.	Time interval since diagnosis (in months)	☐ 12	☐ 23 ☐ 312	☐ >12
14.	Stage of disease	☐ I ☐ II	☐ III	☐ IV
15.	Treatment	☐ Undergoing	☐ Completed	
16.	Type of treatment (if undergoing treatment)	☐ Chemotherapy	☐ Radiotherapy	☐ Combination therapy
17.	Still using medicine	☐ Yes	☐ No	
18.	Food habits	☐ Vegetarian	☐ Non- Vegetarian	
19.	Following any diet	☐ Yes	☐ No	
20.	Which physical activity are you practicing?	☐ Yoga	☐ Exercise	☐ None
21.	If practicing	☐ Once in While	☐ Daily	

Out of the 888 unique articles that are retrieved from PubMed, only 314 articles met the criteria on screening the title and abstract. The full-text screening identified 144 studies and met the target criteria to be included in the review. The data were extracted into a purpose-built, structured template. The data extraction form included publication date, article title, study design, sample size, data source, cancer

TABLE 16.4
QoL Score Calculation
Using EORTC QLQ-C30

Year	Number of Papers
2014	2
2015	33
2016	34
2017	34
2018	36
2019	5

type, QoL instruments, factors affecting QoL, intervention type and duration, outcome measures assessed, copyright/publisher and indexed for. Descriptive statistics were calculated using Microsoft Excel.

16.3.2 Results

16.3.2.1 Articles per Year

The number of included articles grew from 2 in 2014 to 33 in 2015. There is a negligible increase in the papers published from 2015 onwards and a tremendous decrease in 2019. Figure 16.2 shows the number of papers published during the last 5 years.

16.3.2.2 Participants

The study participants in all these studies were the people living with and beyond cancer who had received their cancer diagnosis and were under any type of intervention. Majority of the papers had a smaller sample size which is less than 100. Only one research is conducted with a larger study sample around 2,500. Figure 16.3 shows the number of papers with sample sizes. Sample size selection reflects the intended analysis at the end of the study.

16.3.2.3 Study Design

In more than half of the studies, the data were collected using Randomised Controlled Trial, of which 79% were single-armed, 3% two-armed, and 5% multi-centred. However, few studies used cross-sectional, RENEW, non-randomised design.

16.3.2.4 Data Source

Limitations of all the studies are that they are limited to a particular region and a particular cancer type. Data of study samples are obtained either from cancer registries of their respective countries or from cancer hospitals.

16.3.2.5 Cancer Type

Breast cancer is the most commonly occurring cancer in women and the second most common cancer overall. Although majority of the research studies focussed on

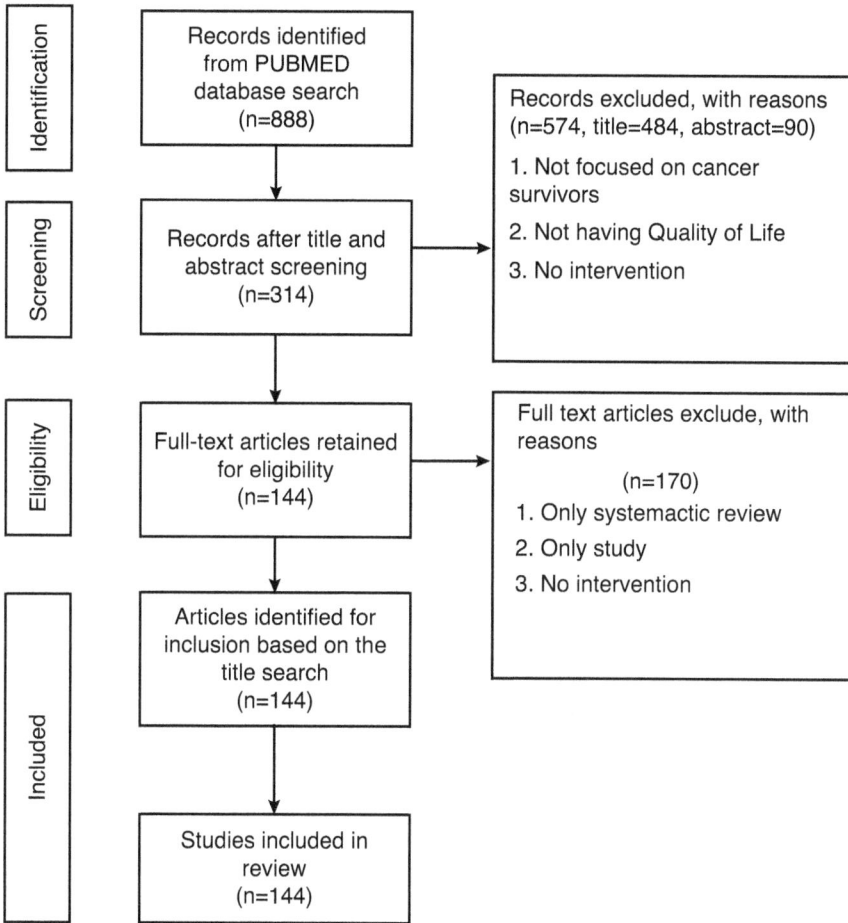

FIGURE 16.2 Number of papers published during the last 5 years.

breast cancer patients, no significant efforts are made to improve the QoL of patients with poorer QoL. Table 16.5 represents the number of articles published under each cancer type (Figure 16.4).

16.3.2.6 QoL Instruments

It is observed that one-fifth of the articles used EORTC QLQ-C30 instruments to measure QoL of cancer survivors in general. Few research studies that focussed on specific type of cancer like breast cancer used the instrument EORTC QLQ-BR23 specific to the breast cancer (Figure 16.5 and Table 16.6).

16.3.2.7 Factors Affecting QoL

Cancer type, pain intensity, fatigue, financial status and health of caregiver were found to be few factors affecting QoL of cancer survivors in general. But none of the studies explicitly revealed the factors considering the meta-analysis of individual patient data.

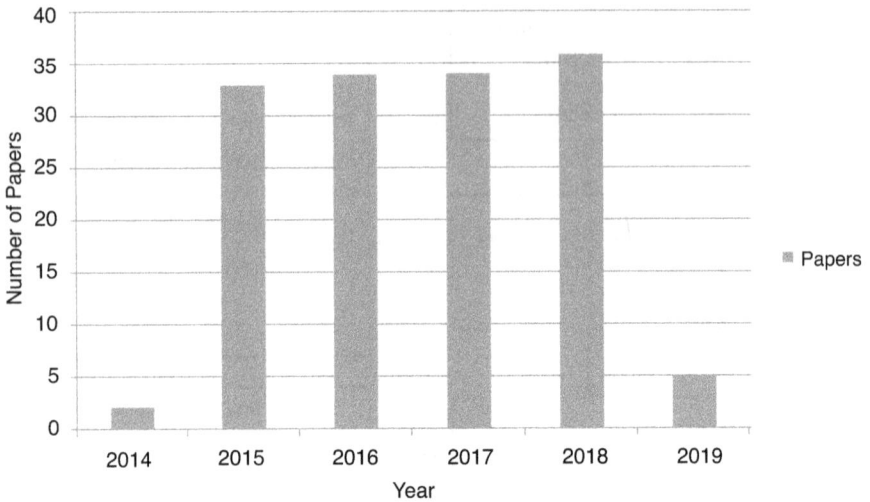

FIGURE 16.3   Number of papers published with the sample size.

TABLE 16.5
The Number of Participants
Involved in Most of the Studies

Participants Range	Number of Participants
0–100	25
101–200	3
201–300	6
301–500	5
501–1,000	4
1,001–2,000	3
2,001–3,000	1

16.3.2.8   Interventions

All the studies reviewed here used distinct, new interventions to improve QoL of cancer survivors. The studies include different cancer types, cancer stages, follow-up times and outcome measures, and they were of variable quality. However, despite these differences, the studies illustrate that these interventions are well received by patients and can be beneficial to them. Majority of the interventions led to a positive impact on QoL. However, determining which intervention was the most successful is difficult. It is observed that many research studies focussed on physical activity which is a non-pharmacological intervention and found considerable improvement in physical domain. However, there are remarkably few papers on digital intervention. But this study reveals that none of the studies developed customised interventions based on the individual patient meta-analysis data. So research is needed to develop personalised intervention strategies (Figure 16.6 and Tables 16.7–16.9).

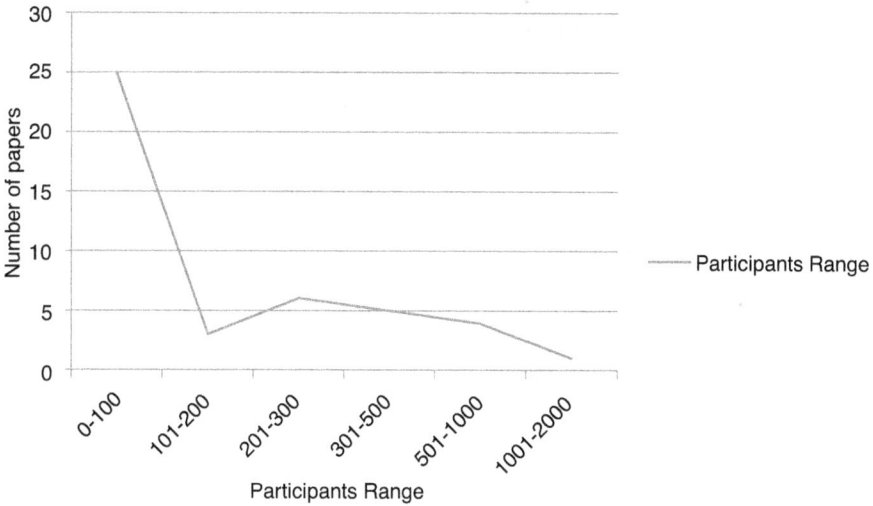

FIGURE 16.4 Number of papers published under each cancer type.

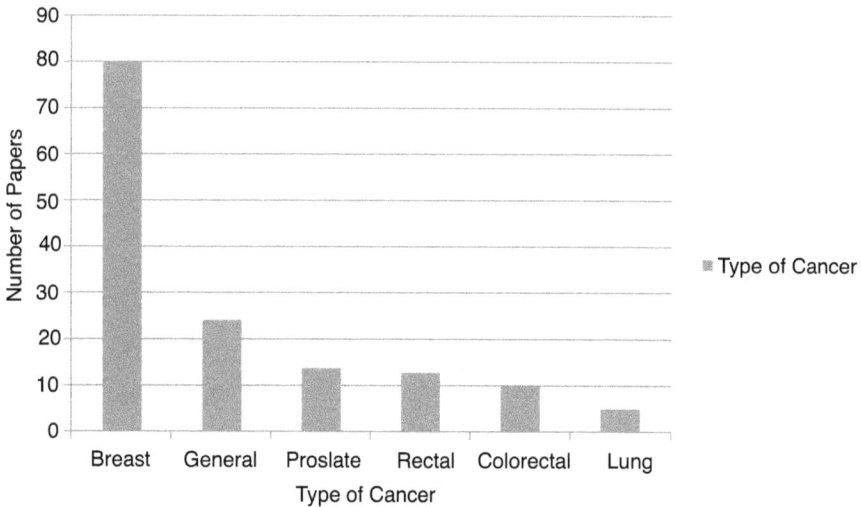

FIGURE 16.5 Number of papers utilising specific type of QoL instruments.

16.3.2.9 Outcome Measures

All the research studies aimed to analyse one or multiple outcomes of overall QoL: five physical domains and eight symptom domains. But no systematic procedure was observed.

To conclude, many interventions were implemented without any proof of support. So we believe that a CI framework should be developed which details step-by-step procedure and provides an evidence-based intervention method.

TABLE 16.6
The Number of Papers Published with Respect to Cancer Type

Type of Cancer	Number of Paper
Breast	80
General	24
Prostate	14
Rectal	13
Colorectal	10
Lung	5
Gynaecologic	5
Endometrial	3
Childhood	3
Cervical	2
Bladder	2
Ovarian	2
Testicular	1
Lymphoma	1
Adolescent	1
Brain	1
Oral	1

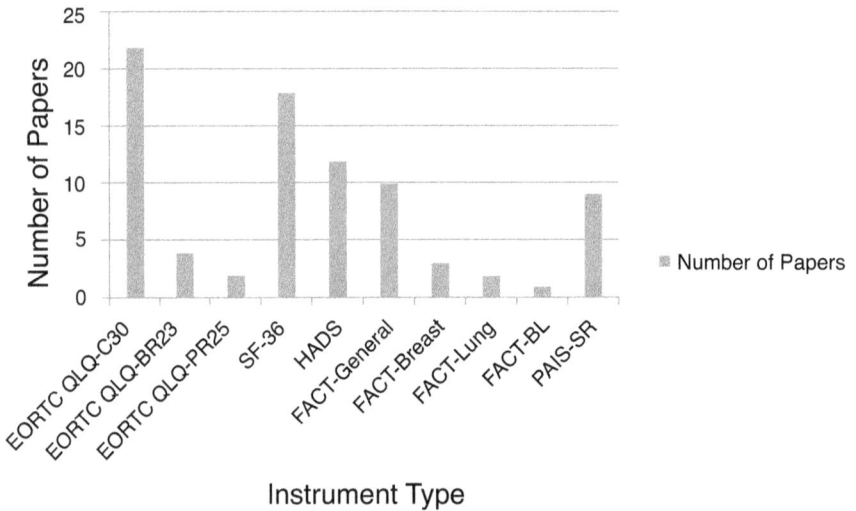

FIGURE 16.6 Number of papers employing particular type of intervention.

16.4 COMPUTATIONAL INTELLIGENCE FRAMEWORK

The aim of this chapter is to formulate a CI framework for recommending personalised health management intervention strategy to cancer survivors so as to improve QoL.

TABLE 16.7

The Number of Papers Published with Different Types of Instruments

Instrument Type	Number of Papers
EORTC QLQ-C30	22
EORTC QLQ-BR23	4
EORTC QLQ-PR25	2
SF-36	18
HADS	12
FACT-General	10
FACT-Breast	3
FACT-Lung	2
FACT-BL	1
PAIS-SR	9

TABLE 16.8

The Number of Papers Published with Different Intervention Types

Non-Pharmacological Categories	Intervention Type	Papers
Social and physical activity	Exercise interventions	30
	Physical activity interventions	20
	Massage	2
Psychological	Education class	12
	Cognitive-related	7
	Counselling	3
	Life style behavioural	3
Activity or mindfulness	Weight loss	9
	Telephone	7
	Home-based	6
	Theoretical	5
	Self-management	3
	Evidence-based	3
	Group-based	3
	Diet-related	2
	Nursing care service	2
Digital intervention	Web-based	8
	Online-based	7
Yoga interventions with meditation	Yoga	3
Overall	Multidisciplinary rehabilitation	9

TABLE 16.9

The citations of various papers with different intervention types

Non-Pharmacological Categories	Intervention Type	Papers	Citations
Social and physical activity	Exercise interventions	30	[1–3,5–10,11,12,13,14,15–18,19,20,21,22,23,24,25,26,27,28,29,30,31]
	Physical activity interventions	20	[5,32,33,34,35,36,37,38,39,40,41,42,43,44,45,46,47,48,49,50]
	Massage	2	[51,52]
Psychological	Education class	12	[53,54,55,56,57,58,59,60,61,62,63,64]
	Cognitive-related	7	[15,39,42,61,65,66,67]
	Counselling	3	[68,69,70]
	Life style behavioural	3	[57,71,72]
Activity or mindfulness	Weight loss	9	[49,73,74,75,76,77,78,79,80]
	Telephone	7	[16,44,81,82,83,84,85]
	Home-based	6	[13,24,35,47,68,86]
	Theoretical	5	[25,51,54,56,87]
	Self-management	3	[24,88,89]
	Evidence-based	3	[44,90,91]
	Group-based	3	[34,41,92]
	Diet-related	2	[14,93]
	Nursing care service	2	[40,94]
Digital intervention	Web-based	8	[28,77,95,96,97,98,99,100]
	Online-based	7	[92,101,102,103,104,105,106]
Yoga interventions with meditation	Yoga	3	[55,107,108]
Overall	Multidisciplinary rehabilitation	9	[18,30,37,43,48,58,109,110,111]

Formulating a personalised health management strategy that recommends intervention programmes towards the improvement of QoL in cancer survivors is an interesting and challenging task. Many research studies have used Randomised Controlled Trial study design and implemented effective intervention strategies on different kinds of cancer patients using different interventions to improve QoL. But the challenges are how to collect, process and analyse increasing amounts of data to deliver the tailored treatment plan and the best outcomes of patients by extracting useful information.

This framework presents a step-by-step procedure to build the prediction model that can recommend the effective intervention strategy tailed to meet the needs of individual cancer patients. This patient-centric cascading prediction model consists of three stages: (1) To identify the factors that affect QoL of cancer survivors; (2) To subsequently identify QoL risk groups that require intervention; and (3) To develop interventions programmes that address the QoL predictors towards the improvement of QoL in cancer survivors.

Numerous randomised controlled trials have evaluated the effects of various interventions on QoL of cancer survivors, with generally small sample sizes and modest effects.

16.4.1 Data Collection

Whenever a patient is diagnosed with cancer, the scores in all five functional domains, eight symptoms and overall QoL should be assessed using QLQ-C30. The calculated scores of these 14 attributes shall be compared with EORTC reference values in the respective domain. Study samples containing sociodemographic data and QoL measurement data shall be collected from the cancer registries and from the literature.

16.4.1.1 From Cancer Registries

A cancer registry is a systematic collection of data about cancers and tumours. Cancer registries are present in 89% of countries with a national cancer control policy. Any patient diagnosed with cancer should be enrolled in a cancer registry. There are two types of cancer registries: (1) population-based cancer registry and (2) hospital cancer registry. Population-based cancer registries monitor the frequency of cancer cases, while hospital cancer registries aim at the improvement of cancer therapies. Data can be sent from a hospital cancer registry to a population-based registry.

16.4.1.2 From the Literature

Literature of cancer studies is present in representative databases like PubMed, ScienceDirect, Scirus, ISI Web of Knowledge, Nursing and Allied Health Literature, PsycINFO, Cochrane Central and EMBASE. Different studies have chosen participants with different eligibility criteria. Many of these studies were either single arm or had small sample sizes.

Lack of data-sharing infrastructure to link institutions, heterogeneity and incompleteness of data, and competition across hospitals resulted in data drought. Hence, data are siloed in individual organisations. It is important to obtain as many relevant studies as possible, as loss of studies can lead to bias. Instead of extracting aggregate data from study reports or from authors, the original research data are sought directly from the investigators of these studies. Data have to be collected from multiple organisations to avoid data biases. Individual patient data meta-analyses allow for adequate statistical analysis of intervention effects and moderators of such effects.

16.4.2 Data Preparation

For developing prediction models using supervised learning, large amounts of patients' data i.e., cancer outcomes data and patient outcomes data is needed. These obstacles need to be addressed. Guidelines have to be developed to support findable, accessible, interoperable and repeatable data use. Oncospace, a multi-institutional big data platform in the field of radiation oncology, consists of database and web-based analysis tools for planning, data import, and outcome predictions.

Individual patient's data from cancer registries and literature must be anonymised by respective authorities, and a unique patient identification number can be provided for further communication and data queries. All these data must be sent to the database for preprocessing. The original data can be archived for backup purpose and can also be examined for consistency and completeness.

Meta-analysis of individual patient's data can be used to identify moderators and intervention methods. Each patient data must be stored in a standard format in the form of electronic health records and should be used for training the model.

16.4.3 STUDY DESIGN

Currently, there is no standard study design. However, the following are the study designs employed in research studies:

- Meta-analysis
- Randomised controlled trial
- Cohort study
- Case-control study
- Cross-sectional study sampling

16.5 DEVELOPMENT OF PREDICTION MODEL

Cascading, an ensemble learning approach, is based on the concatenation of several models. Cascading models use the information collected from the output from a given model as additional information for the next model in the cascade. A two-stage, patient-centric cascading prediction model shall be developed to recommend personalised intervention strategy to the patients so as to improve QoL. Such a model should have a positive impact on the QoL and well-being of patients.

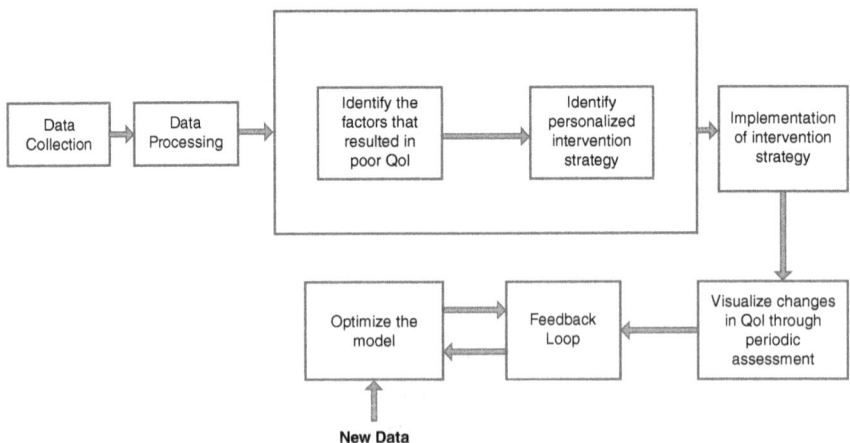

FIGURE 16.7 A Computationally Intelligence Framework for cascading 2-stage prediction model

16.5.1   BUILDING THE MODEL

16.5.1.1   Stage I

Symptoms experienced by cancer patients differ from patient to patient and with many common symptoms overlapping. Majority of the patients face psychological symptoms. So machine learning models should be built which can predict the factors affecting the QoL. The purpose of this phase is to predict the factors that affect QoL of cancer survivors.

Many research studies performed descriptive-analytical cross-sectional methods to estimate mean, std, and QoL, cognitive performance, emotional performance, factors affecting QoL using EORTC QLQ-C30, and sociodemographic and other cancer-specific questionnaires like BR-, CV-, etc. The model should be trained with the existing data in cancer registries. These data are given to the prediction model I for training.

Prediction model stage 1 shall predict the moderators, mediators and predictor factors that affect QoL. The main outcome measures can be overall QoL or specific QoL domains (for example, physical, psychological, functional and social well-being). Currently, the reviews do not have standard outcome selection or reporting convention.

Moderating factors are the ones which influence the strength of a relationship between intervention and QoL. Moderators should also be identified using evidence-based models of why a particular intervention might be more effective for some sub-groups rather than the others. Potential moderators might be demographic, clinical and personal factors. Potential mediators of the intervention effects on QoL shall also be predicted. Mediating factors [112] explain the relationship between intervention and QoL. They play a crucial role in identifying interventions, and they provide information for making interventions more effective. Stepwise Logistic Regression (SLR) analysis along with machine learning techniques like Decision Trees, Random Forest, SVM and ANN may be used for identifying the factors affecting QoL. Data analysis can be performed using Pearson's correlation, independent t, analysis of variance, linear multi-variate regression, etc. Different algorithms like SLR, DT, RF and SVM can be tried to model input variables and identify the factors affecting QoL. Once the risk groups are identified, the recommendation model can use the individual patient meta-analysis data to predict the factors affecting QoL and risk groups with poor QoL. For each type of intervention, we will build prediction models identifying predictors of intervention success.

16.5.1.2   Stage II

Effective interventions to improve QoL of cancer survivors are essential. The studies reveal that no single intervention can be recommended to those patients with poor QoL as interventions were not targeting poorer QoL. Better targeted interventions may result in significant improvement in QoL.

The purpose of this stage is to predict personalised effective intervention programme towards the improvement of QoL in cancer survivors, which supports evidence-based decision making. A particular intervention strategy should specify the type and activities of intervention, duration, frequency of QoL assessment, symptoms to be assessed, etc. To realise such targeted interventions, we must determine which interventions presently available work for which patients and what are the underlying mechanisms (i.e., the moderators and mediators of various interventions). Symptoms of the patients should be recorded as the treatment progresses, and after the intervention, these symptoms should be checked and compared with the predictions generated by machine learning algorithms. The model should validate the different combinations of interventions and therefore predict the right intervention strategy.

16.5.2 MODEL EVALUATION

Cross-validation methods can be employed to obtain a good estimate of the performance. Cross-validation approach splits the whole data several times into different train and test sets, and returns the averaged value of the predicted scores. Mean accuracy, mean area under the curve (AUC), mean squared error (MSE) and adjusted-for-chance mutual information index (AMI) can be calculated for each of the models.

Given the large number of patients diagnosed with cancer and significant amount of data generated during cancer treatment, there is a specific interest in the application of AI to improve oncology care. An online tool should be developed which utilises the three-stage patient-centric cascading prediction model. The aim of the tool shall be to facilitate the accessibility of prediction model for critical evaluation.

16.5.3 DATA VISUALISATION

The measured QoL scores can be visualised in the form of bar graphs or line graphs. Separate graphs showing QoL values for different functional and symptom domains would provide better insights to the consultants. If there is an improvement in QoL scores, then the same intervention strategy can be continued. But if the improvement in QoL score is not as expected, then the qualitative feedback can be collected from the patient and this can be given as input to the model so as to try another intervention strategy.

Along with this, governments should establish National Cancer Control Programme with an aim to reduce the incidence and mortality due to cancer and also to improve QoL of cancer survivors. It should also provide health education and take preventive measures by disseminating appropriate cancer information to the public, screening and diagnosis of cancer, prompt and effective management of cancer, rehabilitation and palliative care.

16.6 CONCLUSION

Majority of the research studies revealed that all the interventions are well received by patients and have a positive impact on QoL. All the studies were on cancer survivors in general. None of the studies developed customised interventions based on the individual patient meta-analysis data. Future research should focus on providing evidence for the effectiveness of interventions, of which activity promotion and management of concurrent symptoms are the most promising. Research is needed to develop personalised intervention strategies. Future research should harness the potential of AI and machine learning algorithms in cancer treatment and subsequently prevent cancer deaths drastically.

REFERENCES

1. Ward, J.S. and Barker, A., 2013. Undefined by data: a survey of big data definitions. arXiv preprint arXiv:1309.5821.
2. Bezdek, J.C., 1994. What is computational intelligence? (No. CONF-9410335-). USDOE Pittsburgh Energy Technology Center, PA (United States); Oregon State Univ., Corvallis, OR (United States). Dept. of Computer Science; Naval Research Lab., Washington, DC (United States); Electric Power Research Inst., Palo Alto, CA (United States); Bureau of Mines, Washington, DC (United States).
3. Kempowsky-Hamon, T., Valle, C., Lacroix-Triki, M., Hedjazi, L., Trouilh, L., Lamarre, S., Labourdette, D., Roger, L., Mhamdi, L., Dalenc, F., & Filleron, T. (2015). Fuzzy logic selection as a new reliable tool to identify molecular grade signatures in breast cancer–the INNODIAG study. *BMC Medical Genomics, 8*(1), 3.
4. Martín, J. C. & Viñán, C. S. (2017). Fuzzy logic methods to evaluate the quality of life in the regions of Ecuador. *Quality Innovation Prosperity, 21*(1), 61–80.
5. Atanasova, I. & Krupka, J. (2013). Architecture and design of expert system for quality of life evaluation. *Informática Económica, 17*(3), 28.
6. Takehira, R., Murakami, K., Katayama, S., Nishizawa, K., & Yamamura, S. (2011). Artificial neural network modeling of quality of life of cancer patients: Relationships between quality of life assessments, as evaluated by Patients, Pharmacists, and Nurses. *International Journal of Biomedical Science: IJBS, 7*(4), 255.
7. Oztekin, A., Al-Ebbini, L., Sevkli, Z., & Delen, D. (2018). A decision analytic approach to predicting quality of life for lung transplant recipients: A hybrid genetic algorithms-based methodology. *European Journal of Operational Research, 266*(2), 639-651.
8. Holmes, S. & Dickerson, J. (1987). The quality of life: design and evaluation of a self-assessment instrument for use with cancer patients. *International Journal of Nursing Studies, 24*(1), 15-24.
9. Burckhardt, C. S. & Anderson, K. L. (2003). The Quality of Life Scale (QOLS): reliability, validity, and utilization. *Health and Quality of Life Outcomes, 1*(1), 60.
10. EORTC QLQ-C30 Scoring Manual Version 3, 2001, Third Edition.
11. Dieli-Conwright, C. M., Courneya, K. S., Demark-Wahnefried, W., Sami, N., Lee, K., Sweeney, F. C., ... Mortimer, J. E. (2018). Aerobic and resistance exercise improves physical fitness, bone health, and quality of life in overweight and obese breast cancer survivors: a randomized controlled trial. *Breast Cancer Research: BCR, 20*(1), 124. https://doi.org/10.1186/s13058-018-1051-6.
12. Pisu, M., Demark-Wahnefried, W., Kenzik, K. M., Oster, R. A., Lin, C. P., Manne, S., ... Martin, M. Y. (2017). A dance intervention for cancer survivors and their partners (RHYTHM). *Journal of Cancer Survivorship: Research and Practice, 11*(3), 350–359. https://doi.org/10.1007/s11764-016-0593-9.
13. Melam, G. R., Buragadda, S., Alhusaini, A. A., & Arora, N. (2016). Effect of complete decongestive therapy and home program on health- related quality of life in post mastectomy lymphedema patients. *BMC Women's Health, 16*, 23. https://doi.org/10.1186/s12905-016-0303-9.
14. Kim, N.-H., Song, S., Jung, S.-Y., Lee, E., Kim, Z., Moon, H.-G., ... Lee, J. E. (2018). Dietary pattern and health-related quality of life among breast cancer survivors. *BMC Women's Health, 18*(1), 65. https://doi.org/10.1186/s12905-018-0555-7.
15. Riedl, D., Giesinger, J. M., Wintner, L. M., Loth, F. L., Rumpold, G., Greil, R., ... Holzner, B. (2017). Improvement of quality of life and psychological distress after inpatient cancer rehabilitation: Results of a longitudinal observational study. *Wiener Klinische Wochenschrift, 129*(19–20), 692–701. https://doi.org/10.1007/s00508-017-1266-z.

16. Thomson, C. A., Crane, T. E., Miller, A., Garcia, D. O., Basen-Engquist, K., & Alberts, D. S. (2016). A randomized trial of diet and physical activity in women treated for stage II-IV ovarian cancer: Rationale and design of the Lifestyle Intervention for Ovarian Cancer Enhanced Survival (LIVES): An NRG Oncology/Gynecologic Oncology Group (GOG-225) Study. *Contemporary Clinical Trials, 49*, 181–189. https://doi.org/10.1016/j.cct.2016.07.005.

17. Rueegg, C. S., Kriemler, S., Zuercher, S. J., Schindera, C., Renner, A., Hebestreit, H., ... von der Weid, N. X. (2017). A partially supervised physical activity program for adult and adolescent survivors of childhood cancer (SURfit): study design of a randomized controlled trial [NCT02730767]. *BMC Cancer, 17*(1), 822. https://doi.org/10.1186/s12885-017-3801-8.

18. Leensen, M. C. J., Groeneveld, I. F., Heide, I. van der, Rejda, T., van Veldhoven, P. L. J., Berkel, S. van, ... de Boer, A. G. E. M. (2017). Return to work of cancer patients after a multidisciplinary intervention including occupational counselling and physical exercise in cancer patients: a prospective study in the Netherlands. *BMJ Open, 7*(6), e014746. https://doi.org/10.1136/bmjopen-2016-014746.

19. Penttinen, H., Utriainen, M., Kellokumpu-Lehtinen, P.-L., Raitanen, J., Sievanen, H., Nikander, R., ... Saarto, T. (2019). Effectiveness of a 12-month exercise intervention on physical activity and quality of life of breast cancer survivors; five-year results of the BREX-study. *In Vivo (Athens, Greece), 33*(3), 881–888. https://doi.org/10.21873/invivo.11554.

20. Farris, M. S., Kopciuk, K. A., Courneya, K. S., McGregor, S. E., Wang, Q., & Friedenreich, C. M. (2017). Associations of postdiagnosis physical activity and change from prediagnosis physical activity with quality of life in prostate cancer survivors. *Cancer Epidemiology, Biomarkers & Prevention: A Publication of the American Association for Cancer Research, Cosponsored by the American Society of Preventive Oncology, 26*(2), 179–187. https://doi.org/10.1158/1055-9965.EPI-16-0465.

21. Strunk, M. A., Zopf, E. M., Steck, J., Hamacher, S., Hallek, M., & Baumann, F. T. (2018). Effects of Kyusho Jitsu on physical activity-levels and quality of life in breast cancer patients. *In Vivo (Athens, Greece), 32*(4), 819–824. https://doi.org/10.21873/invivo.11313.

22. Meneses, K., Gisiger-Camata, S., Benz, R., Raju, D., Bail, J. R., Benitez, T. J., ... McNees, P. (2018). Telehealth intervention for Latina breast cancer survivors: A pilot. *Women's Health (London, England), 14*, 1745506518778721. https://doi.org/10.1177/1745506518778721.

23. Li, Z., Geng, W., Yin, J., & Zhang, J. (2018). Effect of one comprehensive education course to lower anxiety and depression among Chinese breast cancer patients during the postoperative radiotherapy period - one randomized clinical trial. *Radiation Oncology (London, England), 13*(1), 111. https://doi.org/10.1186/s13014-018-1054-6.

24. Gokal, K., Munir, F., Ahmed, S., Kancherla, K., & Wallis, D. (2018). Does walking protect against decline in cognitive functioning among breast cancer patients undergoing chemotherapy? Results from a small randomised controlled trial. *PloS One, 13*(11), e0206874. https://doi.org/10.1371/journal.pone.0206874.

25. Gripsrud, B. H., Brassil, K. J., Summers, B., Soiland, H., Kronowitz, S., & Lode, K. (2016). Capturing the experience: Reflections of women with breast cancer engaged in an expressive writing intervention. *Cancer Nursing, 39*(4), E51–60. https://doi.org/10.1097/NCC.0000000000000300

26. Vallance, J. K., Friedenreich, C. M., Lavallee, C. M., Culos-Reed, N., Mackey, J. R., Walley, B., & Courneya, K. S. (2016). Exploring the feasibility of a broad-reach physical activity behavior change intervention for women receiving chemotherapy for breast cancer: A randomized trial. *Cancer Epidemiology, Biomarkers & Prevention:*

A Publication of the American Association for Cancer Research, Cosponsored by the American Society of Preventive Oncology, 25(2), 391–398. https://doi.org/10.1158/1055-9965.EPI-15-0812.

27. Badr, H., Lipnick, D., Diefenbach, M. A., Posner, M., Kotz, T., Miles, B., & Genden, E. (2016). Development and usability testing of a web-based self-management intervention for oral cancer survivors and their family caregivers. *European Journal of Cancer Care*, 25(5), 806–821. https://doi.org/10.1111/ecc.12396.

28. Galiano-Castillo, N., Cantarero-Villanueva, I., Fernandez-Lao, C., Ariza-Garcia, A., Diaz-Rodriguez, L., Del-Moral-Avila, R., & Arroyo-Morales, M. (2016). Telehealth system: A randomized controlled trial evaluating the impact of an internet-based exercise intervention on quality of life, pain, muscle strength, and fatigue in breast cancer survivors. *Cancer*, 122(20), 3166–3174. https://doi.org/10.1002/cncr.30172.

29. Smith, S. K., O'Donnell, J. D., Abernethy, A. P., MacDermott, K., Staley, T., & Samsa, G. P. (2015). Evaluation of Pillars4life: a virtual coping skills program for cancer survivors. *Psycho-Oncology*, 24(11), 1407–1415. https://doi.org/10.1002/pon.3750.

30. Timmerman, J. G., Tonis, T. M., Dekker-van Weering, M. G. H., Stuiver, M. M., Wouters, M. W. J. M., van Harten, W. H., ... Vollenbroek-Hutten, M. M. R. (2016). Co-creation of an ICT-supported cancer rehabilitation application for resected lung cancer survivors: design and evaluation. *BMC Health Services Research*, 16, 155. https://doi.org/10.1186/s12913-016-1385-7.

31. Bober, S. L., Recklitis, C. J., Michaud, A. L., & Wright, A. A. (2018). Improvement in sexual function after ovarian cancer: Effects of sexual therapy and rehabilitation after treatment for ovarian cancer. *Cancer*, 124(1), 176–182. https://doi.org/10.1002/cncr.30976.

32. Irwin, M. L., Cartmel, B., Harrigan, M., Li, F., Sanft, T., Shockro, L., ... Ligibel, J. A. (2017). Effect of the LIVESTRONG at the YMCA exercise program on physical activity, fitness, quality of life, and fatigue in cancer survivors. *Cancer*, 123(7), 1249–1258. https://doi.org/10.1002/cncr.30456.

33. Carlson, L. E., Zelinski, E. L., Speca, M., Balneaves, L. G., Jones, J. M., Santa Mina, D., ... Vohra, S. (2017). Protocol for the MATCH study (Mindfulness and Tai Chi for cancer health): A preference-based multi-site randomized comparative effectiveness trial (CET) of Mindfulness-Based Cancer Recovery (MBCR) vs. Tai Chi/Qigong (TCQ) for cancer survivors. *Contemporary Clinical Trials*, 59, 64–76. https://doi.org/10.1016/j.cct.2017.05.015.

34. Dominic, N. A., Thirunavuk Arasoo, V. J., Botross, N. P., Riad, A., Biding, C., & Ramadas, A. (2018). Changes in health- related quality of life and psychosocial well-being of breast cancer survivors: Findings from a group- based intervention program in Malaysia. *Asian Pacific Journal of Cancer Prevention: APJCP*, 19(7), 1809–1815. https://doi.org/10.22034/APJCP.2018.19.7.1809.

35. Cases, M. G., Fruge, A. D., De Los Santos, J. F., Locher, J. L., Cantor, A. B., Smith, K. P., ... Demark-Wahnefried, W. (2016). Detailed methods of two home-based vegetable gardening intervention trials to improve diet, physical activity, and quality of life in two different populations of cancer survivors. *Contemporary Clinical Trials*, 50, 201–212. https://doi.org/10.1016/j.cct.2016.08.014.

36. Golsteijn, R. H. J., Bolman, C., Volders, E., Peels, D. A., de Vries, H., & Lechner, L. (2018). Short-term efficacy of a computer-tailored physical activity intervention for prostate and colorectal cancer patients and survivors: a randomized controlled trial. *The International Journal of Behavioral Nutrition and Physical Activity*, 15(1), 106. https://doi.org/10.1186/s12966-018-0734-9.

37. Riedl, D., Giesinger, J. M., Wintner, L. M., Loth, F. L., Rumpold, G., Greil, R., ... Holzner, B. (2017). Improvement of quality of life and psychological distress after inpatient cancer rehabilitation: Results of a longitudinal observational study. *Wiener Klinische Wochenschrift*, 129(19–20), 692–701. https://doi.org/10.1007/s00508-017-1266-z.

38. Gopalakrishna, A., Longo, T. A., Fantony, J. J., Harrison, M. R., & Inman, B. A. (2017). Physical activity patterns and associations with health-related quality of life in bladder cancer survivors. *Urologic Oncology, 35*(9), 540.e1–540.e6. https://doi.org/10.1016/j.urolonc.2017.04.016.

39. Lee, Y.-H., Lai, G.-M., Lee, D.-C., Tsai Lai, L.-J., & Chang, Y.-P. (2018). Promoting physical and psychological rehabilitation activities and evaluating potential links among cancer-related fatigue, fear of recurrence, quality of life, and physiological indicators in cancer survivors. *Integrative Cancer Therapies, 17*(4), 1183–1194. https://doi.org/10.1177/1534735418805149.

40. Aktas, D., & Terzioglu, F. (2015). Effect of home care service on the quality of life in patients with gynecological cancer. *Asian Pacific Journal of Cancer Prevention: APJCP, 16*(9), 4089–4094. https://doi.org/10.7314/apjcp.2015.16.9.4089.

41. Kalter, J., Buffart, L. M., Korstjens, I., van Weert, E., Brug, J., Verdonck-de Leeuw, I. M., … May, A. M. (2015). Moderators of the effects of group-based physical exercise on cancer survivors' quality of life. *Supportive Care in Cancer: Official Journal of the Multinational Association of Supportive Care in Cancer, 23*(9), 2623–2631. https://doi.org/10.1007/s00520-015-2622-z.

42. Cox, M., Carmack, C., Hughes, D., Baum, G., Brown, J., Jhingran, A., … Basen-Engquist, K. (2015). Antecedents and mediators of physical activity in endometrial cancer survivors: Increasing physical activity through steps to health. *Health Psychology: Official Journal of the Division of Health Psychology, American Psychological Association, 34*(10), 1022–1032. https://doi.org/10.1037/hea0000188.

43. Raz, D. J., Sun, V., Kim, J. Y., Williams, A. C., Koczywas, M., Cristea, M., … Ferrell, B. (2016). Long-term effect of an interdisciplinary supportive care intervention for lung cancer survivors after surgical procedures. *The Annals of Thoracic Surgery, 101*(2), 493–495. https://doi.org/10.1016/j.athoracsur.2015.07.031

44. Sun, V., Crane, T. E., Slack, S. D., Yung, A., Wright, S., Sentovich, S., … Thomson, C. A. (2018). Rationale, development, and design of the Altering Intake, Managing Symptoms (AIMS) dietary intervention for bowel dysfunction in rectal cancer survivors. *Contemporary Clinical Trials, 68*, 61–66. https://doi.org/10.1016/j.cct.2018.03.010.

45. Sloan, J. A., Cheville, A. L., Liu, H., Novotny, P. J., Wampfler, J. A., Garces, Y. I., … Yang, P. (2016). Impact of self-reported physical activity and health promotion behaviors on lung cancer survivorship. *Health and Quality of Life Outcomes, 14*, 66. https://doi.org/10.1186/s12955-016-0461-3.

46. Moghaddam Tabrizi, F., & Alizadeh, S. (2018). Family intervention based on the FOCUS program effects on cancer coping in Iranian breast cancer patients: a randomized control trial. *Asian Pacific Journal of Cancer Prevention: APJCP, 19*(6), 1523–1528. https://doi.org/10.22034/APJCP.2018.19.6.1523.

47. Lyons, K. D., Bruce, M. L., Hull, J. G., Kaufman, P. A., Li, Z., Stearns, D. M., … Hegel, M. T. (2019). Health through activity: Initial evaluation of an in-home intervention for older adults with cancer. *The American Journal of Occupational Therapy: Official Publication of the American Occupational Therapy Association, 73*(5), 7305205070p1-7305205070p11. https://doi.org/10.5014/ajot.2019.035022.

48. James, E. L., Stacey, F. G., Chapman, K., Boyes, A. W., Burrows, T., Girgis, A., … Lubans, D. R. (2015). Impact of a nutrition and physical activity intervention (ENRICH: Exercise and Nutrition Routine Improving Cancer Health) on health behaviors of cancer survivors and carers: a pragmatic randomized controlled trial. *BMC Cancer, 15*, 710. https://doi.org/10.1186/s12885-015-1775-y.

49. Patterson, R. E., Marinac, C. R., Natarajan, L., Hartman, S. J., Cadmus-Bertram, L., Flatt, S. W., … Kerr, J. (2016). Recruitment strategies, design, and participant characteristics in a trial of weight-loss and metformin in breast cancer survivors. *Contemporary Clinical Trials, 47*, 64–71. https://doi.org/10.1016/j.cct.2015.12.009.

50. Braam, K. I., van der Torre, P., Takken, T., Veening, M. A., van Dulmen-den Broeder, E., & Kaspers, G. J. L. (2016). Physical exercise training interventions for children and young adults during and after treatment for childhood cancer. *The Cochrane Database of Systematic Reviews*, *3*, CD008796. https://doi.org/10.1002/14651858.CD008796.pub3.

51. Sterba, K. R., Armeson, K., Franco, R., Harper, J., Patten, R., Kindall, S., ... Zapka, J. (2015). A pilot randomized controlled trial testing a minimal intervention to prepare breast cancer survivors for recovery. *Cancer Nursing*, *38*(2), E48–56. https://doi.org/10.1097/NCC.0000000000000152.

52. Brown, J. C., Troxel, A. B., Ky, B., Damjanov, N., Zemel, B. S., Rickels, M. R., ... Schmitz, K. H. (2016). A randomized phase II dose-response exercise trial among colon cancer survivors: Purpose, study design, methods, and recruitment results. *Contemporary Clinical Trials*, *47*, 366–375. https://doi.org/10.1016/j.cct.2016.03.001.

53. Kampshoff, C. S., Chinapaw, M. J. M., Brug, J., Twisk, J. W. R., Schep, G., Nijziel, M. R., ... Buffart, L. M. (2015). Randomized controlled trial of the effects of high intensity and low-to-moderate intensity exercise on physical fitness and fatigue in cancer survivors: results of the Resistance and Endurance exercise After ChemoTherapy (REACT) study. *BMC Medicine*, *13*, 275. https://doi.org/10.1186/s12916-015-0513-2.

54. Hirschey, R., Kimmick, G., Hockenberry, M., Shaw, R., Pan, W., Page, C., & Lipkus, I. (2018). A randomized phase II trial of MOVING ON: An intervention to increase exercise outcome expectations among breast cancer survivors. *Psycho-Oncology*, *27*(10), 2450–2457. https://doi.org/10.1002/pon.4849.

55. Carson, J. W., Carson, K. M., Olsen, M. K., Sanders, L., & Porter, L. S. (2017). Mindful Yoga for women with metastatic breast cancer: design of a randomized controlled trial. *BMC Complementary and Alternative Medicine*, *17*(1), 153. https://doi.org/10.1186/s12906-017-1672-9.

56. Jefford, M., Gough, K., Drosdowsky, A., Russell, L., Aranda, S., Butow, P., ... Schofield, P. (2016). A Randomized Controlled Trial of a Nurse-Led Supportive Care Package (SurvivorCare) for Survivors of Colorectal Cancer. *The Oncologist*, *21*(8), 1014–1023. https://doi.org/10.1634/theoncologist.2015-0533.

57. Lee, C. F., Ho, J. W. C., Fong, D. Y. T., Macfarlane, D. J., Cerin, E., Lee, A. M., ... Cheng, K.-K. (2018). Dietary and physical activity interventions for colorectal cancer survivors: A randomized controlled trial. *Scientific Reports*, *8*(1), 5731. https://doi.org/10.1038/s41598-018-24042-6.

58. Cheng, K. K. F., Lim, Y. T. E., Koh, Z. M., & Tam, W. W. S. (2017). Home-based multidimensional survivorship programmes for breast cancer survivors. *The Cochrane Database of Systematic Reviews*, *8*, CD011152. https://doi.org/10.1002/14651858.CD011152.pub2.

59. Caperchione, C. M., Sabiston, C. M., Stolp, S., Bottorff, J. L., Campbell, K. L., Eves, N. D., ... Fitzpatrick, K. M. (2019). A preliminary trial examining a "real world" approach for increasing physical activity among breast cancer survivors: findings from project MOVE. *BMC Cancer*, *19*(1), 272. https://doi.org/10.1186/s12885-019-5470-2.

60. Shobeiri, F., Masoumi, S. Z., Nikravesh, A., Heidari Moghadam, R., & Karami, M. (2016). The impact of aerobic exercise on quality of life in women with breast cancer: A randomized controlled trial. *Journal of Research in Health Sciences*, *16*(3), 127–132.

61. Stagl, J. M., Bouchard, L. C., Lechner, S. C., Blomberg, B. B., Gudenkauf, L. M., Jutagir, D. R., ... Antoni, M. H. (2015). Long-term psychological benefits of cognitive-behavioral stress management for women with breast cancer: 11-year follow-up of a randomized controlled trial. *Cancer*, *121*(11), 1873–1881. https://doi.org/10.1002/cncr.29076.

62. Conlon, B. A., Kahan, M., Martinez, M., Isaac, K., Rossi, A., Skyhart, R., ... Moadel-Robblee, A. (2015). Development and evaluation of the curriculum for BOLD (Bronx Oncology Living Daily) healthy living: a diabetes prevention and control program for

underserved cancer survivors. *Journal of Cancer Education: The Official Journal of the American Association for Cancer Education*, *30*(3), 535–545. https://doi.org/10.1007/s13187-014-0750-7.

63. Carter, S. J., Hunter, G. R., McAuley, E., Courneya, K. S., Anton, P. M., & Rogers, L. Q. (2016). Lower rate-pressure product during submaximal walking: a link to fatigue improvement following a physical activity intervention among breast cancer survivors. *Journal of Cancer Survivorship: Research and Practice*, *10*(5), 927–934. https://doi.org/10.1007/s11764-016-0539-2.

64. Buffart, L. M., Newton, R. U., Chinapaw, M. J., Taaffe, D. R., Spry, N. A., Denham, J. W., … Galvao, D. A. (2015). The effect, moderators, and mediators of resistance and aerobic exercise on health-related quality of life in older long-term survivors of prostate cancer. *Cancer*, *121*(16), 2821–2830. https://doi.org/10.1002/cncr.29406.

65. Diggins, A. D., Hearn, L. E., Lechner, S. C., Annane, D., Antoni, M. H., & Whitehead, N. E. (2017). Physical activity in Black breast cancer survivors: implications for quality of life and mood at baseline and 6-month follow-up. *Psycho-Oncology*, *26*(6), 822–828. https://doi.org/10.1002/pon.4095.

66. van der Spek, N., Vos, J., van Uden-Kraan, C. F., Breitbart, W., Cuijpers, P., Holtmaat, K., … Verdonck-de Leeuw, I. M. (2017). Efficacy of meaning-centered group psychotherapy for cancer survivors: a randomized controlled trial. *Psychological Medicine*, *47*(11), 1990–2001. https://doi.org/10.1017/S0033291717000447.

67. Maheu, C., Lebel, S., Courbasson, C., Lefebvre, M., Singh, M., Bernstein, L. J., … Sidani, S. (2016). Protocol of a randomized controlled trial of the fear of recurrence therapy (FORT) intervention for women with breast or gynecological cancer. *BMC Cancer*, *16*, 291. https://doi.org/10.1186/s12885-016-2326-x.

68. Rogers, L. Q., Courneya, K. S., Anton, P. M., Hopkins-Price, P., Verhulst, S., Vicari, S. K., … McAuley, E. (2015). Effects of the BEAT Cancer physical activity behavior change intervention on physical activity, aerobic fitness, and quality of life in breast cancer survivors: a multicenter randomized controlled trial. *Breast Cancer Research and Treatment*, *149*(1), 109–119. https://doi.org/10.1007/s10549-014-3216-z.

69. Iyer, N. S., Osann, K., Hsieh, S., Tucker, J. A., Monk, B. J., Nelson, E. L., & Wenzel, L. (2016). Health behaviors in cervical cancer survivors and associations with quality of life. *Clinical Therapeutics*, *38*(3), 467–475. https://doi.org/10.1016/j.clinthera.2016.02.006.

70. Carter, S. J., Hunter, G. R., Norian, L. A., Turan, B., & Rogers, L. Q. (2018). Ease of walking associates with greater free-living physical activity and reduced depressive symptomology in breast cancer survivors: pilot randomized trial. *Supportive Care in Cancer: Official Journal of the Multinational Association of Supportive Care in Cancer*, *26*(5), 1675–1683. https://doi.org/10.1007/s00520-017-4015-y

71. Conroy, D. E., Wolin, K. Y., Blair, C. K., & Demark-Wahnefried, W. (2017). Gender-varying associations between physical activity intensity and mental quality of life in older cancer survivors. *Supportive Care in Cancer: Official Journal of the Multinational Association of Supportive Care in Cancer*, *25*(11), 3465–3473. https://doi.org/10.1007/s00520-017-3769-6.

72. Bernard-Davila, B., Aycinena, A. C., Richardson, J., Gaffney, A. O., Koch, P., Contento, I., … Greenlee, H. (2015). Barriers and facilitators to recruitment to a culturally-based dietary intervention among urban Hispanic breast cancer survivors. *Journal of Racial and Ethnic Health Disparities*, *2*(2), 244–255. https://doi.org/10.1007/s40615-014-0076-5.

73. Winkels, R. M., Sturgeon, K. M., Kallan, M. J., Dean, L. T., Zhang, Z., Evangelisti, M., … Schmitz, K. H. (2017). The women in steady exercise research (WISER) survivor trial: The innovative transdisciplinary design of a randomized controlled trial of exercise and weight-loss interventions among breast cancer survivors with lymphedema. *Contemporary Clinical Trials*, *61*, 63–72. https://doi.org/10.1016/j.cct.2017.07.017.

74. Sheng, J. Y., Sharma, D., Jerome, G., & Santa-Maria, C. A. (2018). Obese breast cancer patients and survivors: Management considerations. *Oncology (Williston Park, N.Y.), 32*(8), 410–417.

75. Christifano, D. N., Fazzino, T. L., Sullivan, D. K., & Befort, C. A. (2016). Diet quality of breast cancer survivors after a six-month weight management intervention: Improvements and association with weight loss. *Nutrition and Cancer, 68*(8), 1301–1308. https://doi.org/10.1080/01635581.2016.1224368.

76. Demark-Wahnefried, W., Colditz, G. A., Rock, C. L., Sedjo, R. L., Liu, J., Wolin, K. Y., … Ganz, P. A. (2015). Quality of life outcomes from the Exercise and Nutrition Enhance Recovery and Good Health for You (ENERGY)-randomized weight loss trial among breast cancer survivors. *Breast Cancer Research and Treatment, 154*(2), 329–337. https://doi.org/10.1007/s10549-015-3627-5.

77. Dittus, K. L., Harvey, J. R., Bunn, J. Y., Kokinda, N. D., Wilson, K. M., Priest, J., & Pratley, R. E. (2018). Impact of a behaviorally-based weight loss intervention on parameters of insulin resistance in breast cancer survivors. *BMC Cancer, 18*(1), 351. https://doi.org/10.1186/s12885-018-4272-2

78. Fazzino, T. L., Klemp, J., & Befort, C. (2018). Late breast cancer treatment-related symptoms and functioning: associations with physical activity adoption and maintenance during a lifestyle intervention for rural survivors. *Breast Cancer Research and Treatment, 168*(3), 755–761. https://doi.org/10.1007/s10549-017-4603-z.

79. DuHamel, K., Schuler, T., Nelson, C., Philip, E., Temple, L., Schover, L., … Carter, J. (2016). The sexual health of female rectal and anal cancer survivors: results of a pilot randomized psycho-educational intervention trial. *Journal of Cancer Survivorship: Research and Practice, 10*(3), 553–563. https://doi.org/10.1007/s11764-015-0501-8.

80. Kitson, S., Ryan, N., MacKintosh, M. L., Edmondson, R., Duffy, J. M., & Crosbie, E. J. (2018). Interventions for weight reduction in obesity to improve survival in women with endometrial cancer. *The Cochrane Database of Systematic Reviews, 2*, CD012513. https://doi.org/10.1002/14651858.CD012513.pub2

81. Grimmett, C., Simon, A., Lawson, V., & Wardle, J. (2015). Diet and physical activity intervention in colorectal cancer survivors: a feasibility study. *European Journal of Oncology Nursing: The Official Journal of European Oncology Nursing Society, 19*(1), 1–6. https://doi.org/10.1016/j.ejon.2014.08.006.

82. Reeves, M. M., Terranova, C. O., Erickson, J. M., Job, J. R., Brookes, D. S. K., McCarthy, N., … Eakin, E. G. (2016). Living well after breast cancer randomized controlled trial protocol: evaluating a telephone-delivered weight loss intervention versus usual care in women following treatment for breast cancer. *BMC Cancer, 16*(1), 830. https://doi.org/10.1186/s12885-016-2858-0

83. Ashing, K. T., & Miller, A. M. (2016, February). Assessing the utility of a telephonically delivered psychoeducational intervention to improve health-related quality of life in African American breast cancer survivors: a pilot trial. *Psycho-Oncology*. England. https://doi.org/10.1002/pon.3823.

84. Wenzel, L., Osann, K., Hsieh, S., Tucker, J. A., Monk, B. J., & Nelson, E. L. (2015). Psychosocial telephone counseling for survivors of cervical cancer: results of a randomized biobehavioral trial. *Journal of Clinical Oncology: Official Journal of the American Society of Clinical Oncology, 33*(10), 1171–1179. https://doi.org/10.1200/JCO.2014.57.4079.

85. Hirschey, R., Docherty, S. L., Pan, W., & Lipkus, I. (2017). Exploration of exercise outcome expectations among breast cancer survivors. *Cancer Nursing, 40*(2), E39–E46. https://doi.org/10.1097/NCC.0000000000000362.

86. Lahart, I. M., Metsios, G. S., Nevill, A. M., Kitas, G. D., & Carmichael, A. R. (2016). Randomised controlled trial of a home-based physical activity intervention in breast cancer survivors. *BMC Cancer, 16*, 234. https://doi.org/10.1186/s12885-016-2258-5.

87. Sohl, S. J., Dietrich, M. S., Wallston, K. A., & Ridner, S. H. (2017). A randomized controlled trial of expressive writing in breast cancer survivors with lymphedema. *Psychology & Health, 32*(7), 826–842. https://doi.org/10.1080/08870446.2017.1307372.

88. Reb, A., Ruel, N., Fakih, M., Lai, L., Salgia, R., Ferrell, B., … Sun, V. (2017). Empowering survivors after colorectal and lung cancer treatment: Pilot study of a Self-Management Survivorship Care Planning intervention. *European Journal of Oncology Nursing: The Official Journal of European Oncology Nursing Society, 29*, 125–134. https://doi.org/10.1016/j.ejon.2017.06.003.

89. Lawn, S., Zrim, S., Leggett, S., Miller, M., Woodman, R., Jones, L., … Koczwara, B. (2015). Is self-management feasible and acceptable for addressing nutrition and physical activity needs of cancer survivors? *Health Expectations: An International Journal of Public Participation in Health Care and Health Policy, 18*(6), 3358–3373. https://doi.org/10.1111/hex.12327.

90. Peoples, A. R., Garland, S. N., Perlis, M. L., Savard, J., Heckler, C. E., Kamen, C. S., … Roscoe, J. A. (2017). Effects of cognitive behavioral therapy for insomnia and armodafinil on quality of life in cancer survivors: a randomized placebo-controlled trial. *Journal of Cancer Survivorship: Research and Practice, 11*(3), 401–409. https://doi.org/10.1007/s11764-017-0597-0.

91. Meneses, K., Gisiger-Camata, S., Schoenberger, Y.-M., Weech-Maldonado, R., & McNees, P. (2015). Adapting an evidence-based survivorship intervention for Latina breast cancer survivors. *Women's Health (London, England), 11*(2), 109–119. https://doi.org/10.2217/whe.14.65

92. Wakefield, C. E., Sansom-Daly, U. M., McGill, B. C., McCarthy, M., Girgis, A., Grootenhuis, M., … Cohn, R. J. (2015). Online parent-targeted cognitive-behavioural therapy intervention to improve quality of life in families of young cancer survivors: study protocol for a randomised controlled trial. *Trials, 16*, 153. https://doi.org/10.1186/s13063-015-0681-6.

93. Orchard, T. S., Andridge, R. R., Yee, L. D., & Lustberg, M. B. (2018). Diet quality, inflammation, and quality of life in breast cancer survivors: A cross-sectional analysis of pilot study data. *Journal of the Academy of Nutrition and Dietetics, 118*(4), 578–588. e1. https://doi.org/10.1016/j.jand.2017.09.024.

94. Taylor, K., Joske, D., Bulsara, M., Bulsara, C., & Monterosso, L. (2016). Protocol for Care After Lymphoma (CALy) trial: a phase II pilot randomised controlled trial of a lymphoma nurse-led model of survivorship care. *BMJ Open, 6*(5), e010817. https://doi.org/10.1136/bmjopen-2015-010817.

95. Kanera, I. M., Willems, R. A., Bolman, C. A. W., Mesters, I., Verboon, P., & Lechner, L. (2017). Long-term effects of a web-based cancer aftercare intervention on moderate physical activity and vegetable consumption among early cancer survivors: a randomized controlled trial. *The International Journal of Behavioral Nutrition and Physical Activity, 14*(1), 19. https://doi.org/10.1186/s12966-017-0474-2.

96. Willems, R. A., Mesters, I., Lechner, L., Kanera, I. M., & Bolman, C. A. W. (2017). Long-term effectiveness and moderators of a web-based tailored intervention for cancer survivors on social and emotional functioning, depression, and fatigue: randomized controlled trial. *Journal of Cancer Survivorship: Research and Practice, 11*(6), 691–703. https://doi.org/10.1007/s11764-017-0625-0.

97. Howell, C. R., Krull, K. R., Partin, R. E., Kadan-Lottick, N. S., Robison, L. L., Hudson, M. M., & Ness, K. K. (2018). Randomized web-based physical activity intervention in adolescent survivors of childhood cancer. *Pediatric Blood & Cancer, 65*(8), e27216. https://doi.org/10.1002/pbc.27216.

98. Kanera, I. M., Bolman, C. A. W., Willems, R. A., Mesters, I., & Lechner, L. (2016). Lifestyle-related effects of the web-based Kanker Nazorg Wijzer (Cancer Aftercare Guide) intervention for cancer survivors: a randomized controlled trial.

Journal of Cancer Survivorship: Research and Practice, *10*(5), 883–897. https://doi.org/10.1007/s11764-016-0535-6.

99. Trinh, L., Arbour-Nicitopoulos, K. P., Sabiston, C. M., Berry, S. R., Loblaw, A., Alibhai, S. M. H., … Faulkner, G. E. (2018). RiseTx: testing the feasibility of a web application for reducing sedentary behavior among prostate cancer survivors receiving androgen deprivation therapy. *The International Journal of Behavioral Nutrition and Physical Activity*, *15*(1), 49. https://doi.org/10.1186/s12966-018-0686-0.

100. Wu, A. W., White, S. M., Blackford, A. L., Wolff, A. C., Carducci, M. A., Herman, J. M., & Snyder, C. F. (2016). Improving an electronic system for measuring PROs in routine oncology practice. *Journal of Cancer Survivorship: Research and Practice*, *10*(3), 573–582. https://doi.org/10.1007/s11764-015-0503-6.

101. Frensham, L. J., Parfitt, G., & Dollman, J. (2018). Effect of a 12-Week Online Walking Intervention on Health and Quality of Life in Cancer Survivors: A Quasi-Randomized Controlled Trial. *International Journal of Environmental Research and Public Health*, *15*(10). https://doi.org/10.3390/ijerph15102081

102. Corbett, T., Walsh, J. C., Groarke, A., Moss-Morris, R., & McGuire, B. E. (2016). Protocol for a pilot randomised controlled trial of an online intervention for post-treatment cancer survivors with persistent fatigue. *BMJ Open*, *6*(6), e011485. https://doi.org/10.1136/bmjopen-2016-011485

103. Atema, V., van Leeuwen, M., Oldenburg, H. S. A., Retel, V., van Beurden, M., Hunter, M. S., & Aaronson, N. K. (2016). Design of a randomized controlled trial of Internet-based cognitive behavioral therapy for treatment-induced menopausal symptoms in breast cancer survivors. *BMC Cancer*, *16*(1), 920. https://doi.org/10.1186/s12885-016-2946-1

104. Ferguson, R. J., Sigmon, S. T., Pritchard, A. J., LaBrie, S. L., Goetze, R. E., Fink, C. M., & Garrett, A. M. (2016). A randomized trial of videoconference-delivered cognitive behavioral therapy for survivors of breast cancer with self-reported cognitive dysfunction. *Cancer*, *122*(11), 1782–1791. https://doi.org/10.1002/cncr.29891.

105. Abrahams, H. J. G., Gielissen, M. F. M., Donders, R. R. T., Goedendorp, M. M., van der Wouw, A. J., Verhagen, C. A. H. H. V. M., & Knoop, H. (2017). The efficacy of Internet-based cognitive behavioral therapy for severely fatigued survivors of breast cancer compared with care as usual: A randomized controlled trial. *Cancer*, *123*(19), 3825–3834. https://doi.org/10.1002/cncr.30815.

106. Hummel, S. B., van Lankveld, J. J. D. M., Oldenburg, H. S. A., Hahn, D. E. E., Broomans, E., & Aaronson, N. K. (2015). Internet-based cognitive behavioral therapy for sexual dysfunctions in women treated for breast cancer: design of a multicenter, randomized controlled trial. *BMC Cancer*, *15*, 321. https://doi.org/10.1186/s12885-015-1320-z.

107. Cramer, H., Rabsilber, S., Lauche, R., Kummel, S., & Dobos, G. (2015). Yoga and meditation for menopausal symptoms in breast cancer survivors-A randomized controlled trial. *Cancer*, *121*(13), 2175–2184. https://doi.org/10.1002/cncr.29330.

108. Janelsins, M. C., Peppone, L. J., Heckler, C. E., Kesler, S. R., Sprod, L. K., Atkins, J., … Mustian, K. M. (2016). YOCAS(c)(R) yoga reduces self-reported memory difficulty in cancer survivors in a nationwide randomized clinical trial: Investigating relationships between memory and sleep. *Integrative Cancer Therapies*, *15*(3), 263–271. https://doi.org/10.1177/1534735415617021.

109. Chen, X., Gong, X., Shi, C., Sun, L., Tang, Z., Yuan, Z., … Yu, J. (2018). Multi-focused psychosocial residential rehabilitation interventions improve quality of life among cancer survivors: a community-based controlled trial. *Journal of Translational Medicine*, *16*(1), 250. https://doi.org/10.1186/s12967-018-1618-0.

110. Schmitt, J., Lindner, N., Reuss-Borst, M., Holmberg, H.-C., & Sperlich, B. (2016). A 3-week multimodal intervention involving high-intensity interval training in female cancer survivors: a randomized controlled trial. *Physiological Reports*, *4*(3). https://doi.org/10.14814/phy2.12693.

111. McEwen, S. E., Davis, A. M., Jones, J. M., Martino, R., Poon, I., Rodriguez, A. M., & Ringash, J. (2015). Development and preliminary evaluation of a rehabilitation consult for survivors of head and neck cancer: an intervention mapping protocol. *Implementation Science: IS, 10,* 6. https://doi.org/10.1186/s13012-014-0191-z.

112. MacKinnon, D. *Introduction to Statistical Mediation Analysis.* Routledge: UK, 2012.

17 Major Depressive Disorder Detection Using Data Science and Wearable Connected Devices

Umar Khalid Farooqui and Archana Sahai
Amity University

17.1 INTRODUCTION

In recent years, mental illnesses like stress and depression have become widespread, important public health issues, and they have a significant impact on society. It affects people of all ages, whether they are male or female, live in cities or rural areas, are educated or not, or are employed or not. Every year, a sizeable number of people in developing nations like India commit suicide as a result of depression. The development of technology and sensors has a significant impact on daily life. With the help of Internet of Things (IoT) and other technologies, numerous researchers are attempting to identify and treat depression at an early stage.

So, let us quickly understand what depression is and how it could be classified.

A prolonged sense of sadness and a loss of interest in things and activities you once found enjoyable are symptoms of the mood disorder depression. Moreover, it may make it difficult to think, remember, eat, and sleep. Depressive disorders are categorised as follows in the Diagnostic and Statistical Manual of Mental Disorders, Fifth Edition (DSM-5) by the American Psychiatric Association:

- Major depressive disorder, also known as clinical depression, is characterised by feelings of sadness, hopelessness, or worthlessness on the majority of days for at least 2 weeks, along with additional symptoms like difficulty sleeping, a lack of interest in previously enjoyed activities, or a change in appetite. This is both the most severe and most prevalent type of depression.
- Mild to moderate depression that lasts for at least 2 years is referred to as persistent depressive disorder (PDD). Compared to major depressive illness, the symptoms are less severe. PDD was once referred to as dysthymia by medical professionals.

DOI: 10.1201/9781003359951-20

- Disruptive mood dysregulation disorder (DMDD): In children, DMDD results in repeated episodes of extreme irritability and persistent irritability. Often, symptoms appear around the age of 10.
- Premenstrual dysphoric disorder (PMDD): With PMDD, you have mood symptoms such as severe irritability, anxiety, or depression in addition to premenstrual syndrome (PMS) symptoms. As your period begins, these symptoms usually go away within a few days, but they can occasionally be so bad that they seriously affect your daily life.
- Depression brought on by a different medical condition: Many medical problems might alter your body in ways that lead to depression. Hypothyroidism, heart disease, Parkinson's disease, and cancer are a few examples. When the underlying illness is successfully treated, depression typically gets better as well.

Out of these, major depressive disorder is a notable one that can take on a variety of forms, including:

Seasonal depression is a type of major depressive disorder that normally develops in the fall and winter and subsides in the spring and summer.

Depression during pregnancy and postpartum: Depression that occurs during pregnancy is referred to as prenatal depression. Depression that appears within four weeks following childbirth is known as postpartum depression. They are known as "major depressive disorder (MDD) with peripartum onset" according to the DSM.

Atypical depression: This ailment, also referred to as MDD with atypical features, has symptoms that are a little different from those of "typical" depression. The primary distinction is a brief mood elevation in response to good things happening (mood reactivity). Increased hunger and susceptibility to rejection are two additional significant symptoms.

In addition to manic or hypomanic episodes, individuals with bipolar illness also experience depressive episodes.

Further, major depression disorder leads to suicidal tendencies. In recent days, suicide has become a serious problem. To save lives, early detection and prevention of suicide attempts should be addressed. Current suicide-detecting techniques are clinical techniques based on the interaction between the targeted population and social workers or expert people using machine learning, deep learning, or feature engineering techniques for automatic detection based on social media information online.

Because they are particularly severe in wealthy nations and emerging economies, mental health concerns including anxiety and depression are causing growing worry in modern society.

In the United States, suicide is one of the top 10 causes of mortality for all age groups and the second greatest cause of death for those between the ages of 15 and 34 (Figure 17.1) [1–4]. While there have been significant improvements in lowering the death toll from diseases like cancer and HIV/AIDS, the suicide rate has remained essentially unchanged over the past 50 years.

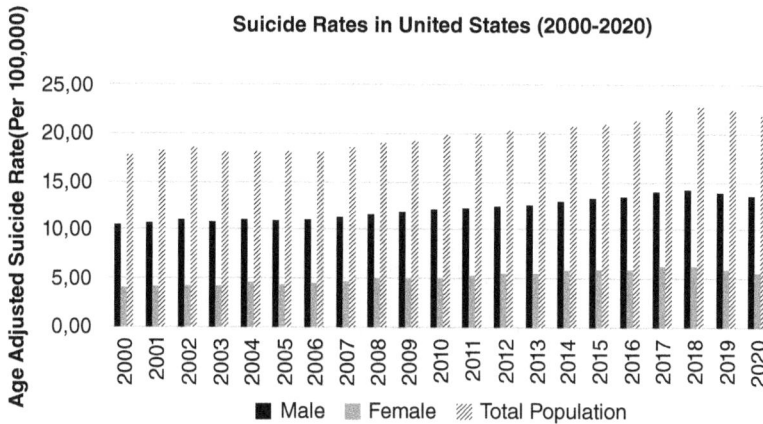

FIGURE 17.1   Suicide facts and figures. Source: Data courtesy of CDC, National Institute of Mental Health USA [5].

We do know that the vast majority of persons who die by suicide have depression or another curable brain disorder [6], despite the fact that some people may think of suicide as being less of a "medical" condition than illnesses like cancer or HIV/AIDS.

People having suicidal thoughts have been treated for these underlying problems with medicine and psychotherapy, and novel drugs (such as ketamine 5) that are researched and produced expressly to address suicidality may help prevent suicide. Additionally, we might be able to develop a technology that can digitally identify high-risk people and suicidal behaviour through the collection and analysis of vast volumes of demographic and/or patient data.

This use of technology has not escaped the attention of important decision-makers. With the objective of "Using Data to Strengthen Mental Health Awareness and Suicide Prevention," the White House recently sponsored the Partnerships for Suicide Prevention event. The White House also hosted a "hackathon" for suicide prevention that took place in five cities and included data scientists, creators, designers, and experts in cutting-edge technology from numerous organisations. They were tasked with working together to create tools, technologies, or data analysis for preventing suicide [7].

Severe mental diseases can cause suicidal thoughts and even suicide attempts if they are not well treated. Some statements posted online are incredibly upsetting and encourage bad behaviours like cyberstalking and cyberbullying. Due to the regular social cruelty that such subpar information engages in, the results can be severe and harmful, leading to rumours or even mental trauma. According to a study, there is a link between online bullying and suicide [1].

Suicidal thoughts, plans, and attempts are frequently preceded by death by suicide, making it a complicated issue [8].

We may have new ways to identify suicidal behaviour, prevent suicide, and save lives thanks to significant advancements in wearable technology and data science.

17.2 DETECTING SUICIDAL BEHAVIOUR

Assessing a patient's risk of suicide is one of a psychiatrist's primary responsibilities. Despite the fact that there are numerous known risk factors for suicide, psychiatrists must use their clinical judgement to determine suicide risk. We have not yet developed an automated approach to model and predict suicide risk objectively; therefore, this rating carries some degree of subjectivity. The risk of suicide can fluctuate quickly and is influenced by biological, social, and psychological factors. A person may be persistently at high risk of suicide, but a more immediate life event may quickly raise that risk further. Suicide risk also has both chronic and acute components.

People who are often at low risk may suddenly find themselves in a situation where they are at high risk. Even neutral or good life experiences can lead to stress and a higher risk of suicide, according to some, despite the fact that stress and bad life events have been related to suicide. Effects of positive life events on suicide risk seem to be multifaceted and poorly understood at the moment [9].

There are various signs that indicate suicidal tendencies. However, the major sign is depression [10]. Bipolar disorder and major unipolar depression are examples of endogenous affective disorders that need to be treated using a medical model because they are brought on by an imbalance in the body's neurochemical, endocrine, and neurobiological systems. Reactive major depression, on the other hand, and adjustment disorders are brought on by outside forces like the unexpected death of a loved one, failure in business or in the workplace, etc. A number of psychiatric disorders, including premenstrual syndrome (PMS), seasonal affective disorder (SAD), and post-traumatic stress disorder (PTSD), also share depressive symptoms as a core component.

In the clinical setting, standardised symptom rating scales and structured or semi-structured diagnostic interviews are used to identify and evaluate the risk of suicide in vulnerable individuals. However, a lot of suicidal behaviours strike high-risk people suddenly and without warning. In order to assess the risk of suicide, it may be helpful to identify sudden changes in suicidal ideation (the desire to die or to be dead). However, current clinical practice is insufficient to identify these sudden changes in suicidal ideation, which occur over time [11].

A wearable sensor system that can continuously track physiological or behavioural signals that reveal suicidal ideation is required (this is your IoT framework). When elevated levels of suicidal ideation are discovered, this non-invasive system can assist clinicians in identifying them and providing information for the efficient management and treatment of suicide risk. The M-SID framework is designed to support the theory put forth by Indic et al. in Refs. [12,13], which links a person's vitality and psychomotor activity to their emotion and disturbed affect. It has been demonstrated that data of heart rate and activity can provide information about suicidal ideation [14].

An overview of the suggested method is presented in Figure 17.2. The M-SID framework assists healthcare professionals in evoking the emotion in real time while allowing them to analyse a pattern of suicidal ideation using the physiological signals of the individual. This M-SID wearable is designed as a portable, intelligent monitoring system that uses a battery management system to supply power. Using

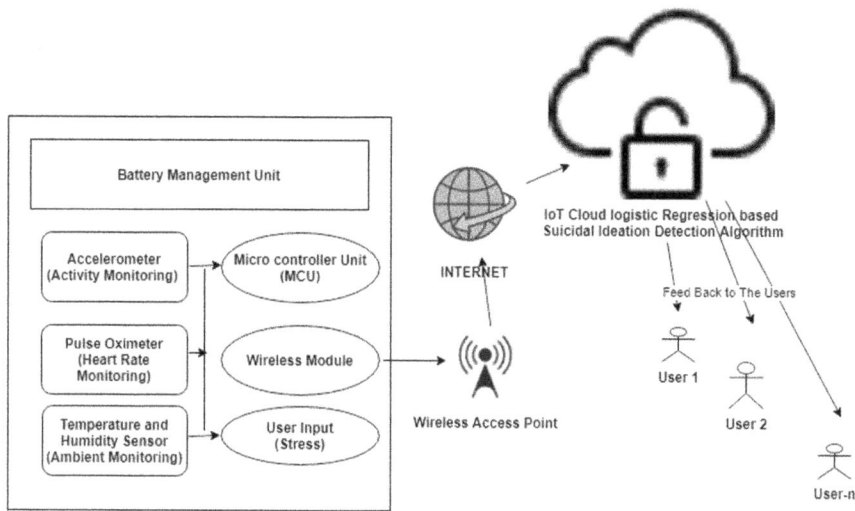

FIGURE 17.2   Basic block diagram of M-SID framework to detect suicidal tendency [12].

a microcontroller unit, the activity and pulse oximeter sensors are wirelessly connected to the Internet. The microcontroller unit's processor and memory allow decisions to be made at the edge.

Further, various researchers have proposed different solutions to detect suicidal tendencies. In the following sections, we examine some of the prominent work and methods used to detect suicidal thoughts.

17.2.1   Use of Computerised Speech and Facial Emotions Analysis

Suicidal thoughts may also be discernible through speech and facial expressions. Researchers can identify changes in the speech patterns of people who are sad and/ or having suicidal thoughts using computerised speech analysis. Individuals with suicidal thoughts may experience changes in their speech frequency, either shifting from high to low frequency or vice versa. Additionally, studies have indicated that depressed individuals have a narrower speech acoustic range [12].

Others are employing computerised real-time facial emotion monitoring to identify tiny alterations in the facial expressions of those who are considering suicide. For instance, researchers from the University of Massachusetts and Affective are trying to pinpoint individuals who are considering suicide based on how they react to several video vignettes. Along with taking body data like skin conductance and heart rate, they also analyse facial expressions using a sophisticated computer program. They want to show that persons with suicidal thoughts and those without have different emotional and physical responses to various movies. Such studies could help us create a quick method to evaluate suicidality, particularly in high-risk people who might not verbally report the existence of suicidal thoughts or plans [13].

A mental state examination (MSE), a methodical yet arbitrary way of evaluating and recording a person's physical and mental presentation, is a skill that psychiatrists and other mental health providers often get training in [14]. One of the most crucial diagnostic and assessment tools for a clinical psychiatrist is the history, along with the MSE [14]. The MSE includes a review of a person's speech, behaviour, mental processes, and mood. The author suggests creating a digital mental state examination (dMSE) that evaluates a patient's presentation using a variety of technologies, such as speech analysis, motion tracking, and natural language processing, to generate quantitative, objective data that may be superior to the subjective reporting of the conventional clinical MSE.

17.2.2   USE OF SMART PHONE AND COMPUTER THERAPY

There is some developing evidence to support the use of computerised suicide prevention methods; automated cognitive behavioural therapy is one such example (CBT). This specific form of talk therapy can be delivered through automated CBT without a human therapist present. Unfortunately, while there is considerable evidence to support the use of such strategies in affective disorders, there is less support for their use in suicide [15].

Smartphone applications (apps) for suicide have even less study accessible. Smartphones and other wearable technology may be the best platforms for diagnosing and even treating mental health issues, according to some. In a recent examination of smartphone apps that deal with suicide, the research team at Brain Power came to some unsettling results. Most apps consisted primarily of symptom checklists or resource listings. Some apps included information that could be potentially dangerous or could make someone's condition worse. Generally speaking, these apps did not use the sophisticated hardware and built-in sensors of smartphones to collect behavioural data or to include any sizeable data analysis.

A suicide prediction system was developed by researchers using information from mood-focused smartphone apps and genetic blood tests (biomarkers) [16].

Up to 13% of apps include potentially harmful content, and more than half lack interactive aspects [15]. Similar issues were raised by a recent systemic evaluation of applications for self-harm and suicide [17]. This implies that we are just beginning to use technologies that can be distributed through an app platform. As major technology corporations become more interested in having an impact on healthcare, this may alter.

17.2.3   USE OF WEARABLE IoT (WIoT)

The infrastructure that connects multiple sensors for tracking human aspects like behaviour, health, wellness, and other data is known as wearable IoT (WIoT). Medical information can be transmitted to doctors using a variety of tiny wearable body area sensors (WBAS) and Internet-connected gateways, where data are gathered, controlled, and monitored. The concept of WIoT has been expanded by the authors in Ref. [18], who also specified its architectural components and the support for big data and the cloud. WIoT has the potential to revolutionise healthcare through early disease identification, more affordable treatment options, and effective ways to remotely monitor patients and their treatments [19].

17.2.3.1 Association Rule Mining and the Apriori Algorithm

Accurately predicting depression is currently a big concern; hence in Ref. [20], the authors offer a model for depression prediction considering 500 individuals with various indicators of depression and the Apriori algorithm and association rule mining.

17.2.3.2 EEG Signal Processing

The author of Ref. [21] employed EEG signal processing to predict depression levels. They processed a model using the connections between sleep and depression. Those who are depressed frequently experience insomnia. Insomnia and hypersomnia are among the sleep problems that affect three-quarters of depressed people. The signs of drinking and sleep difficulties have a significant negative effect on the quality of life, which raises the risk of suicide. They found that the ANFIS findings were marginally superior to the classifier results.

17.2.3.3 Linear Predictive Coding (LPC)

Suicide and depression are growing to be serious health issues. The authors in Ref. [22] have created a model of an emotional speech recognition system employing the Tamil language using linear predictive coding (LPC) and a parameters-based approach. 90% of the highest recognition rate could be achieved with LPC algorithms.

17.2.3.4 I-Vector Technique and Fuzzy Membership Functions

The I-Vector method and fuzzy membership functions were chosen in Ref. [23] to identify the levels of depression in 20 patients. Accuracy, balanced classification rate, peak signal-to-noise ratio, F-measure, and specificity are used to differentiate between the algorithms. To increase accuracy, prior processing is now required to remove any silence that may be present in the audio signals. Fuzzy membership functions were substantially more effective (approx. 97% accurate).

17.2.3.5 T-Bots (Therapy Chatbot)

The authors were able to develop a system to suggest treatments for lowering depression levels using T-bots (Therapy Chatbots), which allow for the detection of a person's level of depression [24]. In this study, a treatment chatbot that can also serve as a buddy or well-wisher is deployed. Those who are depressed don't want to go to therapy, but they can talk to a charming virtual therapist instead of being sad by themselves. T-bots are quite helpful and can act as a personal assistant or psychiatrist, fight depression, provide feedback, determine the severity of the depression, indicate the intensity of therapy, etc.

17.2.3.6 Brief Description

The user will initially put on the fitness band and pair it with their Android phone. The wearable (health band) device provides input in the form of signal data. The hc-05 Bluetooth module is then used to transmit the inputted data to the smartphone. A variety of sensors are included in the wearable gadget, including temperature, heart rate, and MPU 6050 sensors. The system allows the user to view the sensor data that were received from the wearable device. User position (sitting, standing) and the diagnosis of depressive illnesses are the two key judgements that are made using the Support Vector Machine (SVM) algorithm. A notification is given to the carer (guardian) via an Android app if the person is experiencing depression. The Android app is additionally useful for seeing the entire details of an individual.

17.3 GENERAL METHODS AND CLASSIFICATION USED IN PREDICTING THE SUICIDAL TENDENCY

Due to an increase in suicide rates in recent years, suicide detection has attracted the interest of numerous experts and has received substantial research from a variety of angles. The research methods used to study suicide also cover a wide range of disciplines and approaches, such as clinical approaches including patient–clinician interaction [1] and automatic detection using user-generated content (mostly text) [2,3]. The use of machine learning techniques for automatic detection is very common.

Clinical techniques, such as face-to-face interviews and self-reports, are used in traditional suicide detection. A widely used five-item questionnaire was created by Venek et al. [14] as a hierarchical approach to the evaluation of suicidal risks based on the patient's behaviour to assess suicidal risk intentions.

In a dyadic interview, Scherer [25] looked at the prosodic speech patterns and voice quality to distinguish between suicidal and non-suicidal young people.

Other clinical approaches identify functional magnetic resonance imaging-based neuronal representations of words associated with death and life [26], investigate resting heart rate from converted sensory signals [26], and analyse event-related instigators from converted EEG signals [27].

The journey is followed by various other methods used by different authors for finding suicidal tendencies.

Below we summarise a few known methods of finding suicidal tendencies.

17.3.1 CONTENT ANALYSIS

Social media posts by users on websites disclose a wealth of information, including their preferred languages. Using exploratory data analysis on user-generated content, it is possible to gain insight into the linguistic habits and suicide attempters' language use. The thorough study includes topic modelling inside posts on suicide, statistical linguistic aspects, and lexicon-based filtering.

To allow keyword filtering and phrase filtering, a lexicon and glossary of words connected to suicide have been carefully created. Kill, suicide, feeling alone, despondent, and self-injury are all suicide-related words and phrases.

To enable keyword filtering [24,28] and phrase filtering [29], suicide-related keyword dictionaries and lexicons were manually created. Kill, suicide, feeling alone, despondent, and self-injury are all suicide-related words and phrases. Using an annotated Twitter dataset, Vioul'es et al. [30] constructed a point-wise mutual information symptom lexicon. Gunn and Lester [31] examined tweets 24 hours before a suicide attempt's passing. Data from the same platform's language usage were examined by Coppersmith et al. [32].

Strongly negative emotions, anxiety, and hopelessness, as well as other social factors like family and friends, may all play a role in suicidal ideation. Ji et al.'s [33] analysis of suicide-related content using topic modelling and word cloud visualisation revealed that personal and social problems are discussed in relation to suicide. The Twitter social network's connection and communication graphs were examined by Colombo and others [34].

Google Trends analysis for suicide is one of the additional strategies and tactics for risk assessment, response bias evaluation through linguistic hints [35], a hybrid human–machine technique for linguistic examination of the influence of social support risk of suicidal thoughts [36], monitoring of social media content, and study of speech patterns [37].

17.3.2 FEATURE ENGINEERING

The purpose of text-based suicide classification is to ascertain whether candidates have suicidal thoughts through their posts. Natural language processing (NLP) and machine learning techniques have both been used in this area.

1. **Tabular Features**: Tabular data for the detection of suicidal thoughts include survey responses and structured statistical data collected from websites. Such organised data can be utilised directly as characteristics for regression or classification.

   In order to categorise suicide and control groups based on user characteristics and social behaviour variables, Masuda et al. [38] used logistic regression.

   The authors discovered that in a Japanese Social Networking Service (SNS), factors including homophily, local clustering coefficient, and community number have a more significant impact on suicidal ideation. In order to evaluate suicide risk factors, Chattopadhyay [39] employed the Pierce Suicidal Intent Scale (PSIS) and carried out regression analysis. Surveys are a reliable source of tabular features. The International Personality Disorder Examination Screening Questionnaire and the Holmes-Rahe Social Readjustment Rating Scale were utilised by Delgado-Gomez et al. [36]. An application of a multilayer feed-forward neural network was suggested by Chattopadhyay [40].

2. **General Text Features**: Extraction of features from unstructured text is another use of feature engineering. N-gram features, knowledge-based features, syntactic features, context features, and class-specific features make up the majority of the features [41].

   Nine suicidal subjects were the focus of a set of keywords developed by Abboute et al. [42] for vocabulary feature extraction. A lexicon of terms related to suicidal content was created by Okhapkina et al. [43]. They introduced singular value decomposition (SVD) for matrices and term frequency-inverse document frequency (TF-IDF) matrices for communications. Mulholland and Quinn [44] built a classifier to predict the propensity of a lyricist's suicide by extracting language and syntactic factors.

   SVM [41], artificial neural networks (ANN) [45], and conditional random field (CRF) [46] are models for detecting suicidal thoughts with feature engineering. Tai et al. [45] chose characteristics such as a candidate's family history, religious beliefs, family status, history of mental illness, and self-harm behaviour. The effectiveness of several multivariate approaches was compared by Pestian et al. [47] using word counts, POS, concepts, and readability scores as performance indicators. The classification techniques of logistic regression, random forest, gradient boosting decision tree, and XGBoost were also contrasted by Ji et al. [48] in their comparison. Machine learning techniques that have been proven by Braithwaite et al. [49] can successfully identify high suicide risk.

3. **Affective Characteristics**: Both computer scientists and mental health experts have placed a lot of emphasis on affective characteristics, which are among the most noticeable variations between those who attempt suicide and healthy people.

Manual emotion categories were utilised by Liakata et al. [46] and included anger, grief, hopefulness, happiness, peace, fear, pride, abuse, and forgiveness. To develop fine-grained sentiment analysis, Wang et al. [41] merged characteristics of factual (2 categories) and emotive aspects (13 categories).

Similar to this, Pestian et al. [47] recognised the following emotions: abuse, anger, blame, fear, guilt, hopelessness, sadness, forgiveness, happiness, peace, hopefulness, love, pride, thanks, instructions, and information. In order to examine the collected emotional characteristics in suicide blogs and to identify suicidal intentions from a blog stream, Ren et al. [50] suggested a complex emotion topic model.

The eight primary emotions of joy, love, expectancy, surprise, worry, sorrow, rage, and hatred were specifically explored by the writers, along with emotion accumulation, emotion variance, and emotional transition.

17.3.3 Deep Learning

Deep learning has shown considerable success in a variety of fields, including computer vision, natural language processing, and medical diagnosis. It is a crucial technique for automatic suicidal ideation identification and suicide prevention in the field of suicide research. Without using complex feature engineering techniques, it can successfully learn text features automatically. Some people also incorporate extracted features into deep neural networks at the same time.

The multilayer perceptron (MLP) was fed psycholinguistic characteristics and word recurrence by Nobles et al. [51]. Deep neural networks (DNNs) are widely used. Recurrent neural networks, convolutional neural networks (RNNs), and representations of bidirectional encoders (BERT) from transformers, as depicted in Figure 17.3a–c. Typically, natural language text is integrated into spread-out vector space with well-liked word embedding approaches like GloVe [52] and word2vec [53]. Shing et al. employed user-level CNN with the filter size to encrypt user posts 3, 4, and 5 enduring momentary. The memory (LSTM) network, a well-liked RNN version, is used to encrypt text sequences, which is then processed for categorisation using layers that are fully connected [33].

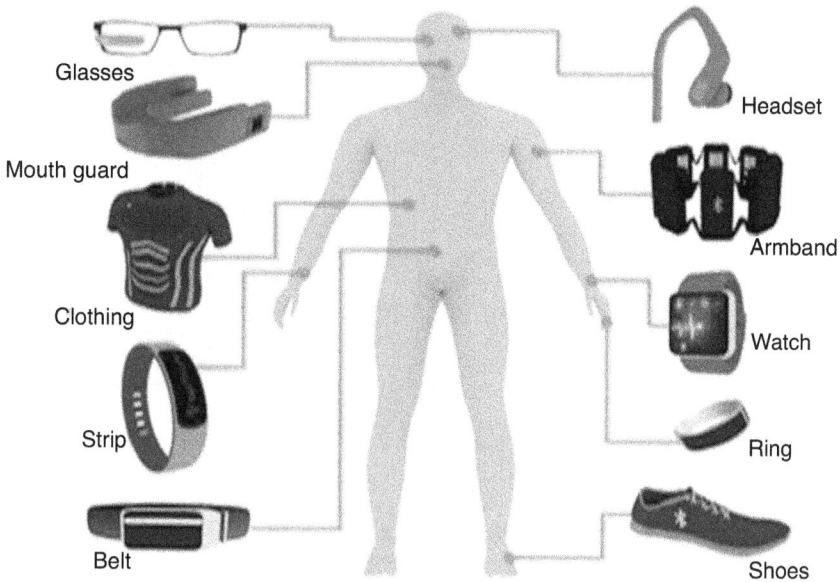

FIGURE 17.3 Wearable devices [54].

New techniques for detecting suicidal thoughts integrate DNNs with other cutting-edge learning paradigms. Ji et al. uses a model to recognize suicidal intent in private chat rooms, model aggregation algorithms for CNNs and LSTMs, the two types of neural networks, were proposed. Decentralised training, on the other hand, depends on coordinators in chat rooms labelling user posts for supervised training, which can only be used in the most basic cases. Using unsupervised or semi-supervised learning techniques could be a preferable option.

By predicting the gender of users as an auxiliary task, Benton et al. [56] predicted suicide attempts and mental health using neural models within the framework of multitask learning. In order to increase performance, Gaur et al. [57] integrated other knowledge bases and an ontology relevant to suicide into a text representation.

Coppersmith et al. [58] created a deep learning model for word embedding using GloVe, self-attention, and bidirectional LSTM for sequence encoding a system for gathering the most informative subsequence. A combination of LSTM, CNN, and RNN for detecting suicidal thoughts. Likewise, Tadesse et al. [59] used LSTM-CNN as a model. Ji and others [48] suggested a careful relation network using LSTM and encoding text and risk indicators using topic modelling (Figure 17.4).

The 2019 CLPsych Shared Task [53] included some well-known architectures based on DNN. A number of well-known deep learning models, including CNN, LSTM, and Synthesis of Neural Networks (NeuNetS) was used [55]. Using a hier-archically organised dual-context paradigm, Matero et al. suggested a dual-context model using hierarchically AttentiveRNN and BERT alert [60].

The so-called hybrid technique is another sub-direction. It integrates representa-tion and minor feature engineering acquiring skills. Chen et al. uses language of

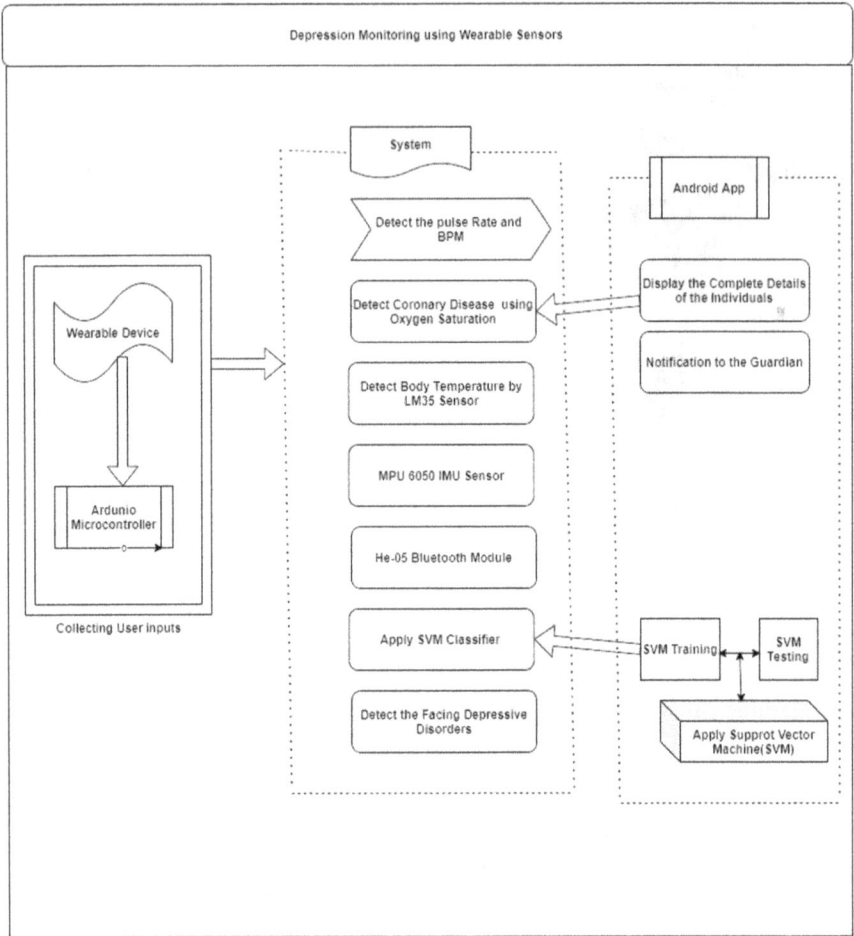

FIGURE 17.4 System architecture and flow of IoMT [16].

suicide is represented by a hybrid behavioral model and categorization framework. The D-CNN model proposed by Zhao et al. uses word embedding characteristics and external tabular data as classification criteria for individuals who attempt suicide.

Sometimes artificial intelligence–powered programmes pick up statistical cues but are unable to discern people's intentions.

Additionally, a lot of brain models are difficult to comprehend. Ref. [57] examines various ways to identify suicidal thoughts from the standpoint of AI, machine learning, and specific domain applications with social implications. Figure 17.5 displays the categorisation from these two angles.

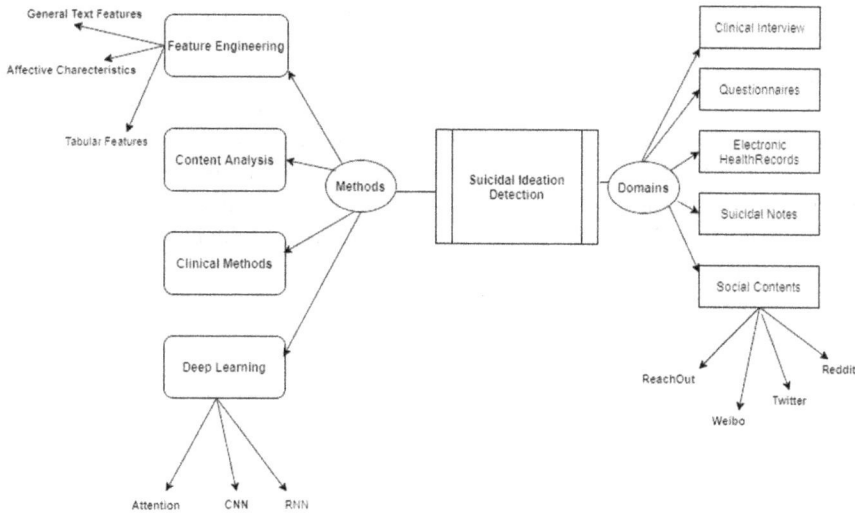

FIGURE 17.5   Classification of suicidal ideation detection [17].

17.4 SUMMARY

The rise in popularity of machine learning has aided studies on the identification of suicidal ideation from multi-modal data and offered a viable means of timely early warning. Deep learning for autonomous feature learning and feature extraction from the text are the main topics of the current study. Numerous canonical NLP features, including TF-IDF, themes, syntactic, emotional, and readability qualities, as well as deep learning models like CNN and LSTM, are frequently used by researchers. These techniques, in particular DNNs with autonomous feature learning, increased the predictive performance and early success in understanding suicidal intention. Some techniques, however, might just teach statistical indications and a lack of common sense. For knowledge-aware suicide risk assessment, the most recent work [61] included external knowledge using knowledge bases and suicide ontologies. It made an impressive advancement in knowledge-aware detection.

17.5 CONCLUSION

In today's society, there is still much that needs to be done to prevent suicide caused by fatal levels of depression. A crucial and successful method of preventing suicide is the early identification of suicidal ideation (major depression disorder). An investigation of current treatment and methods used for the detection and prevention of MDD is from a comprehensive viewpoint that includes clinical techniques such as patient–physician communication and medical signal detection; lexicon-based textual content analysis word cloud visualisation and filtering; feature engineering, encompassing textual, emotive, and tabular characteristics; and representation learning based on deep learning.

It is possible to detect depression by combining certain technology and sensors. In the more recent IoT, sensors with Android applications can be used to build a model that could assist individuals in identifying depression so that they can visit doctors and/or psychiatrists as needed. In some circumstances, mobile applications are created that automatically display the specifics of a person's health band data. Also, the mobile application notifies the person's carer of emergencies.

The system can be improved further by automatically playing cheerful tunes, inspirational words, or video links on a laptop or mobile device based on the SVM rate.

Future suicidal thought detection will most likely take place mostly through online social content. In order to identify online messages containing suicidal ideation and to prevent suicide, it is crucial to create new techniques that can bridge the gap between clinical mental health detection and automatic machine detection.

The majority of this field's research has been done by psychologists using statistical analysis, and computer scientists using feature engineering- and deep learning-based machine learning. We outlined present activities and additionally suggested new potential tasks based on recent research. Finally, we explored some of the limits of the current literature and suggested a number of new lines of inquiry, including the use of cutting-edge learning strategies, interpretable intention understanding, temporal detection, and proactive conversational intervention.

REFERENCES

[1] S. Hinduja and J. W. Patchin. (2010), Bullying, cyberbullying, and suicide. *Archives of Suicide Research*, vol. 14, no. 3, pp. 206–221.
[2] J. Joo, S. Hwang, and J. J. Gallo. (2016), Death ideation and suicidal ideation in a community sample who do not meet criteria for major depression. *Crisis*, vol. 37, no. 2, pp. 161–165.
[3] M. J. Vioul'es, B. Moulahi, J. Az´e, and S. Bringay. (2018), Detection of suicide-related posts in twitter data streams. *IBM Journal of Research and Development,* vol. 62, no. 1, pp. 7:1–7:12.
[4] https://www.nimh.nih.gov/health/statistics/suicide
[5] https://www.cdc.gov/suicide/suicide-data-statistics.html
[6] A. J. Ferrari, R. E. Norman, G. Freedman, A. J. Baxter, J. E. Pirkis,M. G. Harris, A. Page, E. Carnahan, L. Degenhardt, T. Vos et al. (2014), The burden attributable to mental and substance use disorders as risk factors for suicide: findings from the global burden of disease study. *PloS One*, vol. 9, no. 4, p. e91936.
[7] J. Lopez-Castroman, B. Moulahi, J. Az´e, S. Bringay, J. Deninotti,S. Guillaume, and E. Baca-Garcia. (2019), Mining social networks to improve suicide prevention: a scoping review. *Journal of Neuroscience Research*, vol. 98, no. 4, pp. 616–625.
[8] C. M. McHugh, A. Corderoy, C. J. Ryan, I. B. Hickie, and M. M. Large. (2019), Association between suicidal ideation and suicide: meta-analyses of odds ratios, sensitivity, specificity and positive predictive value. *The British Journal of Psychiatry Open*, vol. 5, no. 2.
[9] E. Basca-Garcia, M. M. Perez-Rodriguez, M. A. Oquendo, K. M. Keyes, D. S. Hasin, B. F. Grant, and C. Blanco, "Estimating risk for suicide attempt: Are we asking the right questions? passive suicidal behavior as a marker for suicidal behave "Depression and suicide," https://caps.ucsc.edu/resources/depression.html.

[10] E. Basca-Garcia, M. M. Perez-Rodriguez, M. A. Oquendo, K. M. Keyes,D. S. Hasin, B. F. Grant, and C. Blanco. (2011), Estimating risk for suicide attempt: Are we asking the right questions? passive suicidal behavior as a marker for suicidal behavior. *Journal of Affective Disorders*, vol. 134, pp. 327–332.

[11] P. Sundaravadivel, P. Salvatore, and P. Indic. (2020), "M-SID: An IoT-based Edge-intelligent Framework for Suicidal Ideation Detection," In IEEE 6th World Forum on Internet of Things (WF-IoT), New Orleans, LA, USA, 2020, pp. 1–6, https://doi.org/10.1109/WF-IoT48130.2020.9221279.

[12] P. Indic, P. Salvatore, C. Maggini, S. Ghidini, G. Ferraro, R. J. Baldessarini, and G. Murray. (2011), Scaling behavior of human locometer activity amplitude: association with bipolar disorder. *PLOS One*, vol. 6, no. 5, p. e20650.

[13] P. Indic, G. Murray, C. Maggini, M. Amore, T. Meschi, L. Borghi,R. J. Baldessarini, and P. Salvatore. (2012), Multi-scale motility amplitude associated with suicidal thoughts in major depression. *PLOS One,* vol. 7, no. 6, p. e38761.

[14] V. Venek, S. Scherer, L.-P. Morency, J. Pestian et al. (2017), "Adolescent suicidal risk assessment in clinician-patient interaction," *IEEE Transactions on Affective Computing*, vol. 8, no. 2, pp. 204–215.

[15] A. Birkhofer, G. Schmidt, and H. Forstl. (2005), Heart and brain - the influence of psychiatric disorders and their therapy on the heart rate variability. *Fortschritte Der Neurologie Psychiatrie*, vol. 73, pp. 192–205.

[16] M. Vijay, and K. Joshi. (2020), Depression monitoring using wearable sensors. *International Journal of Emerging Technologies and Innovative Research*, Vol. 7, no. 3, pp. 599–604.

[17] S. Ji, S. Pan, X. Li, E. Cambria, G. Long, and Z. Huang. (2021), Suicidal ideation detection: a review of machine learning methods and applications. *IEEE Transactions on Computational Social Systems*, vol. 8, no. 1, pp. 214–226, https://doi.org/10.1109/TCSS.2020.3021467.

[18] N. Dey, et al., eds. (2018) *Internet of Things and Big Data Analytics Toward Next-Generation Intelligence.* Springer International Publishing.

[19] H. M. Mallikarjun, and H. N. Suresh. (2014), "Depression level prediction using EEG signal processing," In International Conference on Contemporary Computing and Informatics (IC3I). IEEE.

[20] J. Saha, et al. (2016) "A framework for monitoring of depression patient using WBAN," In Wireless Communications, Signal Processing and Networking (WiSPNET), International Conference on. IEEE.

[21] L. Jena, and N. K. Kamila. (2014), "A model for prediction of human depression using apriori algorithm," In Information Technology (ICIT), 2014 International Conference on Industrial Technology (ICIT). IEEE.

[22] U. K. Farooqui, and A. K. Bharti. (2020), A privacy preserving upload model for crowd sourced health care monitoring system. *International Journal of Engineering and Advanced Technology (IJEAT)*, vol. 1, no. 1, pp. 3337–3346.

[23] B. Rani. (2016), "Detecting depression: a comparison between I-Vector technique and fuzzy membership functions." In 2016 International Conference on Inventive Computation Technologies (ICICT). Vol. 1. IEEE.

[24] J. Ramos, J. Hong, and A. K. Dey. (2014), "Stress recognition - a step outside the lab," In Proceedings of the International Conference on Physiological Computing Systems. pp. 107–118.

[25] S. Scherer, J. Pestian, and L.-P. Morency. (2013), "Investigating the speech character-istics of suicidal adolescents," In IEEE International Conference on Acoustics, Speech and Signal Processing. IEEE.

[26] D. Sikander, M. Arvaneh, F. Amico, G. Healy, T. Ward, D. Kearney, E. Mohedano, J. Fagan, J. Yek, A. F. Smeaton et al. (2016), "Predicting risk of suicide using resting state heart rate," In 2016 Asia-Pacific Signal and Information Processing Association Annual Summit and Conference (APSIPA). IEEE, pp. 1–4.

[27] N. Jiang, Y. Wang, L. Sun, Y. Song, and H. Sun. (2015), "An erp study of implicit emotion processing in depressed suicide attempters," In 7th International Conference on Information Technology in Medicine and Education (ITME). IEEE, pp. 37–40.

[28] K. D. Varathan, and N. Talib. (2014), "Suicide detection system based on twitter," In Science and Information Conference (SAI). IEEE, pp. 785–788.

[29] J. Jashinsky, S. H. Burton, C. L. Hanson, J. West, C. Giraud-Carrier,M. D. Barnes, and T. Argyle. (2014), "Tracking suicide risk factors through twitter in the us," *Crisis*, vol. 35, no. 1, pp. 51–59.

[30] M. J. Vioul'es, B. Moulahi, J. Az´e, and S. Bringay. (2018), "Detection of suicide-related posts in twitter data streams," *IBM Journal of Research and Development*, vol. 62, no. 1, pp. 7:1–7:12.

[31] J. F. Gunn. and D. Lester. (2015), "Twitter postings and suicide: an analysis of the postings of a fatal suicide in the 24 hours prior to death," *Suicidologi*, vol. 17, no. 3.

[32] G. Coppersmith, R. Leary, E. Whyne, and T. Wood. (2015), "Quantifying suicidal ideation via language usage on social media," In Joint Statistics Meetings Proceedings, Statistical Computing Section, JSM.

[33] S. Ji, C. P. Yu, S.-f. Fung, S. Pan, and G. Long. (2018), "Supervised learning for suicidal ideation detection in online user content," *Complexity*.

[34] G. B. Colombo, P. Burnap, A. Hodorog, and J. Scourfield. (2016), "Analysing the connectivity and communication of suicidal users on twitter," *Computer Communications*, vol. 73, pp. 291–300.

[35] H. Y. Huang and M. Bashir. (2016), "Online community and suicide prevention: investigating the linguistic cues and reply bias," In Proceedings of the CHI Conference on Human Factors in Computing Systems.

[36] D. Delgado-Gomez, H. Blasco-Fontecilla, F. Sukno, M. S. Ramos-Plasencia, and E. Baca-Garcia. (2012), "Suicide attempters classification: Toward predictive models of suicidal behavior," *Neurocomputing*, vol. 92, pp. 3–8.

[37] M. E. Larsen, N. Cummins, T. W. Boonstra, B. O'Dea, J. Tighe, J. Nicholas, F. Shand, J. Epps, and H. Christensen. (2015), "The use of technology in suicide prevention," In 37th Annual International Conference of the IEEE Engineering in Medicine and Biology Society (EMBC). IEEE, pp. 7316–7319.

[38] N. Masuda, I. Kurahashi, and H. Onari. (2013), "Suicide ideation of individuals in online social networks," *PloS One*, vol. 8, no. 4, p. e62262.

[39] S. Chattopadhyay. (2007), "A study on suicidal risk analysis," In International Conference on e-Health Networking, Application and Services. IEEE, pp. 74–78.

[40] S. Chattopadhyay. (2012), "A mathematical model of suicidal-intentestimation in adults," *American Journal of Biomedical Engineering*, vol. 2, no. 6, pp. 251–262.

[41] W. Wang, L. Chen, M. Tan, S. Wang, and A. P. Sheth. (2012), "Discovering fine-grained sentiment in suicide notes," *Biomedical Informatics Insights*, vol. 5, no. Suppl 1, p. 137.

[42] A. Abboute, Y. Boudjeriou, G. Entringer, J. Az´e, S. Bringay,and P. Poncelet. (2014), "Mining twitter for suicide prevention," In International Conference on Applications of Natural Language to Data Bases/Information Systems. Springer, pp. 250–253.

[43] E. Okhapkina, V. Okhapkin, and O. Kazarin. (2017), "Adaptation of information retrieval methods for identifying of destructive informational influence in social networks," In 31st International Conference on Advanced Information Networking and Applications Workshops (WAINA). IEEE, pp. 87–92.

[44] M. Mulholland and J. Quinn. (2013), "Suicidal tendencies: The automatic classification of suicidal and non-suicidal lyricists using NLP." In IJCNLP, pp. 680–684.

[45] Y.-M. Tai and H.-W. Chiu. (2007), "Artificial neural network analysis on suicide and self-harm history of taiwanese soldiers," In Second International Conference on Innovative Computing, Information and Control. IEEE, pp. 363–363.

[46] M. Liakata, J. H. Kim, S. Saha, J. Hastings, and D. Rebholzschuhmann. (2012), "Three hybrid classifiers for the detection of emotions in suicide notes," *Biomedical Informatics Insights*, vol. 2012, no. (Suppl. 1), pp. 175–184.

[47] J. Pestian, H. Nasrallah, P. Matykiewicz, A. Bennett, and A. Leenaars. (2010), "Suicide note classification using natural language processing: a content analysis," *Biomedical Informatics Insights*, Vol. 2010, no. 3, p. 19.

[48] S. Ji, C. P. Yu, S.-f. Fung, S. Pan, and G. Long. (2018), "Supervised learning for suicidal ideation detection in online user content," *Complexity.*

[49] S. R. Braithwaite, C. Giraud-Carrier, J. West, M. D. Barnes, and C. L. Hanson. (2016), "Validating machine learning algorithms for twitter data against established measures of suicidality," *JMIR Mental Health*, vol. 3, no. 2, p. e21.

[50] F. Ren, X. Kang, and C. Quan. (2016), "Examining accumulated emotional traits in suicide blogs with an emotion topic model," *IEEE Journal of Biomedical and Health Informatics*, vol. 20, no. 5, pp. 1384–1396.

[51] A. L. Nobles, J. J. Glenn, K. Kowsari, B. A. Teachman, and L. E. Barnes. (2018), "Identification of imminent suicide risk among young adults using text messages," In Proceedings of the 2018 CHI Conference on Human Factors in Computing Systems, pp. 1–11.

[52] J. Pennington, R. Socher, and C. Manning. (2014), "Glove: Global vectors for word representation," In Proceedings of the conference on empirical methods in natural language processing (EMNLP), pp.1532–1543.

[53] T. Mikolov, K. Chen, G. Corrado, and J. Dean. (2013), "Efficient estimation of word representations in vector space," arXiv preprint arXiv:1301.3781.

[54] https://itchronicles.com/mobile/wearable-devices-in-healthcare/

[55] S. Ji, G. Long, S. Pan, T. Zhu, J. Jiang, S. Wang, and X. Li. (2019), "Knowledge transferring via model aggregation for online social care," arXiv preprint arXiv:1905.07665.

[56] A. Benton, M. Mitchell, and D. Hovy. (2017), "Multi-task learning for mental health using social media text," In EACL. Association for Computational Linguistics.

[57] M. Gaur, A. Alambo, J. P. Sain, U. Kursuncu, K. Thirunarayan,R. Kavuluru, A. Sheth, R. Welton, and J. Pathak. (2019), "Knowledgeaware assessment of severity of suicide risk for early intervention," In The World Wide Web Conference. ACM, pp. 514–525.

[58] G. Coppersmith, R. Leary, P. Crutchley, and A. Fine. (2018), "Natural language processing of social media as screening for suicide risk," *Biomedical Informatics Insights*, vol. 10, p. 117822261879286.

[59] M. M. Tadesse, H. Lin, B. Xu, and L. Yang. (2020), "Detection of suicide ideation in social media forums using deep learning," *Algorithms*, vol. 13, no. 1, p. 7.

[60] M. Matero, A. Idnani, Y. Son, S. Giorgi, H. Vu, M. Zamani,P. Limbachiya, S. C. Guntuku, and H. A. Schwartz. (2019), "Suicide risk assessment with multi-level dual-context language and bert," In Proceedings of the Sixth Workshop on Computational Linguistics and Clinical Psychology, pp. 39–44.

[61] S. Ji, S. Pan, X. Li, E. Cambria, G. Long and Z. Huang. (2021), "Suicidal ideation detection: a review of machine learning methods and applications," *IEEE Transactions on Computational Social Systems*, vol. 8, no. 1, pp. 214–226, https://doi.org/10.1109/TCSS.2020.3021467.

Index

abdominal circumference (AC) 200
abnormality 237
accuracy 172–174, 178, 181–183
accuracy score for ADHD 148
accuracy score for dyslexia 148
actuator 247
acute care 227
Adaboost 130
ADHD 142
AD8232 IC 230
AI algorithms 187
AI-based IoMT devices 106
AlexNet 220
analyser-based anomaly identification 239
analytics 19
anomaly detecting 237
antidiabetic 33
anxiety attack 129
anxiety disorder 130
ARM processor 229
artificial intelligence (AI) 3, 8, 10, 14, 16, 83, 92, 109, 172, 184, 256
artificial neural network (ANN) 22, 30, 116
attribute based encryption (ABE) 52–54
attribute based signature (ABS) 53–54

Bayesian networks 244
big data 256, 257, 271
Bi-LSTM 130
biological variance 237
bio-sensors 247
biparietal diameter (BPD) 199
bipolar disorder 126
blockchain 8, 9, 13, 14, 16
    applications 43
    framework 42–44
blood sugar 25
Bluetooth 229, 248
body sensor network 8
brain signals 171
broker 20
BSNs 8, 9, 11

cancer survivors 260, 262, 263, 265, 266, 268, 270, 271, 273, 274
cardiovascular diseases (CVD) 109, 110
cardiovascular system 33
CART 131
Celeb Faces Attributes 212
central detection (CD) 249

CERN 138
China's Manchurian pandemic 211
chronic diseases 227
classification 246
client 21
cloud computing 20–22, 58, 60, 66–69
    cloud servers 22
cloud service provider 58
clustering 29
CNN 10, 178, 181
cognitive assessment tools 76
complications 20, 27, 28, 31–33
computational intelligence 93, 256, 268
continuous health examination 227
convolutional neural network (CNN) 10, 23, 30–34, 154–156, 172, 178, 181, 184, 212
covariance matrix 203
COVID-19 4, 10, 11, 15
CRM 247
crown-rump length (CRL) 199
CSV 249

data
    augmentation 23, 26, 30
    big data 23
    cleaning 23
    filtering 23
    fusion 22, 23
    mining 19, 29, 30
    pre-processing 22
data access 46–49
data collection 146
data management 44–45
data mining 19
data preprocessing 218
data preservation 49–51
decision-making 93
decision support systems 28, 29
decision tree 19, 29, 115, 116
decision tree approach 242
decision tree classifier 101
deep auto encoder (DAN) 155
deep learning (DL) 19, 23, 29, 93, 153–168, 171–173, 175, 177, 179–184
Deep Learning Healthcare Diagnosis System (DL-HDS) 159–160
denial of service (DoS) attack 163
depression 171, 173, 177, 180, 183
DESNET 23, 25

device integrity 167
diabetes 19, 21, 27–29, 31, 33
    diabetes treatment 28
diabetes foot 31
diabetes retinopathy 23, 30–32, 34
diagnosis 237
disease diagnosis 94
disease symptoms 97
disruptive mood dysregulation disorder
        (DMDD) 286
doctor 20, 21, 28
DTBM 116
dyslexia 142

EC 172, 174, 180–181, 183, 184
ECG 92, 115, 228
ECG database 231
ECG-data server 231
edge computing 60
EEG 92, 171–177, 180–184, 291
EEG signal processing 131
elderly patients 14
electrocardiogram 19, 25
electroencephalogram 171, 184
electroencephalography 141
electronic health records (EHRs) 46, 94, 247
electronic medical record (EMR) 46
ensemble learning 110, 116
EO 172, 174, 180–181, 183–184
EORTC 260–262, 264, 265, 268, 269, 271, 273
eye 32, 33
eyes close 174, 181–182, 184
eyes open 174, 181–182, 184

FACT 260, 261, 268, 269
false negative 178, 184
false positive 178, 184
fatality 237
FCMIM 116
feature detection 23
feature engineering 286, 295, 298
feature extraction 172–173, 177, 182, 184
feature reduction 30, 34
feature vector 204
femur length (FL) 199
fetal weight (EFW) 199
5G 8, 10, 16, 17
5G/6G 248
fog 8, 11, 14, 16
framework 256, 257, 259, 267, 268, 270, 272

Gaussian mixture models 244
GBBM 116
generative adversarial network (GAN) 155–157, 244
glucometers 25

glucose 19, 21, 22, 25, 27, 28, 33
glucose monitoring 25
glycaemia 25
Google Colab 180
GPRS 229
graphene 63
graphical user interface (GUI) 21, 26

HbA1c 28
head circumference (HC) 172–173, 182, 184, 199
healthcare analytics 20
healthcare diagnosis system 159–160
healthcare infrastructure 33
healthcare internet of things 20
healthcare monitoring 23, 26
health care records (HCR) 167
healthcare solutions 105
health monitoring system 157–159
healthy control 172–173, 177, 184
heart disease 109, 110, 115, 116
humerus length (HL) 200
hybrid random forest with linear model
        (HRF-LM) 116
hyperglycaemia 32, 33
hyper parameter optimization 116
hypoglycaemia 27

imagenet large-scale visual recognition 220
IMIS 27
India 28
infectious diseases 11, 15
information and communication technology
        (ICT) 247
ingestible sensors 7
insulin 21, 22, 27, 29, 31, 33
    insulin pens 21, 22
intelligence of the computer 102
intelligent health systems 93
intensive care unit 4, 16
internal QC 239
International Diabetes Federation 31
Internet of Medical Things (IoMT) 3–13, 59,
        72–85, 91, 131, 186, 238, 246
Internet of Things (IoT) 3, 4, 8, 12, 13, 16, 19–21,
        26, 27, 33, 73, 91
Interplanetary File System (IPFS) 45–46
intrusion detection system 161–166
IoMT architecture 5, 9
IoMT devices 7, 9, 11–15
IoMT gadgets 131
IoMT gateways 248
IoMT sensor and actuator 248
IoMT sensors 95
IoT-based Remote Monitoring System 228
ischaemia 31, 32

Jupiter Notebook 180

kidney 31, 33
K-Nearest Neighbor (KNN) 30, 98, 132
KNN classifier 101

laboratory errors 238
Large Hadron Collider 138
LASSO 116
latent Dirichlet allocation 244
learning disability 141
linear discriminant analysis 115
logistic regression 172–174, 177, 180–184
long-term evolution (LTE) 248
LR 172–174, 177, 181, 184
LSTM 131

machine learning (ML) 3, 4, 8, 10, 14, 16, 22, 29,
        92, 109, 110, 115, 124, 141, 171–174,
        177, 180–182, 184, 256, 258, 273, 274
MADRS scores 132, 133
major depressive disorder 173, 184
MDA 248
MDD 172–174, 180–184
M2DM 28
Medical CQAs 130
medical decision-making 237
medical devices 73
medicinal care 97
medicines 33
mental health 72–85
MHC 132
mobile applications 28
mobile health 73
MongoDB 231
morbidity 237
MSE 129
M-SID framework 288–289
multi layer perception (MLP) 30, 33

Naïve Bayes 19, 98
Naïve Bayes classifier 101
Naïve Bayesian (NB) 173–174, 183–184
NBTree 131
near-field communications (NFC) 248
nervous system 27, 32
neuroischemic foot 32
neuropathic foot 32
neuropathy 31, 32, 34
NodeMCU 230
Node-RED programming tool 230
noncommunicable diseases 227

occipitofrontal diameter (OFD) 200
OCI- DBN 116

optimisation 33, 34
    Adaptive Particle Swam Optimisation
        (APSO) 30
    Grey Wolf Optimiser 33
over-fitting 30

paediatric care 28
passive therapy 25
patients 3, 4, 6–16, 19–21, 23, 25–31, 33
patient-based real-time quality control
        (PBRTQC) 240
patient wireless node (PWN) 230
performance evaluation 147
peripheral vascular disease 31
persistent depressive disorder (PDD) 126, 285
personal emergency response systems 198
personal health records (PHR) 46
physically unclonable function (PUF) 8
physiologies 237
Pima
    Pima Indians Dataset 29, 30
    Pima People 29, 30, 32
population-based anomaly 244, 245
post-traumatic stress disorder (PTSD) 288
prediction 94
prediction model 270–274
predictive model 29, 30
pregnant women in India 187
premenstrual dysphoric disorder (PMDD)
        126, 286
Prenatal Healthcare System of Remote Mother
        and Fetal Surveillance via IoMT 188
pre-processing, feature extraction and feature
        selection 146
principal component analysis (PCA) 116,
        200, 216
privacy protection 51–54
prognosis 237
proof of concept 242
public health records (PHRs) 94
publisher 20

quality of life 256–259, 262, 265

radio frequency identification (RFID) 8, 11,
        61, 248
Random Forest 29, 34, 98, 115, 116
Random Forest Classifier 101
random under-sampling boosting 33
Real-World Masked Face Dataset 212
rectified linear activation function 213
rectified linear unit 23
recurrent neural networks (RNNs) 19, 155–156
regression 29–31, 246
remote diagnosis 4, 14

remote health consultation 4
remote health monitoring 4
remote monitoring of patient 228
remote patient monitoring 4, 6, 9, 10, 14
ResNET 23, 30, 32
restricted Boltzmann machines (RBMs) 19
results 147
retina 23, 30, 32
retinal fundus 32
RFBM 116
RFRFILM 116
RHM 4, 5, 7–14, 16
RNN 131
ROSE sampling 132
Ruzzo–Tompa algorithm 116

scalability 8, 11–13, 16
security and privacy 13, 14
security threats 163
self-reported anxiety scale (SAS) 132
sensitivity 173–174, 178
sensors 19, 21, 26, 188, 248
service fingerprint extraction 167
sibling intractable function families
        (SIFF) 52
smart home devices 80–81
smart hospitals 247
smart inhaler 7
smart pens 28, 29
smartwatches 76–77
SMOTE 116
social network 23
software-defined networking (SDN) 8
spatial sampling density 246
specificity 173–174, 178
specimen-based approaches 242
statistical features 172–173, 177, 184
suicidal tendencies 75–76

Super Bowl event 212
supervised 96
support vector machine (SVM) 22, 30, 31, 116,
        172–174, 177, 180–184

TASK 172, 174, 180–184
TBots 291
telehealthcare 28
    telediabetology 28
threshold optimisations 242
true negative 178, 184
true positive 178, 184
trust execution environment (TEE) 52
TTP 238

ulcers 32
U-Net 155–157
unsupervised 96

validation 22, 30, 32
variational autoencoders 244
virtual reality systems 81–82

wearable devices 295
wearable ECG sensor 229
wearable IoT (WIoT) 290
wearable sensors 76–78
WHO 171
Wi-Fi module 230
Wi-Fi router 230
wireless access point (WAP) 230
wireless sensor networks (WSNs) 8, 9, 62
World Health Organization (WHO) 33

XGBoost 130, 132

Z-Alizadeh Sani 115, 116
ZigBee 229

For Product Safety Concerns and Information please contact our EU
representative GPSR@taylorandfrancis.com
Taylor & Francis Verlag GmbH, Kaufingerstraße 24, 80331 München, Germany

www.ingramcontent.com/pod-product-compliance
Lightning Source LLC
Chambersburg PA
CBHW060327220326
41598CB00023B/2629

* 9 7 8 1 0 3 2 4 1 8 3 1 5 *